S0-BRH-119

COMPREHENSIVE BIOCHEMISTRY

ELSEVIER SCIENCE PUBLISHERS

1 Molenwerf, P.O. Box 211, Amsterdam

ELSEVIER/NORTH-HOLLAND, INC.

52, Vanderbilt Avenue, New York, N.Y. 10017

With 16 plates and 16 figures

Library of Congress Cataloging in Publication Data
Main entry under title:

Selected topics in the history of biochemistry.

(Comprehensive biochemistry; v. 35. Section VI,
A history of biochemistry)
Includes bibliographies and index.
Contents: I. Personal recollections.
1. Biological chemistry-History-Addresses, essays,
lectures. I. Semenza, G., 1928- . II. Series:
Comprehensive biochemistry; v. 35.
QD415.F54 vol. 35 574.19′2s [574.19′2] 83-20491
[QP511]
ISBN 0-444-80507-9 (U.S.: v. 1)

PRINTED IN THE NETHERLANDS

COMPREHENSIVE BIOCHEMISTRY

COMPREHENSIVE BIOCHEMISTRY

SECTION I (VOLUMES 1–4)
PHYSICO-CHEMICAL AND ORGANIC ASPECTS OF BIOCHEMISTRY

SECTION II (VOLUMES 5–11)
CHEMISTRY OF BIOLOGICAL COMPOUNDS

SECTION III (VOLUMES 12–16)
BIOCHEMICAL REACTION MECHANISMS

SECTION IV (VOLUMES 17–21)
METABOLISM

SECTION V (VOLUMES 22–29)
CHEMICAL BIOLOGY

SECTION VI (VOLUMES 30–34)
A HISTORY OF BIOCHEMISTRY

COMPREHENSIVE BIOCHEMISTRY

ALBERT NEUBERGER

Chairman of Governing Body, The lister Institute of Preventive Medicine, University of London, London (Great Britain)

LAURENS L.M. VAN DEENEN

Professor of Biochemistry, Biochemical Laboratory, Utrecht (The Netherlands)

GIORGIO SEMENZA

Laboratorium für Biochemie, ETH-Zentrum Zurich (Switzerland)

Editor of VOLUME 35

SELECTED TOPICS IN THE HISTORY OF BIOCHEMISTRY PERSONAL RECOLLECTIONS. I.

ELSEVIER SCIENCE PUBLISHERS

AMSTERDAM . OXFORD . NEW YORK

1983

GENERAL PREFACE

The Editors are keenly aware that the literature of Biochemistry is already very large, in fact so widespread that it is increasingly difficult to assemble the most pertinent material in a given area. Beyond the ordinary textbook the subject matter of the rapidly expanding knowledge of biochemistry is spread among innumerable journals, monographs, and series of reviews. The Editors believe that there is a real place for an advanced treatise in biochemistry which assembles the principal areas of the subject in a single set of books.

It would be ideal if an individual or a small group of biochemists could produce such an advanced treatise, and within the time to keep reasonably abreast of rapid advances, but this is at least difficult if not impossible. Instead, the Editors with the advice of the Advisory Board, have assembled what they consider the best possible sequence of chapters written by competent authors; they must take the responsibility for inevitable gaps of subject matter and duplication which may result from this procedure.

Most evident to the modern biochemist, apart from the body of knowledge of the chemistry and metabolism of biological substances, is the extent to which we must draw from recent concepts of physical and organic chemistry, and in turn project into the vast field of biology. Thus in the organization of Comprehensive Biochemistry, sections II, III and IV, Chemistry of Biological Compounds, Biochemical Reaction Mechanisms, and Metabolism may be considered classical biochemistry, while the first and fifth sections provide selected material on the origins and projections of the subject.

It is hoped that sub-division of the sections into bound volumes will not only be convenient, but will find favour among students concerned with specialized areas, and wil permit easier future revisions of the individual volumes. Towards the latter end particularly, the Editors will welcome all comments in their effort to produce a useful and efficient source of biochemical knowledge.

M. Florkin†
Liège/Rochester E.H. Stotz

There is a history in all men's lives.
W. Shakespeare, Henry IV, Pt. 2

History is the essence of innumerable biographies.
T. Carlyle, On History

PREFACE TO VOLUME 35

Perhaps one of the most exiting developments in biological sciences in our times has been their merging with chemistry and physics with the resulting appearance of biochemistry, biophysics, molecular biology, and related sciences. The nearly explosive development of these "newcomers" has led to the almost unique situation that these new biological sciences have come of age at time when their founding fathers, or their scientific sons, are alive and active.

It was therefore an almost obvious idea to ask them to write, for the benefit of both students and senior scientists, personal accounts of their scientific lives. With this idea in mind I have already edited two volumes for John Wiley & Sons, who had, however, a somewhat different format.

The chapters in this and in future volumes are meant to complement, with personal recollections, the History of Biochemistry in the Comprehensive Biochemistry series (Vols. 30-33, by M. Florkin and Vol. 34 (forthcoming), by E. Schoffeniels). In fact, it is hoped that the biographical or autobiographical chapters will convey to the reader lively, albeit at times subjective, views on the scientific scene as well as the social evironment in which the authors have operated and brought about new concepts and pieces of knowledge. The Editor considered it presumptuous to give the authors narrow guidelines or to suggest changes in the chapters he received; he thinks that directness and straightforwardness should be given priority over uniformity. The contributions assembled in this volume will convey the flavour of each author's particular personality; whatever the optical distortion of one chapter, it will be compensated by the views in another.

The development of today's life sciences was acted upon by serious and often tragic historical events. The Editor hopes that this message also will reach the readers, especially the young ones.

It proved an impossible task to group the contributors in a strictly logical manner whether according to subject matter, geographical area, or time. In fact, most contributions cross each of these borders. Nevertheless the Editor hopes that the reader will find these contributions as interesting as he did.

The Editor wants to express his gratitute to all individuals who made this series possible; first of all to the authors themselves, who not only wrote the texts, but also willingly collaborated in suggesting further potential contributors, thereby acting as a kind of "Editorial Board at Large". Thanks are due to Ms. U. Zilian who typed most of the correspondence and prepared the index of names.

Swiss Institute of Technology
Zurich, 1983

Giorgio Semenza

CONTRIBUTORS TO THIS VOLUME

H. BLASCHKO
University Department of Pharmacology, South Parks Road, Oxford OX1 3QT (U.K.)

R.K. CRANE
Department of Physiology and Biophysics, College of Medicine and Dentistry of New Jersey, Rutgers Medical School, Piscataway, NJ 08854 (U.S.A.)

P. DESNUELLE
Centre National de la Recherche Scientifique, Centre de Biochimie et de Biologie Moléculaire, 31, Chemin Joseph-Aiguier, B.P. 71, F13277 Marseille-Cédex 9 (France)

H.-G. HERS
Laboratoire de Chimie Physiologique, Université Catholique de Louvain and International Institute of Cellular and Molecular Pathology, UCL 75.39, 75, Avenue Hippocrate, B-1200 Brussels (Belgium)

H.M.KALCKAR
Department of Chemistry, Boston University, 685 Commonwealth Ave, Boston, MA 02215 (U.S.A.)

W.L. KRETOVICH
A.N. Bach Institute of Biochemistry, Leninsky prospect, 33, Moscow B-71 (U.S.S.R.)

L.F. LELOIR
Instituto de Investigaciones Bioquímicas "Fundacion Campomar" y Facultad de Ciencias Exactas y Naturales, Universidad de Buenos Aires, Buenos Aires (Argentina)

C.H. LI
Laboratory of Molecular Endocrinology, Room 1018, HSE, University of California, San Francisco, CA 94143 (U.S.A.)

A.C. PALADINI
Departamento de Química Biológica, Facultad de Farmacia y Bioquímica, Universidad de Buenos Aires, Buenos Aires (Argentina)

K.O. PEDERSEN
Institute of Physical Chemistry, University of Uppsala, Uppsala (Sweden)

N.W. PIRIE
Rothamstead Experimental Station, Harpenden, Herts., AL5 2JQ (U.K.)

J.H. QUASTEL
TRIUMF, The University of British Columbia, 4004, Wesbrook Mall, Vancouver, BC (V6T 2A3 Canada)

S.E. SEVERIN
Department of Biochemistry, Moscow State University, 117234 Moscow B-234 (U.S.S.R.)

LIST OF PLATES

(Photographs reproduced with permission of authors, publishers, and/or owners)

Section VI

A HISTORY OF BIOCHEMISTRY

CONTENTS

VOLUME 35

A HISTORY OF BIOCHEMISTRY

Selected Topics in the History of Biochemistry
Personal Recollections. I.

Chapter 1. The Isolation of Cori-ester, "the Saint Louis Gateway" to a First Approach of a Dynamic Formulation of Macromolecular Biosynthesis
by HERMANN M. KALCKAR

Chapter 2. The Discovery of Sugar Nucleotides
by Luis F. Leloir and Alejandro C. Paladini

Chapter 3. The Road to Ion-coupled Membrane Processes
by Robert K. Crane

Chapter 4. From Fructose to Fructose-2,6-bisphosphate with a Detour through Lysosomes and Glycogen
by Henri-Gery Hers

Chapter 5. Sir Frederick Gowland Hopkins (1861-1947)
by N. William Pirie

Chapter 6. A Short Autobiography
by Juda Hirsch Quastel

Chapter 7. A Biochemist's Approach to Autopharmacology
by Hermann Blaschko

Chapter 8. The Svedberg and Arne Tiselius. The Early Development of Modern Protein Chemistry at Uppsala
by Kai O. Pedersen

Chapter 11. A.N. Bach, Founder of the Soviet School of Biochemistry
by W.L. KRETOVICH

Chapter 12. Sergei E. Severin: Life and Scientific Activity
by S.E. SEVERIN

G. Semenza (Ed.) Selected Topics in the History of Biochemistry: Personal Recollections (Comprehensive Biochemistry Vol. 35)
© 1983 Elsevier Science Publishers

Chapter 1

The Isolation of Cori-ester, "the Saint Louis Gateway" to a First Approach of a Dynamic Formulation of Macromolecular Biosynthesis

HERMAN M. KALCKAR

Department of Chemistry, Boston University, 685 Commonwealth Ave, Boston, MA 02215 (U.S.A.)

The Coris in post-World War I Europe

Carl Cori was born in Prague in 1896, when Prague was part of the Austro-Hungarian Empire. He came from a background of university professors on both sides of the family. The grandfather on the maternal side, Ferdinand Lippich, was Professor of Theoretical Physics at the German University of Prague, a spirited mathematician and a skilled instrument builder. His father, Carl I. Cori, got his M.D. and Ph.D. in zoology in Prague at the Carl Ferdinand University, and in 1898 became the director of the Marine Biological Station in Trieste on the Adriatic Sea. Trieste was at that time part of the Austro-Hungarian Empire. Carl Cori Sr., was one of the leading zoologists and marine biologists in Central Europe. The famous Austrian zoologist Karl von Frisch in his memoirs recalls with appreciation the field trips on the boat "Adria" into the Adriatic Sea. Young Carl Cori grew up to appreciate the multicultural seaport of Trieste and the Adriatic as well as the Tyrolean Alps where the family spent summers. During World War I he studied medicine at the German branch of the Charles University in Prague. (This university also had a Czech branch.) Later young Cori served in the Medical Corps of the Austrian Army. Professor Carl Cori Sr. was at that time in charge of an extensive anti-malaria program.

Gerty T. Cori's maiden name was Gerty Theresa Radnitz. She, too, was born in Prague in 1896. Gerty passed the "Matura" (final examination) in record time and decided to study medicine, possibly influenced by an uncle who was professor of pediatrics in Prague. She met Carl Cori at the

Plate 1. Carl Ferdinand Cori and Gerty Theresa Radnitz Cori.

Medical University College in Prague where they both, being of Austrian background studied at the German University.

In the spring of 1920 both Carl Cori and Gerty Radnitz graduated as M.D.s and by summer they married. At that time World War I was finished, Austria had lost the war, and the Empire began to disintegrate. Prague had by now become the capital of the newly formed Czechoslovakia. The emphasis at the Carl Ferdinand University in Prague had become the training of young Czech doctors mainly as practitioners, for which there was a great need. The Coris had always entertained a warm interest for Czech culture and this most certainly included the founder of the young republic, Thomas Garrigue Masaryk. Years later they compared Thomas Masaryk with Abraham Lincoln. However, the newly born Czech nation had quite naturally too many acute needs to be able to promote research, at least for the next couple of years. The Coris found it therefore more realistic to try to keep their affiliations with the Austrian Medical Schools.

At the University Medical Clinic in Prague under the chairmanship of Professor R.v. Jaksch, it was possible to pursue clinical research. The first publication of the Coris dealt with complement in various diseases and appeared in an immunological journal [1].

Part of the year 1919 was spent in a no man's land, since the disintegration of the Austrian Empire developed gradually. In the early summer of 1919 Carl Cori studied under the spirited pharmacologist Professor Geza Mansfeld at the Hungarian Elisabeth University in Pozsony. In the fall of 1919 Pozsony was, however, being "transformed" into Bratislava as part of the new Republic of Czechoslovakia. This meant that the Mansfeld Institute had to be moved across the Danube farther south. Another member of the Institute who helped in this surreptitious transfer was Dr. Albert Szent-Györgyi. In spite of this nomadic existence a publication appeared in Pflügers Archiv on the excitability of the sinus nodes of the heart and the effect of adrenalin and vagus stimulation [2]. This study was probably largely an outgrowth of the Mansfeld work.

During most of the year 1921 the Coris were forced to work separately. Gerty Cori had obtained a position at "Karolinen Kinderspital" in Vienna. This was a pediatrics unit (led by Professor W. Knoepfelmacher) which gave an opportunity to conduct some research. Gerty Cori pursued problems concerning the importance of the thyroid gland for temperature regulation. The problems had been subject to discussions by pharmacologists such as Geza Mansfeld and Otto Loewi. Gerty Cori investigated temperature regulation in congenital myxoedema [3]. In 1921 Carl Cori was also able to find a position in Vienna at the Pharmacological Institute of the University of Vienna and at the University Clinic under Professor Hans Eppinger. The then retired Professor H.H. Meyer (of the Meyer–Overton theory of narcosis) encouraged Cori's interest in physiology and pharmacology. At that time Professor Meyer had a visit from Dr. Gaylord, the director of the State Institute for the Study of Malignant Diseases in Buffalo, NY, who was trying to recruit gifted young Central Europeans for his institute. Emphasis was on Austrians rather than on Germans because of the anti-German propaganda in the U.S.A. at the end of World War I. The Coris were interested in coming to America, and Dr. Gaylord was more interested in Carl Cori than Carl realized at that time.

Meanwhile Carl Cori spent some stimulating months in the fall of 1921 with Professor Otto Loewi at the Pharmacology Department of the University of Graz, Austria. Professor Loewi was at that time engaged in his pioneering studies of chemical transmission of nerve impulses for which he later was awarded the Nobel Prize. Nevertheless, conditions in the provincial city of Graz were unfavorable. This struck Gerty Cori in particular when she came on her visits from Vienna.

The Coris in the U.S.A.

In 1922 the Coris emigrated to the United States and took a position at the New York State Institute for the Study of Malignant Diseases in Buffalo, NY. The almost 10 years which the Coris spent at the Institute marked the beginning of their intense and fundamental research on carbohydrate metabolism. Their productivity over these years was remarkable, so much the more because they had to perform some service duties dealing with cancer diagnosis during the first couple of years. The first article about carbohydrate metabolism appeared in November of 1922 [4]. The observations on metabolic disturbances of milk sugar metabolism had mainly indirect importance in that they may have contributed to attract the interest of the Coris on galactose metabolism. In 1925, for instance, the Coris described the rapid absorption of galactose in the intestine, galactose being the most active sugar in this respect [5]. The transport of sugars into cells became later an important topic for the Coris, especially by an extension of their studies on insulin and tissue sugar which also originated from their early studies in Buffalo.

In 1924–25 the impact of Otto Warburg's discovery of the high glycolysis in tumors had reached the United States. The Coris were probably the first to recognize the importance of this discovery. Since they worked in a Cancer Institute and were interested in a perceptive approach to carbohydrate metabolism, it was not surprising that they studied the reports from Berlin (all in the German language) with intense interest.

Warburg's first description of the high aerobic and anaerobic glycolysis in tumors was based on in vitro studies utilizing a new, important technique developed in 1923. The Coris were the first to discover that tumors in the living animal also show an abnormally high formation of lactic acid from glucose, as reported in a series of articles in 1925 [6]. The last article in this series is cited in some detail by Warburg himself. It was thus established that high aerobic glycolysis is also a characteristic of tumors in vivo.

The successful contribution to the problem of tumor growth and lactic acid formation encouraged the Coris to focus more attention on the problem of lactic acid formation and the breakdown and resynthesis of carbohydrate in various tissues.

The Coris' first approach to the problem of the economy of lactic acid formation in the body was not like Meyerhof's, from the point of view of thermodynamics, but more from the angle of classical physiology and the growing field of endocrinology. The Coris studied the influence of the hormones insulin and epinephrine on the release of glucose or lactic acid from glycogen.

The Coris were among the first to demonstrate that epinephrine brought about an increase of lactic acid in muscle. The effects of epinephrine on lactic acid formation were first described in 1925 [7]. In the later study of 1927 [8], adrenalectomized animals were used, especially when insulin effects were subject to study. The many conflicting reports on the effect of these hormones remained uninterpretable without an additional concept. This concept later became known as the *Cori Cycle.* The Coris made it clear that the glycogen of the muscle is not available directly as blood sugar but that it is indirectly available. If epinephrine mobilizes muscle glycogen as well as liver glycogen and raises the blood sugar level, the contributions of the two organs are entirely different. During the years 1929 to 1931 studies by the Coris showed clearly that the action of epinephrine on muscle glycogen is a stimulation of breakdown via hexosemonophosphate to form lactic acid. They showed, furthermore, that lactic acid is an excellent precursor of glycogen in the liver. In a magnificent review by Carl Cori [9] a critical evaluation of previous studies on carbohydrate metabolism in the intact animal as well as the introduction of recent experiments and new concepts directed toward clarifying the topic were discussed in detail. This review is actually closer to a monograph and includes almost 500 bibliographic references.

One of the highlights of this review, based on the previous demonstration of the vigorous formation of glycogen in the liver from D-lactic acid [10], deals with the carbohydrate cycle in the intact animal as follows:

"The demonstration of glycogen formation in the liver from lactic acid establishes an important link between carbohydrate metabolism of liver and muscle. It shows that muscle glycogen is not "locked away" as Macleod once expressed it. Muscle glycogen can be converted to liver glycogen by way of blood lactic acid and thus become available as blood sugar. Since blood sugar is convertible to muscle glycogen, the glucose molecule is capable of a complete cycle in the animal body."

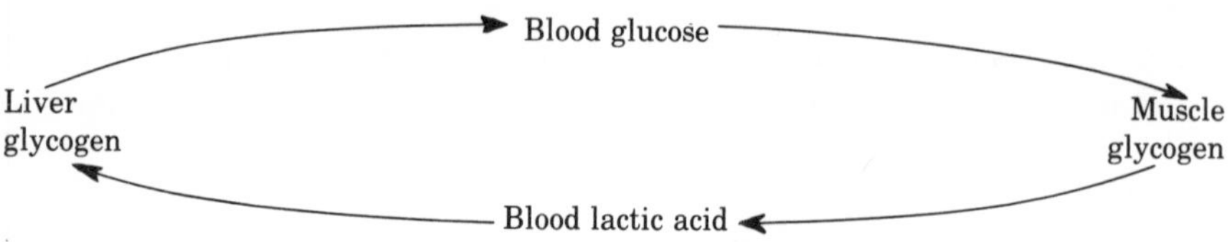

(Quotation and redrawn diagram from C.F. Cori, *Physiol. Rev.*, 11 (1931) 167.

This circulation of carbohydrate material in the animal organism has since been called the Cori cycle. Since epinephrine stimulates lactic acid formation from muscle glycogen as well as glucose formation from liver glycogen, it is not surprising that the previous literature was full of conflicting statements about the effect of that hormone on the levels of liver glycogen.

Research at Washington University, School of Medicine

In 1931 the Coris had developed "a method for the simultaneous determination of hexosemonophosphate (Embden ester) as hexose and as phosphorus in 1 g muscle" [11]. In 1931 Carl and Gerty Cori moved to St. Louis, MO, to be affiliated with Washington University, Medical School and, more precisely, the Department of Pharmacology.

Although the Coris started their St. Louis adventure in the Department of Pharmacology, they had become increasingly interested in biochemistry. The Coris had initiated studies on the phosphorylation of hexoses in isolated muscle. From 1931 to 1936 they were engaged in physiological, pharmacological and biochemical studies of the carbohydrate metabolism of muscle [12]. In accordance with their previous studies on intact animals, they found that physiological amounts of epinephrine stimulated the formation of hexosemonophosphate (Embden ester) from glycogen. This effect was much more pronounced under anaerobic conditions than under aerobic conditions. In the latter case the breakdown of glycogen is greatly decreased. They also found that the combined addition of epinephrine and 2,4-dinitrophenol to isolated muscle brought about a particularly large accumulation of hexosemonophosphate. These seemingly undramatic series of studies from the middle thirties constituted in fact the nucleus for later studies on regulation of metabolism by activation as well as feedback inhibition of metabolic enzymes. The Coris reported some of these studies on isolated muscle at the International Congress of Physiology in Leningrad, 1935.

The discovery of Cori ester and the concept of phosphorolysis

The historical experiment of 1936 [13] which led to the discovery of α-glucose-1-phosphate is worth describing somewhat in detail. Rarely have so few "recipes" meant so much for the future of biochemistry. Frog skeletal muscle was minced with scissors at 0°–2°C. The muscle dispersions were subsequently extracted with water and the "washed" insoluble dispersions were incubated at 20°C in phosphate buffer. Crystalline 5′-adenylic acid was added in varying amounts, increasing from 60 to 120 μg/g muscle [14]. The isolation of the new non-reducing phosphoric ester was accomplished from 400 g frog muscle dispersions, employing as much as 120 mg adenylic acid. The acid-labile "aldose-1-phosphate", suspected to be glucose-1-phosphate, was crystallized as the brucine salt in pure methyl alcohol.

The new hexose ester was the first example of an esterification between orthophosphate and the reducing group of a hexose. An important aspect of the new discovery was the realization that the formation of the 1-ester may be the result of a phosphorolytic split of glycogen.

In a monumental paper which appeared in the 1937 November issue of the *Journal of Biological Chemistry* [15] a full description of α-glucose-1-phosphate as a biologically highly important ester appeared for the first time. The 1-ester which accumulates in Mg^{2+}free muscle extracts in the presence of 5′-adenylic acid was shown upon 10 min acid hydrolysis at 100°C to liberate a hexose with a dextro rotatory power typical of D-glucose. In the same paper a bold and highly successful attempt to obtain the same ester by means of organic chemical synthesis was undertaken. The reducing group was made reactive through bromination; the hydroxyl groups were protected by acetyl groups. After proper identification of the 1-ester formed in muscle extracts in the presence of 5′-adenylic acid as α-glucose-1-phosphate, the question as to its biological origin had to be solved. Soon another development followed.

The fact that the Coris prepared their muscle preparations in the simplest possible way, merely by washing the preparation with chilled distilled water, turned out to be a lucky choice. A few hours incubation of the washed muscle "brei" in phosphate buffer (in the presence of 5′-adenylic acid, as stated) gave them first a chance to collect glucose-1-phosphate. Alternatively, by adding magnesium ions (Mg^{2+}) another enzyme in the muscle extract catalyzed the conversion of 1-ester to a very stable phosphoric ester which the Coris identified as glucose-6-phosphate.

The magnesium-requiring enzyme, catalyzing the conversion from 1-ester to 6-ester, the Coris named phosphoglucomutase [16]. The reaction from 1-ester to 6-ester, although reversible, proceeds greatly in favor of 6-ester formulation, hence the two-step approach was the only way in which some of the basic characteristics of the two enzymes could be disclosed. The second-step enzyme phosphoglucomutase requires magnesium. The first-step enzyme phosphorylase requires an adenine nucleotide, the most effective being 5′-AMP [17].

These experiments also implied that regulation of blood glucose in the liver, thought by prior investigations to be due to the direct hydrolytic action of diastases, may in fact be due to a phosphorolytic fission catalysed by the action of three enzymes, glycogen phosphorylase, phosphoglucomutase and a specific phosphatase, glucose-6-phosphatase in the liver.

In 1938 the Coris showed that the formation of glucose-6-phosphate and free glucose in dialysed liver extracts in the presence of 0.02 M phosphate and 0.001 M 5′-adenylic acid brought about a corresponding decrease of the glycogen levels.

This demonstration raised the highly important question of the feasibility of reversing the phosphorolysis in vitro. In other words, would it be possible to demonstrate glycogen biosynthesis in vitro by means of α-glucose-1-phosphate? This problem was indeed challenging. Physiologists had been told for generations that biosynthesis of macromolecules only takes place in the intact cell, depending on energy metabolism, preferentially respiration.

In collaboration with Dr. Gerhard Schmidt, who had just joined the Cori department in 1939, the first attempt to synthesize a polysaccharide in vitro was initiated [18] and turned out most promising.

Isolation of crystalline muscle phosphorylase catalyzing amylose synthesis

In order to study the biosynthesis of polysaccharides in vitro, it was necessary to try to characterize the nature of the polysaccharides which were formed from glucose-1-phosphate in the presence of the fractionated phosphorylase preparations.

The discovery of α-glucose-1-phosphate had aroused a good deal of excitement in wider circles. The plant biochemist Hanes was able to obtain starch synthesis in vitro with extracts from potato on glucose-1-phosphate as the glucosyl precursor [19]. In 1941 Richard Bear, an X-ray crystallographer at Ames, IA, joined forces with the Coris to study X-ray diagrams of the in vitro generated polysaccharide formed from 1-ester and muscle phosphorylase. Hanes, in collaboration with Astbury and Bell, had tried to obtain X-ray diagrams of naturally occurring amylose and amylodextrins formed from 1-ester incubated with potato preparations. Bear and Cori [20] found only blurred X-ray diagrams from polysaccharides synthesized by the crude liver enzyme. However, the X-ray diagrams of the long unbranched chains synthesized in vitro by muscle phosphorylase turned out to be very clear and strikingly reminiscent of those of plant amylose and amylodextrins.

The first bioengineering of macromolecular biosynthesis stems from the years 1942–1943. The Coris and Dr. Arda Green were able to report that they had been able to purify and crystallize phosphorylase from muscle [21]. By means of synthetic 1-ester made by the Coris and pure crystalline muscle phosphorylase obtained in large amounts, the Coris were able to obtain sufficiently large amounts of the peculiar starch-like polysaccharide to subject it to a chemical analysis along the lines of methylation, a method which is able to give information about the degree of branching in polysaccharides.

The complex enzymatic biosynthesis of the long maltosidic chains was subsequently subjected to a penetrating kinetic analysis. The monumental series of papers which the Coris published together with Arda Green in 1943 represent a feat of bioengineering entirely novel nearly forty years ago [22–24].

The mechanism of enzymatic synthesis of straight and branched polysaccharide was now ripe for a full-fledged investigation. The problem originated essentially from the old observations of 1939 by the Coris and Schmidt that polysaccharide synthesis catalysed by muscle phosphorylase is characterized by a long lag period and by the formation of starch-like polysaccharides that gave a blue color with iodine. In contrast, polysaccharide synthesis catalysed by preparations from liver and heart muscle shows no lag period and gives rise to synthesis of typical glycogen. The lag period characteristic of muscle phosphorylase can be broken either by addition of glycogen in appreciable amounts or by addition of small amounts of extracts of heart muscle or liver. The effect of the extracts could not be ascribed to the varying small traces of glycogen but was ascribed to the presence of another enzyme which might catalyse the formation of branch points (see later).

In the fourth article of the 1943 series of great papers [25], the Coris were indeed on the track of finding a "branching enzyme". By means of some ingenious yet simple experiments they were able to rule out effects from remnants of glycogen in the activating liver or heart muscle fractions. The activating effect of the fractionated extracts looked like an autocatalytical reaction, especially in the absence of any added glycogen. If heated extract was mixed with crystalline muscle phosphorylase *a* and incubated in the presence of 1-ester (5′-adenylate omitted), no reaction occurred; nor did native heart extracts by themselves bring about any polysaccharide synthesis under these conditions. If, however, muscle phosphorylase was permitted to react together with fresh heart extract, 80% of the 1-ester was converted to polysaccharide within one hour. The kinetics of this dramatic reaction was also remarkable in that it showed features pointing toward autocatalysis; a marked lag period of 10–20 min preceded the rapid reaction. The product of the combined action of the two enzymes, i.e., phosphorylase and "phosphorylase starter" was isolated and found to resemble glycogen in that it gave a red-brown iodine color and had activating power on crystalline phosphorylase preparations. This essentially enabled the Coris to formulate their concept of priming in glycogen synthesis.

The interesting question of sequence of action of phosphorylase and the supplementary enzyme was also studied. Amylose, whether natural or formed through the action of crystalline phosphorylase on 1-ester, was not converted to "activating glycogen" by subsequent incubation with heart extracts. Apparently the phosphorylase and the supplementary enzyme must act simultaneously in order to produce an "activating polysaccharide". They showed that branched types of polysaccharides ensue as a result of the action of the two enzymes.

The avalanche of original and fundamental observations of the 1943 publications touched off various new developments in the Cori laboratory which encompassed polymer research studies, enzyme mechanisms and the physiology of carbohydrate metabolism, or more specifically:

(1) The mechanism of enzymatic synthesis of straight and branched polysaccharides.
(2) The nature of conversion of phosphorylase *a* to *b* and the molecular differences between the two types.
(3) The correlation of muscle contraction and recovery and of the effect of hormones to the content of phosphorylase *a* and *b*.

Since 1950 several types of transglycosidation mechanisms have been described. The many different types observed illustrate the high specificity of the enzymes which catalyse these reactions. This high specificity, which is also characteristic of Cori's phosphorylase, came to play an important role in exploring in more depth the mode of action of this enzyme in relation to the polysaccharide.

Studies on the more precise formulation of priming activity of polysaccharides, especially those synthesized enzymatically, were continued in the Cori laboratory by S. Hestrin. Hestrin's extensive studies which were to be so decisive, too, for studies on human hereditary glycogen storage diseases, represented perhaps the most precise application of this approach [26]. The effect of β-amylase and crystalline phosphorylase (in the presence of the excess of phosphate) on enzymatically synthesized polysaccharide was investigated. Two types of dextrin were formed, called "limit dextrins" or "LD". The "LD (amylase)" was a result of digestion of the polysaccharide by splitting off maltosidic units by β-amylase. It whittled down the main chain to one glucose unit distal to the 1,6-glucosidic branch point. The "LD (phosphorylase)" is the limit dextrin formed as a result of exhaustive incubation with phosphorylase. The LD (phosphorylase) contain 2 maltose units distal to the branch point.

Hestrin's formulation of the equilibrium of the phosphorylase reaction as a sum of partial reactions explained why glycogen, within limits of chain length, did not influence the equilibrium of the reaction. One such limit was the LD (phosphorylase) which as Hestrin had shown could only act as a recipient of glucosyl units. *Hence, unlike the other glycogens or amylopectins the concentration of LD (phosphorylase) influences the equilibrium.*

In 1951, G.T. Cori and J. Larner described a new enzyme, "the debrancher" or amylo-1,6-glucosidase [27]. This enzyme which was found in muscle, catalyses the hydrolysis of the 1,6-glucosyl linkages at the branch points. The hydrolysis of the branch points makes it possible for phosphorylase in the presence of excess phosphate (and low concentrations of 1-ester) to split glycogen almost completely.

The branching and the topography of glycogen had been subject to much speculation among chemists. Three models had been proposed. A decision among them could not be reached by analyses using β-amylase since the latter stops at the branch point. The other amylase, α-amylase, splits 1.4 linkages too much at random to be of any use. By 1951 the Coris had the answer: a combination of a splitting of 1.4 linkages by phosphorylase, releasing 1-ester, and a splitting of 1.6 linkages by the glucosidase (the debrancher) to release one molecule of free glucose per branch point, would give a portrait of the topography. *In this way it was found that the number of branch points decreases gradually as glycogen is broken down* [28]. This observation lent strong support to the Meyer model of the tree-like branched structure for glycogen. Further support for this model could be adduced from the early observations by the Coris on the autocatalytic effect of the branching factor on the enzymatic synthesis of glycogen, indicating an *increase of branch points with synthesis.*

In 1951 Leloir isolated and described UDP-glucose [29], a nucleotide which is actually a derivative of Cori-ester, since it is a uridylic derivative of α-glucose-1-phosphate through a pyrophosphate linkage between the phosphate of the 1-ester and that of uridylic acid. UDP-glucose is a more powerful glucosyl donor than the 1-ester, presumably on account of the pyrophosphate linkage to 5′-uridylic acid. Otherwise, as shown by Leloir, the glucosyl group from UDP-glucose is merely forming new terminal maltosidic linkages much in line with the mechanism described by the Coris, i.e. glycogen biosynthesis from UDP-glucose needs, besides the Leloir enzyme, also the branching enzyme of Larner and Cori. The enzyme, UDP-glucose-glycogen synthetase, is clearly different from phosphorylase.

The Coris were successful in using the high specificity phosphorylase in an analysis of certain abnormal polysaccharide structures. Gerty Cori and her coworkers were for instance able to provide a biochemical description of various types of glycogen storage disease in children. Several cases of glycogen storage disease were disclosed to be due to lack of the debrancher enzyme, amylo-1,6-glucosidase. In accordance with this observation, it was found that the glycogen which accumulated had a significantly higher content of end groups, reminiscent of limit dextrin rather than of normal glycogen [30]. It seemed that in this disease glycogen synthesized by UDP-glucose synthetase and branching enzyme could be utilized only as far as the outer branch points.

Another type of glycogen storage disease was found to accumulate starch-like material, amylopectin, in the liver. The polysaccharide was sparingly soluble in cold water and the iodine color was purple. Enzymatic structure analysis was in full accord with these findings in that 1-4 maltosidic chains were found to be longer than in normal glycogen, the values being close to those found for amylopectins. The changes could be explained by a decrease of the "brancher enzyme", 1,4→1,6-transglucosidase (since shown to be the case).

The Coris also discovered the enzymatic defect in the glycogen storage disease called von Gierke's disease. They found that the characteristic enzyme of liver which specifically catalyses the release of glucose from glucose-6-phosphate, the G-6-P phosphatase, is very low or absent in this disease. Consequently, glycogen cannot be broken down to blood glucose which tends to be very low in this disease. 15 years later Carl Cori found this enzyme defect in certain radiation-induced albino mutations in mice described by Dr. S. Glueckson-Waelsch (see later).

Phosphorylases *a* and *b* and their role in metabolic regulation

The discovery of two forms of phosphorylase, "*a*" and "*b*", the latter requiring 5′-adenylic acid to act as a catalyst, was a forerunner of many important developments. Many of these developments were also initiated by the Coris and their coworkers who probed the following questions:

(1) Biochemical characterization of phosphorylase *a* and *b* and the nature of the conversion of *a* to *b*.
(2) Cellular and hormonal regulation of the conversion of phosphorylase *a* to *b*, especially as related to muscular contraction and the action of epinephrine.

In 1956 Madsen and Cori [31] found a derivative of phosphorylase with only one-fourth the molecular weight of phosphorylase *a*. This derivative, which is catalytically inactive, was formed by treatment of phosphorylase *a* with *p*-chloromercuric benzoate (pCMB). The details of this conversion are of great interest. Phosphorylase *a* has 18 available SH groups. If they are titrated with pCMB, catalytic activity disappears. The inactive pCMB derivative was found to have an M_r value of about 100 000 to 125 000, i.e., one-fourth that of phosphorylase *a*.

In 1957 Dr. C.F. Cori and coworkers [32] found that phosphorylase *a* contains 4 mol of pyridoxal-5-phosphate per mol enzyme. Phosphorylase *b* contains 2 mol of pyridoxal-5-phosphate. Fischer et al. [33] have found that phosphorylase *b* treated with sodium borohydride, reducing the aldamine bond of the bound pyridoxal phosphate [34] still remains almost fully active. Precipitation of phosphorylase *a* or *b* with ammonium sulfate at pH 3.5 releases pyridoxal-5-phosphate [35]. The apoenzyme which can be crystallized is catalytically inactive but can be reactivated by addition of pyridoxal phosphate. It is interesting to note that addition of unphosphorylated pyridoxal or 5-deoxypyridoxal will only stabilize the formation of a tetramer but without restoring catalytical activity.

The chemical relation between phosphorylase *a* and *b* has been clarified further through the studies by Krebs and Fischer and their coworkers.

Regulation of phosphorylase in resting and contracting muscle

The work in the late fifties by the Coris' former pupils, Earl Sutherland and Edwin Krebs, and Krebs' later coworker Edmund Fischer, brought to light the existence of regulator enzymes which constitute a whole network of cascade regulators. This line of approach stems originally from some simple observations made by Sutherland and Cori [36] back in 1951. They found that epinephrine plays a crucial role in converting inactive forms of phosphorylase into active forms. This was found to be the case in liver as well as in muscle.

In 1962 Carl Cori and coworkers [40] conclusively demonstrated that the conversion of phosphorylase *b* to *a* is a crucial reaction for the rapid breakdown of glycogen during muscular contraction and following epinephrine administration. During a span of almost 40 years the Cori laboratory had never lost interest in the mechanism of regulation.

In 1945 G.T. Cori had already studied the enzymatic conversion of phosphorylase *a* to *b*. One aspect of this study was the question of the possible functional importance of this conversion for regulation [37]. The problem of turning on and off glycogen breakdown may be approached through a study of the distribution of the two types of phosphorylases, of which *b* is only active if 5′-adenylic acid is present. The outcome of the first attempt to study the distribution of the two phosphorylases during contraction was not too clear. This might be attributed to the fact that the fixation of the muscle was not sufficiently rapid and hence gave rise to strong contractions with concomitant lactic formation.

In 1955 Krebs and Fischer [38] discovered that resting muscle contains largely phosphorylase *b*. They greatly clarified the issue by studying the conversion of phosphorylase *b* to *a* in dialyzed muscle extracts [39]. It was found that the conversion of phosphorylase *b* to *a* required calcium besides magnesium ions and ATP. These studies paved the way for a meaningful attack on the problem of regulation of phosphorylase during muscular contraction.

In the 1962 study from the Cori laboratory [40], it was realized that very special precautions had to be taken in order to obtain a portrait of the situation existing in a resting muscle and during the development of contraction. The entire muscle must be frozen instantaneously and this can be accomplished only by using thin and flat muscles such as the frog's sartorius. The fixation technique was further improved by the use of isopentane at –160°C. In this way the increase in the amount of phosphorylase *a* during an isometric contraction could be measured.

Cori also found that epinephrine in a concentration of 1×10^{-6} M also stimulates the conversion of phosphorylase *b* to *a* in the intact resting muscle. The epinephrine analogue dichloroisoproterenol is a competitive inhibitor which in a concentration of 1×10^{-4} M is able to block epinephrine action but not the action of electric stimulation. Electric stimulation depends directly on the mobilization of calcium ions which activate a specific enzyme, phosphorylase *b* kinase [41], whereas the epinephrine action is mediated by the enzymatic formation of cyclic 3′, 5′-adenylic acid. In this connection it is important to note that Sutherland and coworkers found that dichloroisoproterenol also inhibits the effect of epinephrine on cyclic 3′, 5′-adenylic acid in heart muscle [42].

It is known that the rate of breakdown of glycogen may increase several hundred times during a short tetanus and subsequently reach resting levels during relaxation. *The elegant kinetic analysis of the formation of phosphorylase* a *from* b *indicated that this conversion may actually exert a decisive regulatory control in muscle.*

1947 Nobel Prize for Physiology and Medicine

In December 1947 Carl and Gerty Cori were awarded the 1947 Nobel Prize for Physiology and Medicine "for their discovery of the course of the catalytic conversion of glycogen", as cited from Professor Hugo Theorell's presentation [43] of the prize to the famous couple. Gerty Cori became the third woman (after Mme. Marie Curie and Mme. Irene Joliot Curie) and the first American woman to receive the Nobel Prize. They shared the Nobel honors with Dr. Bernardo Houssay of Argentina, who was awarded the 1947 prize for Physiology and Medicine "for his discovery of the importance of the anterior pituitary hormone for the metabolism of sugar". Dr. Carl Cori replied on behalf of himself and his wife.

"It is our belief that art and sciences can best grow in a society which cherishes freedom and which shows respect for the needs, the happiness, and the dignity of human beings. My wife and I are proud to have been honored by a country that excels in all these qualities, and we are happy to be guests in this beautiful and hospitable city" [43].

The effect of hormones on carbohydrate metabolism in vitro

In the Nobel citation by Professor Theorell another aspect of the work closely related to Dr. Bernardo Houssay's work on the anterior pituitary hormone was mentioned. In 1945–46 the Coris had investigated the rate of hexokinase (glucose phosphorylation by ATP giving glucose-6-phosphate) in crude muscle extracts from diabetic animals and found that the rate of muscle hexokinase could be stimulated by addition of insulin. The Cori group reported that crude muscle extract preparations from alloxan diabetic rats showed a distinct decrease of hexokinase activity as compared with preparations from normal rats and that this activity could be stimulated by addition of 50 μg of insulin [44]. Even more convincing were studies in which small amounts of an amorphous fraction from the adrenal cortex were used as an inhibitor of hexokinase [44, 45]. This type of inhibition could also be overcome by addition of insulin in the amount used on the preparations of hexokinase from diabetic rats.

The studies on the effect of hormones on muscle hexokinase were difficult to reproduce in that only about one-third of the experiments showed a distinct hormone effect. The stimulation by insulin was particularly erratic. The value of these bold attempts to study hormone effects on the molecular basis did not fail, however, to bear fruit in several directions. One direction was towards the molecular level, a lead which commenced with the above-mentioned observations on the effect of epinephrine on glycogen breakdown by Sutherland and Cori in 1951 and culminated with the work by Krebs and Fischer and by Sutherland on the phosphorylase kinase regulation system. The other direction remained more confined to physiological studies on cell membranes of intact organs. Yet in many ways these studies were also sophisticated and they deserve brief mentioning.

Already during 1947, Carl Cori in collaboration with Michael Krahl decided to study glucose uptake in alloxan-diabetic rats by using the Gemmill technique of measuring glucose uptake by the isolated intact diaphragm. This technique formed the basis for a series of important studies from the Cori laboratory. Krahl and Cori [46] were able to show that the glucose uptake of diaphragms from diabetic rats is greatly stimulated by addition of insulin. Moreover, the rate of glucose utilization of diaphragms from diabetic rats is lower than that from normal rats. Many of these observations were confirmed and extended by Park also working in the Cori laboratory. Park and Krahl also succeeded in finding an increased rate of glucose uptake in diaphragms from hypophysectomized animals and in inhibition by pituitary extracts [47].

Around 1949–1950 Rachmiel Levine and his group published a series of papers on the mode of action of insulin. In these papers Levine put the spotlight on the cell membrane as the target for the action of insulin [48]. The somewhat complicated preparation used (eviscerated and nephrectomized animals) was largely composed of skeletal muscle. Galactose was used instead of glucose since galactose is not phosphorylated or metabolized in other ways by this tissue. Hence changes in the distribution between blood and muscle brought about by insulin would have to be attributed solely to an effect on the transport through the membrane of the muscle cell. Levine and coworkers were able to show that injection of insulin (1 unit/kg) markedly lowered the final stationary blood level of galactose as compared with distribution experiments without insulin. This was a strong indication that more galactose had entered into the skeletal muscle in the presence of insulin than in its absence. Subsequently the Levine group conducted investigations on the action of insulin on the transfer of a variety of sugars across cell barriers [49]. Sugars related to galactose and glucose with respect to the first 3 carbons such as L-arabinose or D-xylose responded to insulin the same way as galactose did, i.e., the uptake into muscle was stimulated by insulin. This stimulation was absent if sugars like D-arabinose or L-rhamnose were used.

During the late fifties Carl Cori and his coworkers resumed their work on carbohydrate metabolism of the isolated diaphragm. In a series of papers on tissue permeability, paper number III dealt with the effect of insulin on the pentose uptake by the diaphragm [50]. In accordance with Levine and his group it was shown directly by determination of the intracellular levels of D-xylose that insulin stimulated the penetration of this sugar into the diaphragm. The uptake of 2-deoxyglucose was also strikingly stimulated by insulin [51].

The most recent developments tend to favor the idea that hormones like insulin and epinephrine act on membrane proteins as well as on enzyme proteins. Insulin, for instance, stimulates the uptake of various sugars as well as amino acids into a variety of mammalian cells. It also stimulates various metabolic pathways and among them also certain types of hexokinases (ATP-glucokinases) [52, 53].

Personal data from later years

In 1956 a special tribute was paid the Coris in the form of an anniversary volume entitled *Enzymes and Metabolism: A Collection of Papers Dedicated to Carl F. and Gerty T. Cori on the Occasion of their 60th Birthday* as a special issue of *Biochimica Biophysica Acta.* Nobel prize winner Bernardo Houssay of Buenos Aires wrote an introduction.

> "A vast general culture and a thorough training in Medicine, Physiology and Pharmacology gave them an integrated view of relations between the chemical changes studied *in vitro* with the behavior of isolated tissues and the entire organism under normal or pathological conditions. Their discoveries have permanent historical importance and have led to fundamental advances in our knowledge of cell physiology" [54].

The volume contained many important articles. Papers on the phosphorylase *a* and *b* conversion by Sutherland and his group and on the conversion of *b* to *a* by Krebs and Fischer were included. Arda Green and William McElroy published their elegant work on crystalline firefly luciferase. Leloir and Cardini added new knowledge to the metabolism of glucosamine phosphate. Arthur Kornberg, Sylvia R. Kornberg and E.S. Simms, at that time working in the Microbiology Department at Washington University Medical School, described an interesting case of metaphosphate synthesis in *E. coli.* Ochoa and coworkers Marianne Grunberg-Manago and P.J. Ortiz, described the enzyme polynucleotide phosphorylase, the discovery of which sparked further work on biosynthesis of nucleic acids in vitro. It is well known that Leloir, Kornberg and Ochoa discovered the decisive pathways for biosyntheses of macromolecules in the cell. In this volume, together with Kurt Isselbacher and Elizabeth Anderson, I had the opportunity to describe the nature of an inborn error, galactosemia, involving a disturbance in UDP-galactose formation. This was a topic close to the Coris' interests at that time.

Besides Dr. Houssay's introduction, the articles, and a listing of all the great accomplishments stemming from 30 years of research from the Cori laboratory, the anniversary volume also contains a quotation from the discoverer of DNA, Friedrich Miescher, to the Swiss physiologist, W. Hiss, about the importance of phosphoric acid. It read:

> "I cannot help regarding the chemical dynamics of phosphoric acid towards water, bases and protein as one of the most promising keyholes through which it should be possible to peep into the interior. There one finds those easy transitions between salt-like and ester-like bonds on which everything ultimately depends" (Vogel, 1978).

This citation was indeed a fitting tribute to the accomplishments of the Cori laboratory at Washington University. In 1936 the Coris provided the first substantial demonstration that phosphorolysis may be a process of great importance in biology by the discovery of α-glucose-1-phosphate formation from glycogen in muscle homogenates. About 15 years later a member of the Cori laboratory, Dr. Mildred Cohn, synthesized ^{18}O-labelled phosphate and showed by the strictest scientific criteria in existence that phosphorylase does indeed catalyse a genuine phosphorolysis of glycogen [55].

About the same time an important Ph.D. thesis appeared from the Cori laboratory, sponsored by one of the members of the department, Dr. Sidney Velick. The Ph.D. candidate, Jane Harting, had observed the formation of acetyl phosphate from acetaldehyde and inorganic phosphate catalysed by an enzyme from muscle, glyceraldehyde-3-phosphate dehydrogenase (first crystallized in the Cori laboratory). Harting and Velick as well as Racker first formulated a generalized scheme of phosphorolysis as a part of an oxidation-reduction process. The forerunners of this scheme were Warburg and Negelein, Lipmann and Lynen. Nevertheless the discovery of the formation of acetylphosphate from acetaldehyde catalysed by glyceraldehyde dehydrogenase and the demonstration of the importance of the sulfhydryl groups of the enzyme for this formation was decisive for a modern formulation of a coupling of phosphorylation with an oxidation-reduction. Harting and Velick [56] followed essentially the formulation used by E.R. Stadtman in the description of the enzyme phosphotransacetylase. In this simple way the groundwork was laid for another type of phosphorolytic split, in this case, of an acyl-mercapto linkage. Thus, the Cori laboratory gave birth to several types of phosphorolytic reactions and may well have served as a "priming center" for additional related discoveries. Three former members of the Cori department had the opportunity to discover related types of "dry" fissions [57–59]. The third member [59] and a fourth member [60] described reactions which were decisive biosynthetic reactions, important for the formation of macromolecules.

It was fortunate that Gerty Cori lived to see this anniversary volume. For almost 10 years she had courageously fought a chronic anemia which finally took her life.

Gerty Cori died in the autumn of 1957. Her last paper from the summer of that year dealt, like her first paper of 1921, with problems related to pediatrics and medicine. Her first paper originated from a pediatrics unit and her last paper of 1957 dealt with their work on glycogen deposition diseases in children and appeared in an international pediatrics journal, *Modern Problems in Pediatrics* [61].

Dr. Bernardo Houssay's memorial speech for Gerty Cori stressed her devotion to the medical sciences by these words:

> "Gerty Cori's life was a noble example of dedication to an ideal, to the advancement of science for the benefit of humanity."

This ideal went beyond medicine to the arts and sciences, as also stressed by Dr. Houssay:

> "The Coris have not only carried out personal work of extraordinary originality and significance, they have also inspired and directed one of the most active centers of biochemical research. Their laboratory became the point of attraction of all workers interested in carbohydrate metabolism, and more than sixty first class scientists have published papers which are the result of the stimulating atmosphere of the St. Louis laboratories. One of my former associates, Luis F. Leloir worked for several months with them."*

Dr. Severo Ochoa and I wrote memorial words in *Science* [62]. Several books and articles in biochemistry were dedicated to her memory; one might in particular mention Fruton and Simmond's textbook in General Biochemistry, Second Edition, 1958.

Both Gerty and Carl Cori took a deep interest in science, education, art

---

*B. Houssay, a personal communication to the Cori family and their friends, 1957.

and problems of a free society. Gerty Cori expressed her belief in imaginative science in the following forceful words:

"I believe that in art and science are the glories of the human mind. I see no conflict between them. In the past they flourished together during the great and happy periods of history and those men seem to me shortsighted who think that by suppressing science they will release other creative qualities" [63].

It is impossible to list all the honors which were awarded the Coris besides the Nobel Prize. They were both members of the National Academy of Sciences as well as the Philosophical Society, Benjamin Franklin's illustrious society in Philadelphia. They were awarded honorary degrees from many universities. Among the many distinguished honorary degrees one might cite that presented to Dr. Carl Cori at Cambridge University at the time of the First International Congress of Biochemistry in 1949. The *Doctor Honoris Causa* was cited in Latin and he was presented in Latin to the University of Cambridge and a large international assembly of biochemists — Duco ad vos Biochemiae apud Washingtonienses Professorem Carolum Ferdinandum Cori.

Dr. Carl Cori is also a Foreign Member of the Royal Society, London, and he became a Foreign Member of the Danish Royal Society during the time when Niels Bohr was still president of that society. Among the distinguished honors awarded Dr. Cori from chemists I might cite the Willard Gibbs Medal of the American Chemical Society.

In spite of his reservations about getting involved in a large number of administrative committees, Carl Cori served on a number of important committees. During World War II, he was responsible for a laboratory under contract with the Office of Scientific Research and Development (OSRD), which was concerned with the action of toxic gases on enzyme systems. He has served on the Committee of Overseers at Harvard University, on the Scientific Advisory Committee of the Massachusetts General Hospital and on the Helen Hay Whitney Foundation, Scientific Advisory Council. Gerty Cori served on the first advisory board of the National Science Foundation.

Upon his retirement in 1966, Carl Cori became Visiting Professor of Biological Chemistry, Harvard Medical School and Massachusetts General Hospital. He is more active than ever in research. In 1967 an interesting paper dealing with problems of allosteric interaction of substrates on phosphorylase *a* appeared [64]. This was the last paper from Washington

University Medical School. The first paper to appear from Massachusetts General Hospital and Harvard Medical School was an interesting joint enterprise with Dr. Salome Gluecksohn-Waelsch of the Department of Genetics, Albert Einstein College of Medicine in New York [65]. In this work they discovered a complex type of a highly interesting case of hereditary deficiency in glucose-6-phosphatase in mice. It will be recalled that Gerty and Carl Cori discovered a hereditary defect of this enzyme in man 15 years earlier.

Carl Cori has shown equal enthusiastic activity in teaching. He has given many seminars at the Massachusetts General Hospital, Harvard Medical School and lectures at Harvard University, the Weizmann Institute in Israel and many other places. In the summer of 1967 he was awarded the Walker Ames Professorship as a Visiting Lecturer in the Department of Biochemistry (Dr. Hans Neurath, Chairman), University of Washington, Seattle, WA.

The son of Carl and Gerty, Thomas C. Cori, born in 1936, is the President and Chief Operating Officer of the Sigma Chemical Company. Carl Cori and his wife Anne, who were married in 1960, live an active and interesting life in Cambridge and Ipswich, MA, where they pursue their interests in arts, humanities and gardening.

Acknowledgments

I want to express my gratitude to Dr. Carl Cori for his generous and helpful interest in this essay and his useful suggestions.

The collection of publications from the Cori laboratory over more than 40 years which was brought by Dr. Carl Cori to the Massachusetts General Hospital and which constituted the most unique nucleus of the Gerty T. Cori library was an important factor in gathering the many relevant facts together for this essay.

REFERENCES

1 K. Cori and G. Radnitz, Z.f. Immunitätsforsch., 29 (1920) 445–462.
2 K. Cori, Pflügers Arch. Gesamt. Physiol., 184 (1920) 272–280.
3 G.T. Cori, Z. Gesamt. Exp. Med., 25 (1921) 150–169.
4 G.W. Pucher and K.F. Cori, J. Biol. Chem., 54 (1922) 567–578.
5 C.F. Cori, J. Biol. Chem., 66 (1925) 691–715.
6 C.F. Cori and G.T. Cori, Proc. Soc. Exp. Biol. Med., 22 (1925) 254–255; C.F. Cori and G.T. Cori, J. Biol. Chem., 64 (1925) 11–12; 65 (1925) 397–405.
7 C.F. Cori, J. Biol. Chem., 63 (1925) 253–268.
8 C.F. Cori and G.T. Cori, J. Biol. Chem., 74 (1927) 473–494.
9 C.F. Cori, Physiol. Rev., 11 (1931) 143–275.
10 C.F. Cori and G.T. Cori, J. Biol. Chem., 81 (1929) 389–403.
11 G.T. Cori and C.F. Cori, J. Biol. Chem., 94 (1931) 561–579.
12 See, for instance, A.H. Hegnauer and G.T. Cori, J. Biol. Chem., 105 (1934) 691–703.
13 C.F. Cori and G.T. Cori, Proc. Soc. Exp. Biol. Med., 34 (1936) 702–705.
14 G.T. Cori and C.F. Cori, Proc. Soc. Exp. Biol. Med., 36 (1937) 119–122.
15 C.F. Cori, S.P. Colowick and G.T. Cori, J. Biol. Chem., 121 (1937) 465–477.
16 G.T. Cori, S.P. Colowick and C.F. Cori, J. Biol. Chem., 124 (1938) 543–555.
17 G.T. Cori, S.P. Colowick and C.F. Cori, J. Biol. Chem., 123 (1938) 381–389.
18 C.F. Cori, G. Schmidt and G.T. Cori, Science, 89 (1939) 464–468.
19 C.S. Hanes, Proc. Roy. Soc. London, B, 129 (1940) 174–207.
20 R.S. Bear and C.F. Cori, J. Biol. Chem., 140 (1941) 111.
21 A.A. Green, G.T. Cori and C.F. Cori, J. Biol. Chem., 142 (1942) 447–448.
22 A.A. Green and G.T. Cori, J. Biol. Chem., 151 (1943) 21–29.
23 G.T. Cori and A.A. Green, J. Biol. Chem., 151 (1943) 31–38.
24 C.F. Cori, G.T. Cori and A.A. Green, J. Biol. Chem., 151 (1943) 39–55.
25 G.T. Cori and C.F. Cori, J. Biol. Chem., 151 (1943) 57–63.
26 S. Hestrin, J. Biol. Chem., 179 (1949) 943–955.
27 G.T. Cori and J. Larner, J. Biol. Chem., 188 (1951) 17–29.
28 J. Larner, B. Illingworth, G.T. Cori and C.F. Cori, J. Biol. Chem., 199 (1952) 641–651.
29 L.F. Leloir, in W.D. McElroy and B. Glass (Eds.), A Symposium on Phosphorus Metabolism, Johns Hopkins Press, Baltimore, 1951, pp. 67–93.
30 G.T. Cori, in The Harvey Lectures, Series 48, Academic Press, New York, 1953, pp. 145–171.
31 N.B. Madsen and C.F. Cori, J. Biol. Chem., 223 (1956) 1055–1065.
32 T. Baranowski, B. Illingworth, D.H. Brown and C.F. Cori, Biochim. Biophys. Acta, 25 (1957) 16–21.
33 E.H. Fischer, A.B. Kent, E.R. Snyder and E.G. Krebs, J. Am. Chem. Soc., 80 (1958) 2906–2907.
34 D.H. Brown and C.F. Cori, in P.D. Boyer, H.A. Lardy and K. Myrbäck (Eds.), The Enzymes, Vol. 5, Academic Press, New York, 1961, pp. 207–228.
35 C.F. Cori and B. Illingworth, Proc. Natl. Acad. Sci. USA, 43 (1957) 547–552.
36 E.W. Sutherland and C.F. Cori, J. Biol. Chem., 188 (1951) 531–543. E.W. Sutherland, in W.D. McElroy and B. Glass (Eds.), Phosphorus Metabolism, Johns Hopkins Press, Baltimore, 1951, pp. 53–61.
37 G.T. Cori, J. Biol. Chem., 158 (1945) 333–339.

38 E.G. Krebs and E.H. Fischer, J. Biol. Chem., 216 (1955) 113–120.
39 E.H. Fischer and E.G. Krebs, J. Biol. Chem., 216 (1955) 121–132.
40 W.H. Danforth, E. Helmreich and C.F. Cori, Proc. Natl. Acad Sci. USA, 48 (1962) 1191–1199.
41 E.G. Krebs, D.J. Graves and E.H. Fischer, J. Biol. Chem., 234 (1959) 2867–2873.
42 F. Murad, T.-M. Chi, T.W. Rall and E.W. Sutherland, J. Biol. Chem., 237 (1962) 1233–1238.
43 Presentations of the 1947 Nobel Prize for Physiology and Medicine by Professor Hugo Theorell (Biochemical Department of the Karolinska Institute). Les Prix Nobel en 1947. Stockholm Imprimerie Royale, Norstedt, Stockholm 1949.
44 C.F. Cori, The Harvey Lectures, Series 41 (1945–1946) 253–272. (Lecture delivered May 16, 1946).
45 S.P. Colowick, G.T. Cori and M.W. Slein, J. Biol. Chem., 168 (1947) 583–596.
46 M.E. Krahl and C.F. Cori, J. Biol. Chem., 170 (1947) 607–618.
47 C.R. Park and M.E. Krahl, J. Biol. Chem., 181 (1949) 247–254.
48 R. Levine, M. Goldstein, S. Klein and B. Huddleston, J. Biol. Chem., 179 (1949) 985–986.
49 M.S. Goldstein, W.L. Henry, B. Huddleston and R. Levine, Am. J. Physiol., 173 (1953) 207–216.
50 D.M. Kipnis and C.F. Cori, J. Biol. Chem., 224 (1957) 681–693.
51 D.M. Kipnis and C.F. Cori, J. Biol. Chem., 234 (1959) 171–177.
52 C. Sharma, R. Manjeswar and S. Weinhouse, J. Biol. Chem., 238 (1963) 3840–3845.
53 H.M. Katzen and R.T. Schimke, Proc. Natl. Acad. Sci. USA 54 (1965) 1218–1225.
54 B.A. Houssay, Biochim. Biophys. Acta, 20 (1956) 11–15.
55 M. Cohn, J. Biol. Chem., 180 (1949) 771–781.
56 J. Harting and S.F. Velick, J. Biol. Chem., 207 (1954) 857–865.
57 M.E. Friedkin and H.M. Kalckar, in P.D. Boyer, H. Lardy and K. Myrbäck (Eds.), The Enzymes, Vol. 5, Ch. 15, Academic Press, New York, 1961, pp. 237–255.
58 M. Grunberg-Manago and S. Ochoa, J. Am. Chem. Soc., 77 (1955) 3165–3166.
59 A. Kornberg, I.R. Lehman, M.J. Bessman and E.S. Simms, Biochim. Biophys. Acta, 21 (1956) 197–198.
60 L.F. Leloir and C.E. Cardini, in P.D. Boyer, H. Lardy and K. Myrbäck (Eds.), The Enzymes, Vol. 6, Ch. 19, Academic Press, New York, 1962, pp. 317–325.
61 G.T. Cori, Modern Probl. Pediat., 3 (1957) 344–358.
62 S. Ochoa and H.M. Kalckar, Science, 128 (1956) 16–17.
63 G.T. Cori, This I Believe, Collected Interviews (E.R. Murrow, Ed.).
64 E. Helmreich, M.C. Michaelides and C.F. Cori, Biochemistry, 6 (1967) 3695–3710.
65 R.P. Erickson, S. Gluecksohn-Waelsch and C.F. Cori, Proc. Natl. Acad. Sci. USA, 59 (1968) 437–444.

G. Semenza (Ed.) Selected Topics in the History of Biochemistry: Personal Recollections (Comprehensive Biochemistry Vol. 35)

Chapter 2

The Discovery of Sugar Nucleotides

LUIS F. LELOIR and ALEJANDRO C. PALADINI*

*Instituto de Investigaciones Bioquímicas "Fundación Campomar" y Facultad de Ciencias Exactas y Naturales, Universidad de Buenos Aires, and *Departamento de Química Biológica, Facultad de Farmacia y Bioquímica, Universidad de Buenos Aires, Buenos Aires (Argentina)*

The discovery of sugar nucleotides

Leloir: Partial accounts of various aspects of the research of our group in Buenos Aires have appeared in different articles [1–3]. For Dr. Semenza's collection I thought it might be a good idea to check my statements so as to correct the failures of my memory. For this purpose I asked Dr. Alejandro Paladini, who worked in our group from 1947 to 1951, to give his account at the same time as mine. Dr. Paladini joined the Foundation with a Fellowship. We had to select between several candidates, but there were two who were rather outstanding. One was Dr. Paladini who had some experience and was known to be a promising young man, the other candidate was a girl of great beauty. It was a rather painful decision but I think I made the correct one.

In 1947 the Fundación Campomar was still in its infancy – a rich textile manufacturer, Jaime Campomar, had decided to create a research laboratory and he entrusted the organization and the selection of personnel to me. The first investigator to join the Foundation was Ranwel Caputto who had recently returned from the Biochemical Laboratory of Cambridge, the director of which was Sir Frederick Gowland Hopkins, one of the founders of British biochemistry. A few years before I had also received my initial biochemical training in that same laboratory. Caputto had worked with a well known biochemist, Malcolm Dixon and had succeeded in crystallizing a glyceraldehyde dehydrogenase which turned out to be the same as glyceraldehyde phosphate dehydrogenase. At that time it was quite a feat to crystallize an enzyme. After Caputto several

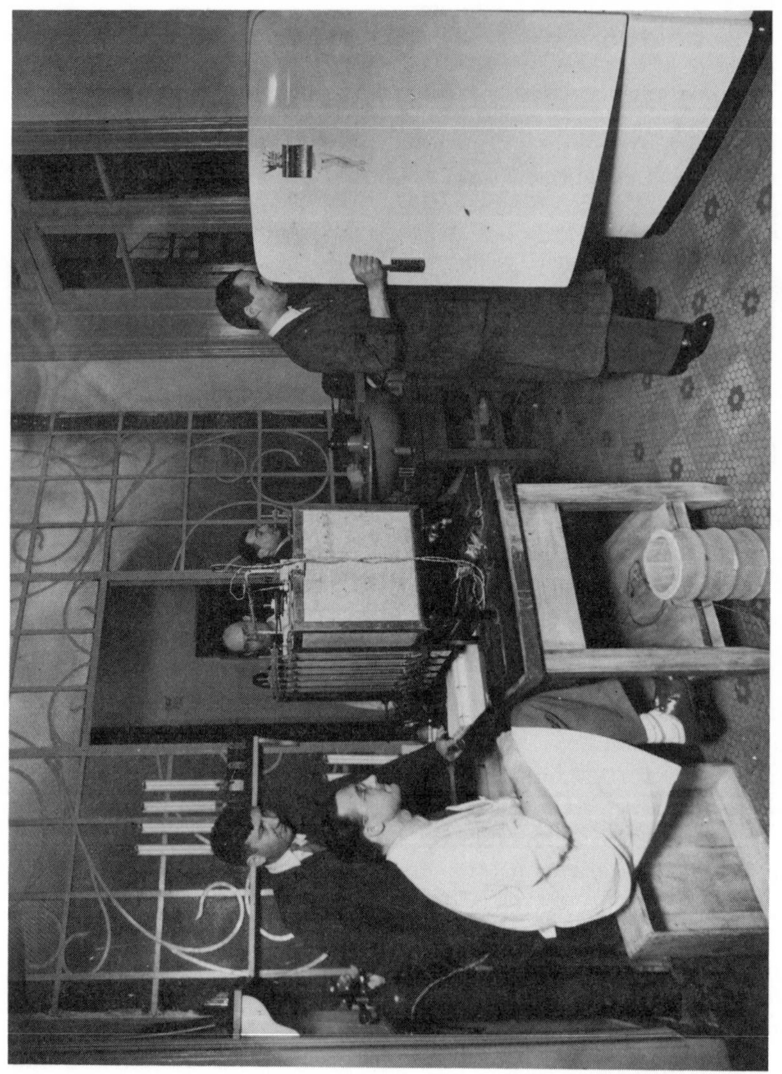

Plate 2. The entrance hall of the laboratory of the Instituto de Investigaciones Bioquímicas "Fundación Campomar" during the years 1947 to 1957. Left to right: R. Caputto, A. Paladini, C. Cardini, R. Trucco and L. Leloir.

other investigators joined the Foundation: Dr. Raul E. Trucco, a microbiologist, Carlos E. Cardini, a biochemist and Naum Mittelman, a biophysical chemist. For political reasons prevailing at that time in Argentina, many of the members of the University had resigned and we were among them. We had a small private laboratory located in an old house (Julián Alvarez 1719, Buenos Aires) which only covered about 100 square meters. The equipment was poor but the place had the advantage that it was very quiet. We did not have any teaching duties nor meetings, so that we could think all day long about research problems. In later years I missed that quiet atmosphere very much.

Paladini: I would like to describe my first impressions when I joined the group. It was quite striking to me, used to the frigid atmosphere of official institutes, to become a member of a joyful lot of people where the head of the laboratory made witty, but friendly remarks all the time, and had the answer "I don't know..." to many daily questions. The equipment of the lab looked luxurious to me but even by the standards of, say, 30 years ago, it was very modest. The most sophisticated piece was a Beckman DU spectrophotometer, run by dry cells which were very difficult to get. We also had a Klett–Summerson photocolorimeter, a microscope and a home-made Warburg respirometer. No refrigerated centrifuge was available for a long time and of course all the initial basic discoveries on sugar nucleotides were made without the help of radioactive materials.

1948–1949 Glucose-diphosphate

Leloir: Our first project was to continue the studies on cell-free acid oxidation which I had initiated with Dr. Juan M. Muñoz when we worked in the Faculty of Medicine in Buenos Aires University. Our studies had been carried out on liver homogenates and now the idea was to study the process in bacteria. I do not remember why, but the project did not progress and we abandoned it. Looking back now it seems that it was a rather good project.

Dr. Caputto had done his thesis work on the mammary gland phosphatase and told us that he had done some preliminary experiments according to which it seemed that gland homogenates would produce lactose from glycogen. We tried to repeat those experiments with the crude methods of that time, one of which consisted in preparing the osazones of the sugars and looking at the crystals under the microscope. What was obtained from glycogen was probably a mixture of glucose, maltose and maltotriose which are the products formed by the action of α-amylase.

In those days it was known from the work of the Coris that glycogen could be formed from glucose-1-phosphate with the enzyme phosphorylase and it had been shown also by Hassid, Doudoroff and coworkers that sucrose could be formed from glucose-1-phosphate and fructose with an enzyme of bacterial origin (from *Pseudomonas*). Consequently the idea that lactose originated from glucose-1-phosphate and galactose was floating in the air. With the idea of looking for a lactose phosphorylase we started studies with a lactose-utilizing yeast (*Saccharomyces fragilis*) which we grew on whey in large milk cans. The search for an enzyme that would reversibly yield glucose-1-phosphate and galactose from lactose resulted only in the finding of a lactase [4]. This led us to the study of galactose utilization. Evidence for the presence of an enzyme catalyzing galactose phosphorylation was obtained and the name galactose kinase was proposed. The reaction product was found to be galactose-1-phosphate [5]. This ester had previously been isolated from the liver of galactose-fed rabbits [6]. The next obvious step was to find out the fate of galactose-1-phosphate. It was then that we stepped into a fertile field of study which compensated our previous series of failures.

Incubation of galactose-1-phosphate with our enzyme preparation led to the formation of glucose-6-phosphate. This reaction could be easily measured by making use of the fact that the latter but not the former reduces alkaline copper solutions. Besides using galactose-1-phosphate as substrate we sometimes used glucose-1-phosphate, both of which were obtained by chemical synthesis. When the enzyme preparations appeared inactive we often tried reactivation with heated yeast extracts which we called by the German expression Kochsaft. We often obtained activations with the Kochsaft but the results were very confusing because we used galactose-1-phosphate in some experiments and glucose-1-phosphate in others. However, after many discussions and apparently contradictory experiments we realized that we were dealing with two thermostable factors, one for each of the following reactions:

galactose-1-phosphate → glucose-1-phosphate → glucose-6-phosphate (1)

At that time the enzyme corresponding to the second step, phosphoglucomutase was well known. It had been discovered by Cori et al. [7, 8] and obtained in crystalline state by Najjar [9]. However, no cofactor for phosphoglucomutase was known to be required. By measuring the activation of yeast phosphoglucomutase we could estimate fairly well the concentration of our thermostable cofactor. The substance was purified by a procedure which consisted mainly in the precipitation with metals. At this point we sent a preliminary letter to *Archives of Biochemistry* [10], titled: *A coenzyme for phosphoglucomutase.* At the end of the report we had written

"Kendall and Stickland [11] obtained an activation of phosphoglucomutase by adding hexose diphosphate but Cori et al. [8] were unable to obtain any effect. The hexose diphosphate preparation of Kendall and Stickland may have been contaminated with the new co-enzyme reported in this paper."

After sending the letter we happened to test fructose-1,6-diphosphate again although it had already given negative results in previous experiments. This time, however, it proved to be active and still worse, it was found that the purified preparations of our cofactor contained as much fructose diphosphate as commercial preparations of the latter! We decided immediately to ask the editors to send our letter back or throw it in the waste-paper basket, but in the meantime we were thinking desperately. Our desperate thinking generated the idea that our cofactor might not be fructose-1,6-diphosphate but glucose-1,6-diphosphate, acting as shown in Fig. 1. The idea was that a transfer of phosphate occurred so that the diphosphate became regenerated at each step.

It was also reasoned that the two di-esters could be distinguished by treatment with hot alkali which would destroy the fructose ester because of the presence of a free-reducing group while the non-reducing glucose ester would remain intact. Most of the time one's hypothesis proves to be false but this time it was correct and thus glucose-1,6-diphosphate was firmly established as a new cofactor [12, 13].

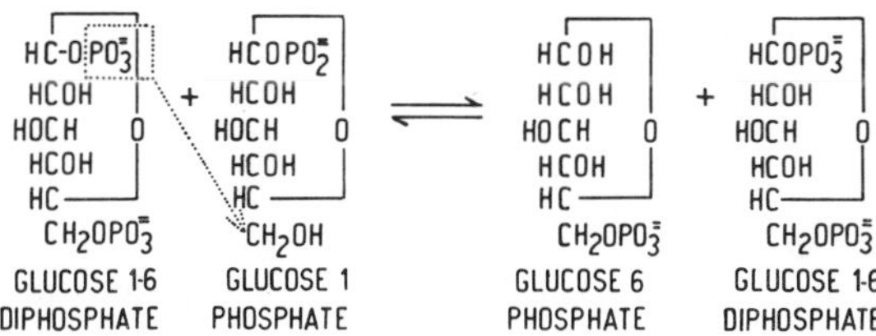

Fig. 1. The mechanism of action of glucose-1,6-diphosphate as cofactor of phosphoglucomutase.

The scheme was confirmed by experiments of Sutherland et al. [14] who studied the reaction with ^{32}P-labeled compounds. The structure of glucose-1,6-diphosphate was confirmed by chemical synthesis [15] and the biosynthesis from glucose-1-phosphate and ATP [16] was described as well as by a transphosphorylation between two molecules of glucose-1-phosphate catalyzed by a bacterial enzyme [17].

1950 UDPG

Once we knew the structure and function of one of our cofactors we could channel our efforts to the study of the other one. It could be estimated by using galactose-1-phosphate as substrate, excess phosphoglucomutase and determining the glucose-6-phosphate formed [18]. After a few purification steps such as precipitation of the mercury salt, adsorption on charcoal etc. it seemed that cofactor activity was related to a substance absorbing at 260 nm. The substance was at first thought to be an adenosine derivative but later was found to contain uridine.

Paladini: Although the spectrum of the cofactor was similar to that of adenosine the changes observed at different pH values could not be accounted for on that basis. The only paper available, at that time, reporting the spectra of purine and some pyrimidine bases and nucleosides was that of Hotchkiss [19] but it did not include uridine. Fortunately, Dr. Caputto while idly looking through a newly arrived volume of the *Journal of Biological Chemistry* found a spectrum identical to that of our compound. It was that of uridine [20]. There was little doubt, thereafter, that the cofactor was a derivative of uracil. This was confirmed in several ways but one of them was accidental: the paper chromatograms in which we attempted to identify uridine in the hydrolysates of the cofactor were dried in a corner of the laboratory, very close to a fluorescent tube. The spectra obtained with the material eluted from the papers gave clear indications of a photochemical modification which, nevertheless, was identical to that obtained with authentic samples of uridine.

Leloir: This was rather exciting, because at that time uridine was known only as a nucleic acid constituent. Uridine-3′-phosphate had been described but uridine-5′-phosphate, UTP and UDP were only detected years later. Besides uridine, our cofactor was found to contain glucose and two phosphates.

In February 1950 we published a note in Nature [21] in which the correct structure was proposed but the position at which the phosphate was linked to the uridine was left open. The glucose was erroneously shown as the β anomer. Soon afterwards we published a full paper on the isolation and structure of UDPG. It was authored by Caputto et al. [22]. At that time we used to rotate the names in each paper so that the last one would become the first in the next publication.

A titration curve of the substance showed only two primary phosphoric acid groups per uridine. Other properties of the substance were as follows: (a) mild acid hydrolysis (15 min at pH 2 at 100°C) liberated a sugar which was identified as glucose by paper chromatography in several solvents, by color reactions and by its fermentability by yeast. Splitting off the glucose unmasked its reducing power and liberated a secondary acid group of phosphoric acid; (b) further acid hydrolysis (15 min in 1 N acid at 100°C) liberated 1 mol of inorganic phosphate and another secondary acid group of phosphoric acid appeared; (c) the substance remaining after splitting off the glucose and one phosphate was identified as uridine-5′-phosphate by comparison with a synthetic specimen kindly supplied by Professor Todd.

Paladini: The first determination of the molecular weight of the cofactor was laboriously made using an excellent Kuhlman microbalance which Leloir had bought many years before the Institute was born. To attain an accuracy of about 0.01 mg I had to apply a very time-consuming estimation of the amplitude of the oscillations. Finally, the acceptable value 630 was obtained instead of the theoretical 566 by measuring the dry weight of a volume of a solution of known absorbance.

When UDPG was run on paper with an alkaline solvent (ethanol—ammonia) it was found to decompose into uridine-5′-phosphate and an unknown substance that had a higher mobility than hexose phosphates. This substance was not easy to identify but after quite a lot of work and thinking we were able to present evidence showing that the decomposition product was the 1,2 cyclic monophosphate of glucose. On acid treatment the cyclic ester gave glucose-2-phosphate and the 1-phosphate which the acid broke down to free glucose [23].

The formation of the cyclic phosphate indicated that the glucose in UDPG is the α anomer because the β has the first two hydroxyl groups in *trans* position thus making difficult the formation of a five-membered ring.

Leloir: Around 1950 research was not as competitive as it is nowadays but nevertheless one study related to ours appeared. Park and Johnson [24] who were investigating the action of penicillin in *Staphylococcus aureus* observed that this substance produced an increase in labile phosphate which turned out to belong to a uridine containing compound.

The state of the field in 1951 was reported in the Symposium on Phosphorus Metabolism organized by W.D. McElroy which took place in Baltimore [25]. Park reported the isolation from penicillin-treated *Staphylococcus* of three uridine containing compounds. The simplest of them had the structure of UDPG but with a sugar which was unknown at that time and that we now know as acetylmuramic acid. Another compound had one alanine residue in addition and the third compound contained one lysine, one glutamic acid and three DL-alanine residues joined to the nucleotide sugar. When my turn came to report our findings I mentioned the discovery of guanosine diphosphate mannose and the alkaline decomposition of UDPG.

1951 UDP-galactose

Another thing I reported at the Phosphorus Symposium was the reversible formation of UDP-galactose from UDP-glucose on incubation of the latter with our yeast extracts. This reaction could be studied thanks to a procedure for the separation of glucose from galactose by paper chromatography developed by Jermyn and Isherwood [26]. The reaction reached an equilibrium when the ratio UDP-glucose/UDP-galactose was about 3. These results were published in a five page paper [27]. It was then assumed that the transformation of galactose-1-phosphate catalyzed by what we used to call galacto-waldenase occurred as follows:

$$\text{galactose-1-P} + \text{UDPG} \longrightarrow \text{glucose-1-P} + \text{UDP-galactose} \quad (2)$$

$$\text{UDP-galactose} \longrightarrow \text{UDP-glucose}$$

$$\text{SUM: Galactose-1-P} \longrightarrow \text{Glucose-1-P}$$

The first reaction: a transfer of uridylic acid, was hypothetical but the formulation was later proved to be correct by Kalckar and coworkers. The equilibrium of the sum of the reactions is reached when the ratio glucose-P to galactose 1-P is also 3 [28].

The difference between galactose and glucose consists only in the stereochemical position of the hydroxyl at position four. The mechanism by which this change occurs was quite mysterious and there were several unproved hypotheses. The mystery was clarified when Kalckar and coworkers discovered that DPN(NAD) was required for the transformation. This was a rather exciting finding. I remember very well that one morning I received a telegram from Denmark. It read: "UDP-galactose epimerase requires DPN" – signed Herman. This finding was a strong indication that the process occurs by an oxidoreduction reaction and further work is consistent with the idea that hydrogen is removed from the carbon at position 4 and a keto intermediate is formed.

Several important contributions from Kalckar and coworkers [29] to the field have been mentioned. When he was asked in what he was working he sometimes answered: I am interested in Place Pigalle and operate along la rue du Dr. Leloir. Not everybody understood the joke right away. When he said Pigalle he meant P-gal (phosphogalactose) and not the well known district of Paris which houses painters' ateliers, writers' cafes, night clubs and other attractions. As to the rue du Dr. Leloir he meant some of the pathways of galactose metabolism and sugar nucleotides which we also studied.

Although our laboratories worked on similar lines there never was any rivalry or resentment among us as often happens. We exchanged information quite freely.

1952 UDP-acetylglucosamine

Paladini: Another member of the family of sugar nucleotides was discovered in our laboratory by sheer serendipity. I was very much involved in finding good solvents for paper chromatography of UDPG and related compounds. It should be remembered, by the way, that the use of ion exchange columns was, at that time, still in its infancy in the world. I assayed many different solvents but obtained the best results with various combinations containing ethanol. As mentioned before ethanol–ammonia decomposed UDPG but we could detect another substance, which was not affected by the alkaline reaction and which was provisionally referred to as UDPX and later identified as UDP-acetylglucosamine [30].

Leloir: The main difficulty in our study of UDPX was that at the time we did not have a sample of acetylglucosamine to compare with our unknown sugar. Knowing how simple it is to *N*-acetylate aminosugars this seems incredible. The procedure consists simply in adding acetic anhydride to the aminosugar at neutral reaction. Once we had identified the sugar moiety it was easy to establish the structure because everything was the same as with UDPG.

This substance was later isolated in larger amounts by ion exchange column chromatography and studied more carefully [30, 31].

Paladini: The use of ion exchange resins to isolate UDPG and other nucleotides was a great advance but to obtain sizeable amounts of these it was necessary to handle big columns and hundreds of liters of eluents. For the collection of the samples an unusual fraction collector was devised. The flasks were flat bottles originally used to store water in refrigerators. Many dozens of them were placed side by side on a shelf two meters high from the floor. Along the shelf, in front of the bottles, ran a sort of toy railroad and the wagon carried the effluent tube from the column. It was placed so as to discharge in the flasks through a basculating siphon. The wagon was pulled by a string attached to a heavy counterpoise and the change of flask was ingeniously achieved by a sort of ratchet mechanism activated by the weight of the siphon operating on nails placed all along the edge of the shelf. The column was suspended high in the air and the eluting solvent in a 50-liter bottle, was handled by rope and pulley, to attain its proper position near the ceiling of the room. Unbelievably, this contraption worked perfectly well for several years.

Charcoal absorption proved to be a good step in the purification of sugar nucleotides. The difficult step was the elution. In connection with this I can lighten this recollection by mentioning that we used to have around the lab all kinds of fancy bottles, that were originally containers of perfumes, shampoos or prescriptions. Many came from Dr. Leloir's home and he insisted in storing reagents in them with the idea that the non-uniformity of the shapes and colors helped to avoid mistakes.

The best eluent from charcoal for the sugar nucleotides landed in a bottle of perfume named Lotus Flower. For years we never used any other name for it. This gave rise to many amusing incidents when explaining our research to some visitor and we inadvertently described the experiment made with the Lotus Flower solution.

GDP-mannose

Leloir: Another compound which like UDP-acetylglucosamine was first detected by paper chromatography was studied with E. Cabib [32]. It was prepared from a yeast extract by precipitation of the charcoal and electrophoresis on starch. The guanosine moiety was easily recognizable because it gives a very typical spectrum with a peak at 260 μm and a shoulder at about 280 μm in neutral or acid solution.

The biosynthesis of UDPG. Experiments had been carried out by Dr. Trucco in 1951 [33] in which he incubated glucose-1-phosphate, UDP and ATP with yeast enzymes. He found that addition of the three substances was required for maximal UDPG formation. The latter was measured as activator of the galactose-1-P → glucose-1-P transformation. Trucco believed that UDPG was formed as follows:

$$\text{UDP} + \text{ATP} \longrightarrow \text{UTP} + \text{ADP} \quad (3)$$
$$\text{UTP} + \text{Glucose-1-P} \longrightarrow \text{UDPG} + \text{inorganic pyrophoshate} \quad (4)$$

Clearer results were later obtained by Munch-Petersen et al. [34] by measuring the reaction from right to left, that is starting from UDPG and inorganic pyrophosphate.

1953-55 UDP-glucuronic acid

Dutton and Storey [35] were studying the reaction by which glucuronosyl residues become attached to phenols and various other compounds in liver, a process often called detoxication [36].

Liver enzymes were found to add glucuronosyl residues to *O*-aminophenol or menthol, faster when a heated extract of liver was added. The active substance was purified and it turned out to be identical to UDPG except that the $-CH_2OH$ of position 6 of the glucose was replaced by COOH (carboxyl). This was the first case in which the donor function of a sugar nucleotide was obtained experimentally and was therefore a very important finding.

The mechanism of formation of UDP-glucuronic acid was elucidated by Strominger et al. [37]. They used an enzyme from liver and DPN (NAD) as oxidant.

1953 Biosynthesis of disaccharides

A donor role of UDPG was suggested by Buchanan et al. [38] and Buchanan [39] to explain some results on the formation of sucrose. The experiments consisted in identifying the products formed by photosynthesizing cells from labeled inorganic phosphate and CO_2. Two-dimensional paper chromatography showed that the UDPG spot decreased with time while sucrose increased. Buchanan [39] suggested that glucose was transferred to fructose phosphate to give sucrose phosphate.

A few years later we were able to obtain an enzyme that catalyzed this reaction. However, we first detected another enzyme which leads to the biosynthesis of trehalose (a disaccharide consisting of two glucoses joined at position 1).

Early in our studies on UDP-glucose we began to suspect that it might have some other function besides its role in the galactose-1-phosphate $\rightleftarrows$ glucose-1-phosphate transformation. In the paper where we reported the isolation of UDP-glucose [22] we commented that

> "The yeast used (for the isolation) was not adapted to galactose and hardly fermented this sugar. This raises the question as to whether UDPG may not have some other function besides accelerating the galactose-glucose transformation."

We often asked one another the question "What is the use of UDPG?" and I think we even thought of instituting a prize for the person who would solve the riddle.

Since we had a method for estimating UDPG we tried measuring its disappearance under different conditions and adding various substances. We found that the addition of glucose-6-phosphate to yeast enzymes increased the disappearance. The change was traced to the formation of trehalose phosphate as follows:

$$\text{UDPG} + \text{glucose-6-phosphate} \rightarrow \text{UDP} + \text{trehalose phosphate} \qquad (5)$$

The results were reported in a preliminary note in the *Journal of the American Chemical Society* [40] and later in a full paper in the *Journal of Biological Chemistry* [41]. Trehalose phosphate may be acted upon by a specific phosphatase to yield trehalose which is a reserve carbohydrate found in yeasts and insects. A similar procedure was then applied to a wheat-germ extract and it was found that UDPG acted as glucose donor to fructose [42] to yield sucrose. Only free sucrose was detected as the reaction product so that we thought that there might be a mistake in the suggestion of Buchanan et al. [38] that sucrose phosphate was the precursor of the free sucrose. However further research [43, 44] proved that the suggestion was probably right. After some steps of purification it became evident that we were dealing with not one but with two enzymes one which yielded free sucrose and another sucrose-phosphate.

$$\text{UDPG} + \text{fructose} \rightarrow \text{sucrose} + \text{UDPG} \quad (6)$$

$$\text{UDPG} + \text{fructose-6-phosphate} \rightarrow \text{sucrose-phosphate} + \text{UDP} \quad (7)$$

These enzymes were then found in other plant tissues and are now believed to be responsible for sucrose formation in the plant kingdom.

Hexosamine metabolism. For some time we wondered which was the role of UDP-acetylglucosamine in tissues and this led us to study various processes in which hexosamines are involved.

An enzyme was detected in extracts of *Neurospora crassa* [45], which catalyzed the reaction.

Hexose-6-phosphate + glutamine → glucosamine-6-phosphate + glutaminic acid (8)

The paper also mentioned some other reactions such as the acetylation of glucosamine-6-phosphate and a reversible transfer of phosphate from position 1 to 6, a reaction similar to that catalyzed by phosphoglucomutase. The mutase was studied in detail by Reissig [46] who also collaborated with Strominger and myself [47] in improving the method of estimation of acetylhexosamines.

An enzyme present in liver which catalyzes the phosphorylation of galactosamine by ATP was also described [48]. This enzyme which was also found in brain and in *Saccharomyces fragilis* was believed to be the same as galactokinase. Some time afterwards enzymes catalyzing the phosphorylation of acetylglucosamine and acetylgalactosamine phosphate were detected [49].

In 1955 Pontis ran an extract of bovine liver through an anion exchange column and separated the UDP-acetylglucosamine fraction. The sugar moiety was found to be a mixture of acetylglucosamine and acetylgalactosamine [50]. With Cardini [51] I tried to detect an enzyme that would interconvert the above mentioned sugar nucleotides. This led us to make our greatest published mistake. What we did was to incubate UDP-acetylglucosamine with a liver extract and then separated the free acetylhexosamines by chromatography on borate-treated paper. We could detect the formation of a substance which had the same mobility on paper as acetylgalactosamine and gave the same color reaction. The results were published in a paper entitled *Enzymic formation of acetylgalactosamine.* After some time Comb and Roseman [52] discovered that what was formed was not acetylgalactosamine but acetylmannosamine. The two hexosamines have the same mobility when chromatographed on borate-treated paper but could be distinguished easily after deacetylation by treatment with ninhydrin followed by paper chromatography. We had carried out this test but either we did not do it correctly or we were blinded by the conviction that the product was acetylgalactosamine.

Another enzyme of hexosamine metabolism which we found with Cardini [53] was one present in kidney and liver which catalyzes the reaction.

$$\text{Glucosamine-6-phosphate} \rightarrow \text{fructose-6-phosphate} + \text{ammonia} \qquad (9)$$

The curious property of this enzyme is that it is activated by acetylglucosamine-6-phosphate.

Biosynthesis of polysaccharides

Leloir: Studies on polysaccharide biosynthesis from sugar nucleotides were initiated by Glaser and Brown [54] who obtained evidence that UDP-acetylglucosamine could act as donor in the formation of hyaluronic acid (a polymer of acetylglucosamine and glucuronic acid) by Rous chicken sarcoma homogenates. Glaser and Brown [55] also obtained incorporation of radioactivity from UDP-acetylglucosamine in chitin (a polymer of acetylglucosamine) with an enzyme preparation from *Neurospora crassa.* Soon afterwards studies on the biosynthesis of various polysaccharides were published by various workers.

An important development was the finding of an enzyme which catalyzes transfer of glucose from UDPG to glycogen. This enzyme is now colloquially known as glycogen synthetase. Until then it was firmly believed that glycogen was formed from glucose-1-phosphate by the action of enzyme phosphorylase. The reaction:

$$\text{Glycogen} + \text{phosphate} \rightarrow \text{glucose-1-phosphate} \qquad (10)$$

is reversible so that synthesis can be achieved at low phosphate concentration. One of the important advances in biochemistry was the "in vitro" synthesis of glycogen by Cori and Cori in 1936. Since then the textbooks made phosphorylase responsible for the synthesis of glycogen. Some inconsistencies in the theory were pointed out by Sutherland [56]. According to him it was difficult to understand why the activation of phosphorylase, for instance with epinephrine, always produced glycogen degradation. The reaction shown in equation 10 will go to the right when the inorganic phosphate concentration is high and to the left when phosphate is low and glucose-1-phosphate high. In tissues the relation of inorganic phosphate/glucose-1-phosphate was always such that glycogen was broken down. This difficulty had been dismissed before by saying that cells have compartments and that at the site of phosphorylase action the concentrations might be such as to favor glycogen synthesis.

After examining the available facts Niemeyer [57] suggested that UDPG might be the donor for glycogen synthesis. Finally the point was settled by the detection of the corresponding enzyme in liver. The reaction was measured by estimating the UDP formed. No labeled compounds were used in the initial experiments [58] but further work with ^{14}C-labeled UDPG confirmed the finding [59].

The activity of the enzyme was rather low so here again we tried the addition of "Kochsaft" and found that it produced an activation. The identity of the active substance was easily established and it turned out to be glucose-6-phosphate and it was found to be very specific as an activator.

Since most of the glycogen synthetase activity was recovered in the high-molecular-weight glycogen after subcellular fractionation [60] it was reasoned that something similar might occur with starch-synthesizing enzymes. It was found that starch granules catalyze transfer of glucose from UDP-glucose to starch [61, 62]. Later Dr. Eduardo Recondo prepared ADP-glucose by chemical synthesis and it proved to act as a much more efficient donor than UDP-glucose [63].

It is now accepted that ADP-glucose is the natural donor for the synthesis of starch in plants and of glycogen in bacteria. The enzyme which leads to its synthesis was first detected in potatoes by Espada [64]. The reaction is as follows:

$$\text{ATP} + \text{glucose-1-P} \rightarrow \text{ADP-glucose} + \text{UDP} \qquad (11)$$

But these findings approach modern times and are not history any more.

Final remarks

After the first members of the sugar nucleotide family were discovered, studies were continued in various parts of the world and many new compounds were described. In the 1963 census [65] more than 40 sugar nucleotides were reported. Furthermore, many enzymatic reactions involving these substances were discovered. Transfer of sugars to many oligo and polysaccharides was detected. It seemed that direct transfer of monosaccharides would explain the biosynthesis of the multitude of polysaccharides that are found in Nature. However, in 1967 another mechanism was uncovered. In studies on the biosynthesis of *Salmonella* lipopolysaccharide by P. Robbins and coworkers [66] and of bacterial peptidoglycan by Strominger and others [67] it was found that lipid intermediates were involved. These intermediates turned out to be derivatives of a polyprenol, undecaprenol, a compound containing 55 carbon atoms and eleven double bonds. Similar studies with animal tissues showed that a related intermediate, dolichol, containing about 100 carbon atoms and 19 double bonds is involved in the glycosylation of some proteins.

The style of research has changed considerably in the years that followed World War II. At first research was carried out by a few highly motivated individuals with rather scant means. Later the research workers became more professional and had the advantage of using more sophisticated equipment and receiving generous government grants. Looking back at the years 1947–1960 they seem to have been very fertile. Those were pleasant years in which we often had the joy of doing a successful experiment and were surrounded by a very stimulating atmosphere.

REFERENCES

1 L.F. Leloir, Biochem. J., 91 (1964) 1–8.
2 L.F. Leloir, in R. Piras and H.G. Pontis (Eds.), Biochemistry of the Glycosidic Linkage, PAABS Symposium, Vol. 2, Academic Press, New York, 1972, pp. 1–18.
3 L.F. Leloir, in E.Y. Lee and E.E. Smith (Eds.), Biology and Chemistry of Eucaryotic Cell Surfaces, Academic Press, New York, 1974, pp. 1–2.
4 R. Caputto, L.F. Leloir and R.E. Trucco, Enzymologia, 12 (1948) 263–276.
5 R.E. Trucco, R. Caputto, L.F. Leloir and N. Mittelman, Arch. Biochem., 18 (1948) 137–146.
6 H.W. Kosterlitz, Biochem. J., 37 (1943) 318–321.
7 G.T. Cori, S.P. Colowick and C.F. Cori, J. Biol. Chem., 123 (1938) 375–380.
8 G.T. Cori, S.P. Colowick and C.F. Cori, J. Biol. Chem., 124 (1938) 543–555.
9 V.A. Najjar, J. Biol. Chem., 175 (1948) 281–290.
10 R. Caputto, L.F. Leloir, R.E. Trucco, C.E. Cardini and A. Paladini, Arch. Biochem., 18 (1948) 201–203.
11 L.P. Kendall and L.H. Stickland, Biochem. J., 32 (1938) 572–584.
12 C.E. Cardini, A. Paladini, R. Caputto, L.F. Leloir and R.E. Trucco, Arch. Biochem., 22 (1949) 87–100.
13 L.F. Leloir, R.E. Trucco, C.E. Cardini, A. Paladini and R. Caputto, Arch. Biochem., 19 (1948) 339–340.
14 E.W. Sutherland, M. Cohn, T. Posternak and C.F. Cori, J. Biol. Chem., 181 (1949) 153–159.
15 L.F. Leloir, O.M. Repetto, C.E. Cardini, A. Paladini and R. Caputto, Asoc. Quim. Argentina, 37 (1949) 187–191.
16 A. Paladini, R. Caputto, L.F. Leloir, R.E. Trucco and C.E. Cardini, Arch. Biochem., 23 (1949) 55–66.
17 L.F. Leloir, R.E. Trucco, C.E. Cardini, A. Paladini and R. Caputto, Arch. Biochem., 24 (1949) 65–74.
18 R. Caputto, L.F. Leloir, R.E. Trucco, C.E. Cardini and A. Paladini, J. Biol. Chem., 179 (1949) 497–498.
19 R.D. Hotchkiss, J. Biol. Chem., 175 (1948) 315–332.
20 J. Ploeser and H.S. Loring, J. Biol. Chem., 178 (1949) 431–437.
21 C.E. Cardini, A. Paladini, R. Caputto and L.F. Leloir, Nature, 191 (1950) 191–193.
22 R. Caputto, L.F. Leloir, C.E. Cardini and A.C. Paladini, J. Biol. Chem., 184 (1950) 333–350.
23 A. Paladini and L.F. Leloir, Biochem. J., 51 (1952) 426–430.
24 J.T. Park and M.J. Johnson, J. Biol. Chem., 179 (1949) 585–592.
25 W.D. McElroy and B. Glass (Eds.), Symposium on Phosphorus Metabolism, Johns Hopkins Press, Baltimore, 1951.
26 M.A. Jermyn and F.A. Isherwood, Biochem. J., 44 (1949) 402–407.
27 L.F. Leloir, Arch. Biochem., 33 (1951) 186–190.
28 L.F. Leloir, C.E. Cardini and E. Cabib, Asoc. Quim. Argentina, 40 (1952) 228–234.
29 H. Kalckar, in G.F. Springer (Ed.), Polysaccharides in Biology, 1957, p. 165.
30 E. Cabib, L.F. Leloir and C.E. Cardini, Cienc. Invest., 8 (1952) 469–471.
31 E. Cabib, L.F. Leloir and C.E. Cardini, J. Biol. Chem., 203 (1953) 1055–1070.
32 E. Cabib and L.F. Leloir, J. Biol. Chem., 206 (1954) 779–790.

33 R.E. Trucco, Arch. Biochem. Biophys., 34 (1951) 482–483.
34 A. Munch-Petersen, H.M. Kalckar, E. Cutolo and E.E.B. Smith, Nature, 172 (1953) 1036–1037.
35 G.J. Dutton and I.D.E. Storey, Biochem. J., 53 (1953) XXXVII.
36 I.D.E. Storey and G.J. Dutton, Biochem. J., 59 (1955) 279–288.
37 J.L. Strominger, H.M. Kalckar, J. Axelrod and E.S. Maxwell, J. Am. Chem. Soc., 76 (1954) 6411–6412.
38 J.G. Buchanan, J.A. Bassham, A.A. Benson, D.F. Bradley, M. Calvin, L.L. Daus, M. Goodman, P.M. Hayes, V.L. Lynch, L.T. Norris and A.T. Wilson, in W.D. McElroy and B. Glass (Eds.), A Symposium on Phosphorus Metabolism, Johns Hopkins Press, Baltimore, 1952, pp. 440–466.
39 J.G. Buchanan, Arch. Biochem. Biophys., 44 (1953) 140–149.
40 L.F. Leloir and E. Cabib, J. Am. Chem. Soc. 75 (1953) 5445–5446.
41 E. Cabib and L.F. Leloir, J. Biol. Chem. 231 (1958) 259–275.
42 L.F. Leloir and C.E. Cardini, J. Am. Chem., Soc., 75 (1953) 6084.
43 C.E. Cardini, L.F. Leloir and J. Chiriboga, J. Biol. Chem., 214 (1955) 149–155.
44 L.F. Leloir and C.E. Cardini, J. Biol. Chem., 214 (1955) 157–165.
45 L.F. Leloir and C.E. Cardini, Biochim. Biophys. Acta, 12 (1953) 15–22.
46 J.L. Reissig, J. Biol. Chem., 219 (1956) 753–767.
47 J.L. Reissig, J.L. Strominger and L.F. Leloir, J. Biol. Chem., 217 (1955) 959–966.
48 C.E. Cardini and L.F. Leloir, Arch. Biochem. Biophys., 45 (1953) 55–64.
49 L.F. Leloir, C.E. Cardini and J.M. Olavarría, Arch. Biochem. Biophys., 74 (1958) 84–91.
50 H.G. Pontis, J. Biol. Chem., 216 (1955) 195–202.
51 C.E. Cardini and L.F. Leloir, J. Biol. Chem., 225 (1957) 317–324.
52 D.G. Comb and S. Roseman, Biochim. Biophys. Acta, 29 (1958) 653–654.
53 L.F. Leloir and C.E. Cardini, Biochim. Biophys. Acta 20 (1956) 33–42.
54 L. Glaser and D.H. Brown, Proc. Natl. Acad. Sci. USA, 41 (1955) 253–260.
55 L. Glaser and D.H. Brown, Biochim. Biophys. Acta, 23 (1957) 449–450.
56 E.W. Sutherland, Ann. N.Y. Acad. Sci., 54 (1951) 693–706.
57 H. Niemeyer, Metabolismo de los Hidratos de Carbono en el Higado, Universidad de Chile, Santiago, 1955, p. 148.
58 L.F. Leloir and C.E. Cardini, J. Am. Chem. Soc., 79 (1957) 6340–6341.
59 L.F. Leloir, J.M. Olavarría, S.H. Goldemberg and H. Carminatti, Arch. Biochem. Biophys., 81 (1959) 508–520.
60 L.F. Leloir and S.H. Goldemberg, J. Biol. Chem., 235 (1960) 919–923.
61 M.A.R. de Fekete, L.F. Leloir and C.E. Cardini, Nature, 187 (1960) 918–919.
62 L.F. Leloir, M.A.R. de Fekete and C.E. Cardini, J. Biol. Chem., 235 (1960) 636–641.
63 E. Recondo and L.F. Leloir, Biochem. Biophys. Res. Commun., 6 (1961) 85–88.
64 J. Espada, J. Biol. Chem., 237 (1962) 3577–3581.
65 E. Cabib, Annu. Rev. Biochem., 32 (1963) 321–354.
66 A. Wright, M. Dankert, P. Fennessey and P.W. Robbins, Proc. Natl. Acad. Sci. USA, 57 (1967) 1798–1803.
67 J. Higashi, J.L. Strominger and C.C. Sweeley, Proc. Natl. Acad. Sci. USA, 57 (1967) 1878–1884.

G. Semenza (Ed.) Selected Topics in the History of Biochemistry: Personal Recollections (Comprehensive Biochemistry Vol. 35)

Chapter 3

The Road to Ion-coupled Membrane Processes

ROBERT K. CRANE

Department of Physiology and Biophysics, University of Medicine and Dentistry of New Jersey, Rutgers Medical School, Piscataway, NJ 08854 (U.S.A.)

Washington College

Nothing has more impressed me in my life than the extent to which the direction of it has been determined by pure chance. The day I left St. Andrews School, Walden Pell advised me, "Do not go into science" and I was bent on following that wise advice because it well matched my lack of interest. I didn't have the money to go up to Princeton to study philosophy so I settled for a tuition scholarship at Lehigh University and the practical purpose of becoming a civil engineer so as to join my father in business. Failing to get a loan for bed and board at Lehigh from a well-to-do cousin, I turned to Congressman Goldsborough of Denton for an appointment to the U.S. Naval Academy at Annapolis. Failing that, I registered at the very last minute for the fall, 1938, semester at Washington College hoping to transfer after one year. I chose Washington College because my older brother, Walter, had entered Washington College the previous year and my mother had taken up residence at Chestertown in order to care for him. Walter was chronically sick, incurably so. He died in 1939. My father had offered him a choice among several colleges of similar size and provincial atmosphere. Whatever he told my father, Walter confided in me that he had actually chosen Washington College because during one of his visits he happened to see cross the campus a strikingly beautiful young woman with shoulder-length blond tresses. I do not doubt that I owe my life in science to Mary Lillian Knotts.

Then, as now, at Washington College, freshmen were required to take one laboratory science course. But which course? The choices were biology

Plate 3. Robert K. Crane

or chemistry or physics. Viewing physics as far too much pain to suffer for a mere curricular requirement, I tossed a mental coin between biology and chemistry. The coin landed for chemistry and, of course, for Kenneth Buxton, the professor of it. He fished well for science, that man. He had me hooked and netted by the end of the first semester. Chemistry, the way he taught it, profoundly interested me.

Not wishing to diminish the lustre of the later and recent development of Washington College, I am still, nonetheless, profoundly impressed with the exceptional quality of the faculty in those years just before World War II. Though comparative anatomy bored me it was not the fault of Julian Corrington. I found his histology and embryology compellingly beautiful. Physics, which I had earlier shunned, came belatedly alive as taught by Jesse Coop and I ended up taking all but one of his courses. I ran out of time. The German I learned from Arthur Davis passed for me my examination at Harvard four years later with less than a month's review. My singular failure was not to arrive at a mastery of calculus. For some reason I understand it easily but cannot put it to use. This failure has led to a continuing deficiency in the level of my understanding of physical chemistry which I have always regretted and, on occasion, suffered from.

The first of what I might call stirrings of interest in research occurred while I was returning in the rain from my brother's funeral. I found myself wanting to study and understand the basis of the kidney function which had failed him. In my last year, Albert Kline arrived at Washington College fresh from his Ph.D. at The Johns Hopkins with S. O. Mast. For himself, Mast claimed that all one needed to do research was "a microscope and a drop of pond water" but his students were thoroughly grounded in biochemistry. Albert taught me no courses but during the many hours we spent together I absorbed the understanding that biochemistry is the route to the solution of biological problems. Albert and I spent most nights between 10 PM and 2 AM at the Kent and Queen Anne's Hospital where we were setting up a clinical chemistry laboratory for the use of Dr. Dick, the chief surgeon. It was probably there in that laboratory or perhaps during the midnight break for a meal with the duty nurses that I started to think about biomedical research in earnest and formed the notion to seek, first, an M.D. and then a Ph.D. as the best way to develop myself. But it was not to be.

Interim

In 1942, the V-12 program by means of which all medical students were paid members of the armed forces had not been started and I was no better off than in 1938. Medical School was out. I turned down offers of assistantships for graduate work in Organic Chemistry at Maryland and Georgia Tech. E.V. McCollum accepted me at the School of Public Health at the Johns Hopkins University but refused to seek a draft deferment. I did not want to waste my chemistry so I entered training for TNT production at the Atlas Powder Company installation at Tamaqua, PA. I did well during the training period so the Reynolds Experimental Laboratory decided after the training was over to keep me on to work for the company instead of shipping me off to Paducah or Weldon Springs. This laboratory served among other purposes as the quality control laboratory for the various acid plants and powder works belonging to the Atlas Powder Company. During the year I stayed, I was taught to be a competent analytical chemist. I also learned to write monthly reports of progress and I had my first taste of research while analyzing ethylene glycol dinitrate vapors in the dynamite mixing houses. These houses are surrounded by heavy earthworks and with good reason. I am told that the mixing house in which I took most of my samples for assay blew up a year or so later.

Albert Kline re-entered my life in 1943 by transmitting an offer to teach chemistry at the Kirksville (Missouri) State Teachers College. Being disillusioned by the modesty of my war effort role at Tamaqua, I accepted the position by a letter to President Walter Ryle in which I carefully omitted mentioning my age, 23. The regular staff was gone off to war and I was alone in the department. I thoroughly enjoyed trying to teach everything: general, qualitative, quantitative, organic and even biochemistry. I also did everything else, from preparing reagents to washing the bottles into which they were put. During the year I tried to get into the V-12 student program, but was now off phase. I also made plans to take some graduate work in biochemistry at the University in Columbia. However, within a few months I began to want to be directly in the war so I went down to St. Louis and applied for a commission in the Navy. The next step was to volunteer for the draft. At the end of the school year I was drafted. I thoroughly enjoyed my time in the Navy. I would have been happy to have made a career of it after my experiences in the South Pacific as a deck officer on the 2100-ton destroyer, Killen, but I was deterred by two facts of Navy life. First, I could not bring myself to foist upon Mildred the sorry lot of a Navy wife which she had not chosen. Second, shoreside duty in peacetime is a frightful bore.

Harvard

My wife urged me to go on to study biochemistry. I now had the money, that is, 3 years eligibility under the GI bill so there was no reason not to. I applied to Harvard, Michigan and Hopkins and was accepted at all three because, thankfully, the graduate record exam did not appear as a requirement until a year or so later. I chose to go to Boston because I was very uncertain of my abilities and wanted to know as quickly as possible whether I should remain in science. I was sure they would let me know. Eric Ball twitted me for years because of the letter I wrote. Having lost, during my time in service, any really positive sense of direction, I described my career goals literally in terms relating to my experiences up to that time. I wanted to do for others, I wrote, in effect what Ken Buxton had done for me. I wanted to teach in a small college and carry on research on the side. I have long wondered why I was accepted. And not solely because of the letter. Years later, I remarked to David Brown, who directed the graduate student program in Carl Cori's department, that the requirements he set for admission would have without question excluded me.

Harvard was a shock. I was 26 years old and up to that time I had met many people of admirable talent but I had yet to meet anyone I considered to be really and truly bright. At Harvard, I met really bright people, faculty and students, by the bunches. My amazement probably accounts for the fact that, while there, I spent seven days and evenings of every week either studying or doing research. I was running scared. I did very well in my course work and I finished my thesis and I was out of there in 3 years, a perhaps record time. It's just as well, because I think I could not have withstood even one more year of that cruel pace.

Shortly after my arrival, Eric Ball put me to work to make crystalline yeast hexokinase by Kunitz' method [1] which had just been published. There were three notable consequences of this project. First, my initial encounter with Baird Hastings came with his making it perfectly clear that he did not like the smell of yeast and that he had little use for any graduate student who let it loose in the hallways. Second, I never got any crystals. This latter experience has repeated itself several times in my career serving to fortify my view of myself more as a physiological chemist than as a biochemist. In my view, a biochemist can crystallize an enzyme. Third, for the required report on the project I reviewed the literature on hexokinase in such a thorough fashion that Eric Ball, being impressed, offered me the chance to work for the summer in the laboratory of G.H.A. Clowes at the Marine Biological Laboratory at Woods Hole.

Woods Hole was a welcome respite from the disciplined studies in Boston. It also offered my first real glimpse into the world of research and resulted in my first publication [2]. More profoundly for my future, it introduced me to M.E. Krahl, a long-time associate of Clowes and the director of this summer laboratory. Krahl was also an Associate Professor in Carl Cori's department. It was Krahl who persuaded Cori to look me over and to take me on in his department 3 years later. Altogether, at that period, I spent two summers at Woods Hole, passing up a third because of the need to write my thesis before joining Lipmann. Of all the remarkable things I saw and heard in those summers, two remain especially sharp in memory. I came head to head with the concept of vitalism in the person of L.V. Heilbrunn who offered the observation after one of my Tuesday evening presentations that what I had measured was not interpretable in biological terms because I had broken the cells. I wanted to rise in response, but Eric Ball held me down. The other memory is of Benjamin Libet and his report on nerve sheath ATPase [3]. My retention of the former is self-explained, my retention of the latter, though it could seem prescient, actually mystifies me. I also gained some confidence in my ability to come to grips with the meaning of experiments. In preparation for going down to work with Clowes and Krahl I studied all of their publications on dinitrophenol effects on sea urchin eggs and arrived at the conclusion that action at a single point in metabolism could explain all of their observations on intact cells. I did not know what this point was but looking for it became moot that very summer with the announcement of the uncoupling effect on aerobic phosphorylation by Farney Loomis (see [4]).

Back at Harvard, Eric Ball specified that whoever wanted to be his graduate student would have to study CO_2 fixation in retina. My thesis project was thus decided. The studies I did on it [5, 6] had little impact on the world but a great deal of impact on me. Eric's approach to graduate student teaching was "sink or swim". If one asked for counsel the response was a routine, "I can give you 3 minutes, fella". In consequence, I was forced to do my own thinking. I was often in conflict with Eric about the meaning of the data and frequently designed my experiments with the view of proving him wrong. I wasn't terribly happy about the arrangement at the time but I can see how the experience taught me to search out the critical, decisive experiment and to do it. Since I have come to consider this to be the litmus test of a researcher, I have belatedly to acknowledge that Eric certainly tried to see to it that I learned how to do research. Which is, after all, what I was there for.

A year with Fritz Lipmann

I joined Fritz Lipmann at the Massachusetts General Hospital in September of 1949, 3 months before the defense of my thesis, 7 months before the award of the Ph.D. Why actually Lipmann offered me the job which Nate Kaplan was leaving I do not know, though I know that my name was put up for it through the good offices of Dave Novelli. As it turned out, I was the maverick in the laboratory. Everyone else was engaged in hot pursuit of the cofactor for acetylation. I was involved with phosphorylation; both by myself and together with Hermann Niemeyer. Hermann found the acceptor effect in mitochondrial phosphorylation in studies of the action of thyroid hormones on mitochondrial activity [7]. I studied the arsenate effect on phosphorylation [8].

If I was goaded by Eric Ball to learn to do research, I was challenged by Fritz Lipmann to learn to think. The year 1949–50 was a rare and remarkable experience. Nearly every afternoon, Fritz would appear in my cubbyhole laboratory, sit down and begin to talk. Lipmann's description for the way to do research is, "Follow your nose". It was the same in those conversations. We wandered everywhere, "nosing" out a trail in the thicket of biochemistry. This habit of Lipmann's caused me to conceive of and bring to fruition the arsenate work while he was absent for two months, there being little chance to complete a day's work when he was home. It also led to my conceptual discovery of the mechanism of the glyceraldehyde-3-phosphate dehydrogenase reaction which I later transmitted to Jane Harting.

In 1950, it was still conventional wisdom that 1,3-diphosphoglycerate was produced by the oxidation of a complex between 3-phosphoglyceraldehyde and inorganic phosphate. The concept, originated by Warburg and carried on by Lipmann [9] had not yet been challenged. In one of our conversations we discussed this reaction. Lipmann restated his view that oxidation in a phosphate system could generate an "energy-rich" bond, oxidation in a water system could not. Knowing there was no experimental basis for Warburg's concept, I posed the contrary view that oxidative formation of an acyl-enzyme followed by phosphorolysis was much more likely especially in view of the fact that in the next room Stadtman and Novelli were studying a nearly ideal model enzymic reaction for my proposed phosphorolysis step; transacetylation from CoA to phosphate.

A decade with Carl Cori

I went to St. Louis in September of 1950 hoping to put my idea to the test. On arrival, however, I learned that Sidney Velick was deeply into a variety of studies of glyceraldehyde-3-phosphate dehydrogenase and that he had a graduate student, Jane Harting, working on the question of mechanism. She was in the laboratory next to me. I explained my concept and she proceeded to formulate experimental tests for it. At one point there was a question of continuing because of a lack of progress. However, I persuaded her that she had not yet made a decisive test, whereupon she continued working and shortly thereafter got evidence that she was on the right track. She ended up with an elegant thesis [10,11]. What was important to me at the time was that my insight had proved to be correct. What was to be important to me later was the uncomplicated understanding of the independence of mechanism and thermodynamics which permitted me to have had this insight.

In my own laboratory, I continued studies of the so-called gel phosphate that I had started in Boston [12]. Carl Cori took an interest in the work and passed on numerous ideas. However, beyond being able to conclude that gel phosphate was inorganic phosphate inside the mitochondrial membrane and having some hints of nucleotide transfer across the membrane there was little sense to be made of the results and they were never published. Only much later did I appreciate that I had made my first foray into membrane transport.

Hexokinase

In mid-1951, Cori suggested that I work on hexokinase. There were two aspects of particular interest. There was Weil-Malherbe's observation of glucose-6-phosphate inhibition [13]. There was also a strong activating effect of muscle extracts which Claude Liebecq had observed but had not explained. Within a couple of months, Alberto Sols arrived and we set out on $2^1/_2$ years of fruitful collaboration.

For 30 years my work has more or less steadily developed from the beginnings Alberto and I made together, that is, during all this time I have been studying, in a sense, the "first step" of glucose metabolism. Of course, in the meantime the concept of what is the "first step" has changed. Before Rachmiel Levine discovered that insulin acts at the cell membrane to enhance glucose transport, "first step" meant the hexokinase reaction; afterward, it meant transport. Since the effect on transport was published in 1949 [14], why then did we start work on hexokinase as the "first step" two years later? Well, as everyone knows it usually takes a few years for such a revolutionary concept as Levine's to be appreciated, tested by others and accepted. Moreover, in the case of the Cori laboratory acceptance was naturally rather more difficult to achieve than elsewhere. In that laboratory there had developed over a quarter of a century a strong, continuing commitment to a concept that insulin and other polypeptide hormones acted intracellularly, directly on key enzymes in metabolism [15]. During the early 50's Carl Cori was reluctant to permit, indeed one may fairly say he was strongly opposed to, the investment of effort in studies of membrane transport.

In the four major papers that Sols and I wrote on hexokinase [16–19] seniority of authorship suggests rather faithfully our individual approaches to and tastes in research. I frankly did not see the special utility of the extensive specificity studies [18] at the time we began them. Working alone, I would probably not have taken that route, but of course I joined in enthusiastically as Sols joined in those issues which were more attractive to me. It was a good thing I did. Though the substrate specificity studies were not the chief contribution of our work on hexokinase, I depended heavily on them and used our same collection of glucose analogs to make several substantial points when I later studied glucose transport.

What I have just written brings to light that among my several shortcomings as a researcher, one that has caused me much trouble is my tendency to do an experiment only when I have in mind a specific question to answer or an hypothesis to test. This means conversely almost never to experiment without one or the other. I shy from collecting data, "just to see what happens" and have to prod myself into it. Yet much experience has taught me that, when done by an astute observer, to collect data is frequently the faster route to those things you want to know and often enough the only route to those things you did not expect. Some others, I have found, "collect data" rather routinely and some have their favorite things to try. For Sols it was, and perhaps still is, specificity [20]. Carl Cori always wanted to know the effect of temperature.

Another important aspect of doing research, or at least of having it understood when published, that I ought to have learned early, but have not even yet learned despite the years, is that of the title of the paper. The message, it seems, must be in the title else it is almost certain to be overlooked or, if not overlooked, ignored. The messages from our work with hexokinase and glucose-6-phosphate were news in their time but have only lately been generally regarded that way, perhaps because of the titles [16, 19].

Metabolite regulation of glucolysis

In our first paper [16] we solved the problem of the activating effect of muscle extracts. It was due to their content of phosphofructokinase. We also went on to show, convincingly, that the cellular concentration of a metabolite could regulate glycolysis and in this specific case, that the level of glucose-6-phosphate determined in part the balance between glycogenolysis and glucolysis, turning the latter off when the former was active. Serving both as substrate for phosphofructokinase, the first "irreversible" step of glycolysis, and as an inhibitor of hexokinase, glucose-6-phosphate provided for a steady-state relationship between them in a model two enzyme system; that is, the substrate-inhibited rate of the first enzyme equaled the substrate-stimulated rate of the second.

Metabolite regulation of glycogenolysis by glucose-6-phosphate, among other compounds, was to come later and a role for glucose-6-phosphate in the regulation of glycogen synthesis was to be found when synthesis was finally determined to have a route other than the back reactions of glycogenolysis (e.g. [21]). At the time Sols and I were together, it was dictum in the Cori Laboratory that glycogen was synthesized in vivo by mass action reversal of the phosphorylase reaction as could be demonstrated in vitro (see e.g. [22]). Later also, control of the utilization of glucose-6-phosphate by phosphofructokinase was also indicated to be metabolite regulated [23] following which this area of research blossomed with the elegant studies of Passonneau and Lowry [24, 25]. Even now, there are still new things to be found in this old field. In just the past year α-glucose-1,6-diphosphate which we [19] showed to be a good inhibitor of hexokinase has been demonstrated to be important in the regulation of hexokinase in brain [26].

When the "feed-back" or "end product" inhibition effect in bacteria was described several years after our hexokinase studies [27, 28] the principle of metabolite control seemed to me to be not very different from ours but the elegance of the mechanism in bacteria was compelling. The feed-back of the synthesized metabolite in bacteria is not to inhibit the enzyme forming it as in the hexokinase case but to skip back to inhibit a much earlier one, usually the first, in the sequence. Although the discovery of this effect was attributed for a long time to others it is now widely appreciated that it was clearly described very early by Dische [29].

A regulatory site for glucose-6-phosphate

Our paper on the specificity of glucose-6-phosphate inhibition [19] was devoted to the proof that hexokinase possesses a specific binding site for glucose-6-phosphate in addition to the binding site at which glucose-6-phosphate is formed in the forward reaction or is bound as a substrate in the reverse reaction. This concept of a metabolite specific, independent regulatory site was novel at the time. We knew it was novel and went to a good deal of special care to make certain of our demonstration of it. However, only lately has our description of this in the text, but not the title, been seen by others to be the same thing introduced a good deal later by Monod et al. [30] under the name allosteric inhibition. Part of the delay is certainly owed to the inappropriate use by some of kinetic data to try to decide a physical question (e.g. [31]). Another part, come to light more recently, may be the fact that most allosteric enzymes appear to have subunits, which hexokinase does not, and the argument has been turned around. The new argument goes something like this, if the possession of subunits is the mark of an allosteric enzyme how could we have discovered allosteric inhibition with an enzyme that does not have subunits? Be that as it may, my dear friend Alberto Sols continues to do combat and has apparently finally proven, by a physical method, that we were right all along. At least, this time there is no equivocation in the title, *Brain hexokinase has two spatially discrete sites for binding of glucose-6-phosphate* [32].

Membrane transport

Increasingly during the 50's we found it necessary to invoke the vague concept of "compartmentation" in order to explain the failure of enzymes in cells to behave in the manner predicted from their properties in vitro

[22]. I think it must have been this which led Cori to become re-interested finally in the cell membrane although his focus was not, as it had been in his early years, on "the selective permeability of living membranes" [33]. Now, he wanted to know how these selective events at the cell membrane might influence cellular metabolic rates [22]. Sometime during Richard Field's first post-doctoral year in the department, he was given Cori's blessing to import from Paul Zamecnik's laboratory a dozen or so Swiss mice inoculated with Ehrlich Ascites Tumor and work began. I was only tangentially involved in the preliminary work. However, when Field went back to Boston, Cori asked me to take over the project.

It was great fun. For the next two years, Carl and I discussed yesterday's results or tomorrow's experiment nearly every day. It was very much a joint venture into the unknown for both of us. We toyed with every imaginable idea or concept about carriers and how they might work, from ferryboats, to swinging gates, to revolving doors. One afternoon we discussed the possibility of doing an experiment showing that hexokinase could play the role of carrier for glucose. By next morning, we had each concluded that the experiment could not work because of the expected preponderance of diffusion of free over bound glucose. Our assessment in that case was undoubtedly correct but we, nonetheless, shared in chagrin when the conceptually identical experiment with hemoglobin and oxygen appeared some years later [34, 35]. This latter experiment worked because of the limited solubility of oxygen in water.

On the whole, as I look back on it, it seems to me that most of the ideas for experiments on the ascites tumor cells originated with Carl. While I was busy devising techniques, getting experiments done and analyzing data, he had the leisure to reflect. In any case, the value to my personal growth of that lengthy, close relationship with Carl Cori cannot possibly be overestimated. If any period of my professional life can be judged to be so, that was my period of maturation as a researcher. Erich Heinz has told me that the paper we wrote [36] was used in Frankfort for many years as a "textbook" case of carrier-mediated transport.

Toward the mechanism of glucose active transport

While Cori's interest continued on in the direction of the influence of permeability on metabolic events with Helmreich, Kipnis, Narahara and others, my own diverged toward the question of mechanism. At the phenomenological level facilitated diffusion was, in principle, solved. I wanted to know the mechanism of active transport and how metabolic energy might be transduced to the accumulation of substrate. I decided that the intestine was the tissue to work with because it actively transported glucose and its mucosa was more accessible than that of the kidney. It was important to study glucose, I felt, because at neutral pH glucose carries no charge of any kind in contrast to amino acids which, though they may be neutral in an electrical sense, actually carry at least two charges, one +, the other –. It seemed to me that the presence of these charges could introduce unmanageable complexities into studies of mechanism. I had in mind such well known phenomena as the role of K^+ as an essential activator for pyruvate kinase [37]. It was somewhat wryly that I was later to recall this line of thought as I proceeded to introduce Na^+ into the mechanism of glucose transport. Nonetheless, it looks now that I had reason on my side. An activator effect of Cl^- for animo acid transport has been reported (e.g. [38]).

Fall of the phosphorylation-dephosphorylation hypothesis

In early 1956, we in the Cori laboratory, probably more than others (see [22]), were aware that the long-held hypothesis of phosphorylation-dephosphorylation (Lundsgaard, 1933) had become shaky indeed with the demonstration by Sols [39] that extracts of intestinal mucosa did not phosphorylate Csaky's actively transported analog, 3-O-methyl glucose, and the knowledge that children suffering the genetic absence of kidney glucose-6-phosphatase showed no glucosuria [40]. However, Sols also did not find phosphorylation of galactose which was known to be metabolized by intestinal tissue so there was a question whether he had measured the total phosphorylation specificity of the tissue. It was also possible to invoke the activities of a different kinase and a different phosphatase, e.g., alkaline phosphatase, if the process were viewed as being restricted to the cell membrane [41, 42]. A decisive experiment was needed.

One day, while half-listening to Melvin Cohn trying to persuade me of the advantages of working with bacteria on the basis of the recent successes at the Pasteur Institute [43], I interrupted him with a proposal to use the recently introduced sac technique of Wilson and Wiseman [44] and several of the extensive Sols–Crane collection of glucose analogs to test which hydroxyl groups of glucose were necessary for active transport. Mel agreed with my perception. Within the week, Stephen Krane and I had completed and sent for publication the studies that showed that the hydroxyls at carbon-1 and carbon-6 were dispensable [45], thus disproving the hypothesis of phosphorylation-dephosphorylation. The studies, incidentally, also clearly demonstrated that glucose was not required to mutarotate during transport. This was, of course, of lesser importance because the mutarotase hypothesis [46] could apply, in any case, only to facilitated diffusion. Mutarotation cannot supply the energy requirement of active transport.

Detour

The road was nearly clear now of historical debris. We had virtually isolated the hydroxyl group at carbon-2 as being the only possible site for a covalent reaction to occur and I began to consider the experiments needed to test for reactions of this hydroxyl group when our way was blocked in an unexpected manner. Carl Cori ordered a halt to our work. The reason was that Tom Wilson had been hired and would arrive in the fall.

My reaction was complex. From my earliest days in St. Louis, Cori had profoundly impressed me and I tried to meet his expectations of me, or what I thought were his expectations. Now in my sixth year there, whatever I had actually done, I seemed to be clearly informed that it was in conflict with his firm convictions and stern principles. Cori's attitude in this instance did not square with his oft-repeated claim that a researcher is free to work on whatever he wants, so perhaps I had it wrong about his reasons, perhaps also as to what he actually meant. Nonetheless, I succeeded in arousing in myself a feeling of guilt sufficient to justify stopping work on intestinal transport. I could understand that my knowledge of Wilson's coming could have, probably did, focus my thoughts on the utility of the everted sac method which, then, opened the way for my analysis and resolution of the first part of the question of mechanism.

I wrote to Tom Wilson enclosing a copy of the Crane and Krane [45] manuscript. I offered him the use of the Sols–Crane collection of glucose analogs should he want to pursue specificity studies and hinted at collaboration. Tom responded gracefully, pointing out that his method was published, that a number of people were already using it and that he had no monopoly. He stopped short, however, of suggesting that I go ahead with my studies although there seemed to be no conflict in our research interest. Wilson wanted to, and did, when he arrived in the fall, pursue the search for a phosphorylated intermediate of transport; a search I saw to be futile. In fact, Wilson did not finally give up the phosphorylation-dephosphorylation concept until the completion two years later of his studies with Landau on whether transported glucose went through the intracellular pool of glucose-6-phosphate. It did not [48].

Stopping work on intestinal transport alone accounts for the detour which Steve Krane and I made into a study of galactose uptake by kidney slices [47]. As expected the slice technique was too complicated in terms of the possible membranes involved in sugar transport to be of any value in pursuing the question of mechanism. However, our results in that study added to others done earlier (e.g. [36]) did bring me to the firm conviction, as I wrote to W.D. Lotspeich in September of 1957, that the glycoside, phlorizin, inhibits glucose active transport "primarily and probably solely by preventing entry of sugar into the cell". As will be seen, this understanding of the action of phlorizin was an important link in the chain of evidence leading me to the gradient hypothesis. In 1957, many, including Lotspeich, still believed that phlorizin inhibited sugar transport by acting on intracellular phosphorylation reactions [49, 50].

After Tom arrived, I tried to persuade him to drop his search for the, to my mind, non-existent compound and to collaborate on the further specificity studies needed to nail down the question of the unique need for the hydroxyl at carbon-2. Fairly late in the year after, I think, he had already accepted a post in physiology at Harvard, Tom agreed and we proceeded to the next experiments in the program [51]. We confirmed the uniqueness of the carbon-2 hydroxy group. Altogether it was not a bad bargain. I could now continue the tests for a chemical reaction.

A group transfer reaction with glucose is not the mechanism

In the fall of 1957, with Wilson gone to Harvard and Steve Krane returned for a second year, little time was lost in devising a strategy to test for any possible reactions at the carbon-2 hydroxyl. One of our approaches was to look for ^{18}O exchange in that hydroxyl; there wasn't any. The other was to use an analog of glucose lacking a carbon-bound hydrogen at position 2; it was actively transported. Thus, with the very great and indispensable help of Mildred Cohn and George Drysdale, we succeeded in eliminating all possibilities except esterification which, of course, could occur without ^{18}O exchange [52]. As we cautiously acknowledged in the paper [53] phosphorylation at carbon-2 could not be ruled out but we underscored the point that there was no evidence for a chemical reaction. It was at this point, early 1958, that I stopped believing in any possibility of covalent transduction of energy to glucose as the basis for active transport.

What may not be immediately clear nowadays is the dramatic change in viewpoint which necessarily accompanied, in 1958, rejection of a covalent reaction with glucose as the means of energy transduction, especially when that rejection was made by a biochemist trained by Fritz Lipmann and Carl Cori. Indeed, Lipmann's concept of group transfer reactions had just been hailed as the new truth and adopted for membrane transport by Peter Mitchell under the banner of "chemiosmosis" [54–56].

Energy transduction is at the brush-border membrane

Now, however, it became more imperative than ever that we know at which end of the epithelial cell the gradient of glucose was established, the brush border membrane or the baso-lateral membrane. Was energy transduced to the carrier which phlorizin inhibition revealed to be in the brush border or was it transduced to a carrier at the other end. In my letter to Tom Wilson in early 1956, I had told him of my interest to make this determination by means of autoradiography. However, I didn't get around to giving it a serious try until the fall of 1957. I couldn't get it to work. I turned then to the preparation of single, isolated cells reasoning that if the cells accumulated sugar to high levels it would prove that the gradient was established at the brush border membrane. Of course, a failure to accumulate sugar would prove nothing because such failure could be attributed to leakiness of the membranes torn from their natural surroundings. All attempts failed and to support leakiness as the reason, all my preparations of single cells failed to exclude the dye, nigrosin [57].

There were, of course, many other experiments done to round out our understanding of intestinal sugar transport [49, 50, 58] and it is curious to me now that I would not at that time accept accumulation in our intact villi preparation [58] as proof of a brush border location for energy transduction. Intact villi were prepared as the first step in the isolation of single cells so I was working with them throughout 1958. I suppose I could not bring myself to argue something that was probably true but not demonstrable, at least at that time; that the lamina propria provided no restriction to the diffusion of sugar and that the observed accumulation must have been intracellular.

Certainly my attitude was rigorous as is expected of a good chemist, but it impresses me more now as rigor of such a degree as to approach, if not to attain, rigidity. Mike Krahl used to complain that I was much too literal. I still am for that matter and I doubt that any clearer proof could be found than the fact that I would not accept a brush border locale for energy transduction until it was proved unequivocally in a collaboration with members of Ollie Lowry's department of the floor above, and with his micromethods [59]. Everted sacs were frozen during active transport, sectioned, and samples of cells and lamina propria were taken for assay. At all time periods, from 15 seconds to 15 minutes, the concentration in the cells far exceeded that in the lamina propria. We now knew that the same brush border membrane carrier which phlorizin inhibited also received the energy transduced from metabolism. The question was how.

The idea of Na^+-coupling

We were now in early 1959 and by mid-summer I had decided that energy was transduced by a Na^+ circuit established by an ATP-energized pump. However, to get there I had first to decide that energy was not delivered to the carrier by a covalent reaction. After all, we had not ruled out covalent reactions as a class, we had simply ruled out covalent reactions with the substrate. Covalent reactions with the carrier remained a possibility and as usual, thermodynamics made no restriction on the kind of covalent reaction that could take place.

However, for covalent energy transduction to the carrier, I only seriously considered alternate phosphorylation-dephosphorylation. The clear reason for this restricted view was the existing knowledge about the $Na^+ + K^+$ pump which was easily understood as driven by a reaction of ATP with the transport protein rather than with the substrate [60] though modelling had not advanced much beyond the early concept of Shaw [61]. I no longer remember whether the concept of the cyclic interconversion of phosphorylase *a* to phosphorylase *b* by ATP-phosphorylation and phosphatase-dephosphorylation was actually known in the Cori laboratory in early 1959. If it was it could have served as a model (Fig. 2). I rather think it came just a bit later [62] after I had already committed to Na^+ and could no longer be confused. At any rate, I got rid of phosphorylation-dephosphorylation from my thinking in a rather cavalier, perhaps naïve, way that in today's climate with protein phosphokinases and phosphoprotein phosphatases acting everywhere on intrinsic membrane proteins I would not have dared to do. Nor would I have been able to do it had the cells I was working with been round rather than columnar and less highly specialized in structure. I took another look at epithelial cells in phase

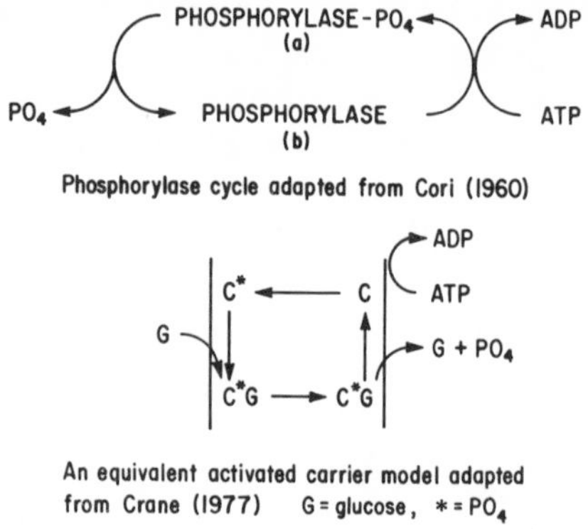

Fig. 2. Model of possible energy transduction to glucose transport by carrier phosphorylation-dephosphorylation.

contrast and electron micrographs and told myself that ATP-molecules, lacking tails, were not going to swim all the way from the mitochondria, through the terminal web and along the length of the narrow channels of the microvilli to the outer portions of the brush border membrane against the currents generated by the substances being absorbed and carry out phosphorylation at anywhere near the rate needed to account for the rate of intestinal absorption.

The obverse of the same proposition gave me immediately to understand that whatever was the driving force, it needed to be readily available at the membrane. Given the state of the art in 1959, the conclusion that the driving force was some kind of ion seemed, at this point, to be inescapable. However, what I knew about ions and membranes was about what I remembered from my physiology course at Harvard in 1947. Added to this was Na^+-K^+-ATPase [60] which I heard from Skou, himself, in a seminar in the physiology department in St. Louis. I also had the impression that it was by then widely accepted that in order to survive all animal cells were required to pump Na^+ out of themselves. This last I checked out with Dan Tosteson. His answer, "yes" pushed me along.

It was very satisfying to get to this point. From the first, I had not liked phosphorylation-dephosphorylation because it seemed energetically wasteful. Tacking active transport onto a process which the cell was obliged to carry out anyway to insure its own survival was consequently very attractive. The cell had once again adapted to the evolutionary burden of its transmembrane ionic profile by making use of it. As for getting membrane work out of an ion gradient, which some to this day find difficult to accept, what was in my mind was no more than the principle that if energy can be transduced from a chemical reaction to create an ion gradient, the reverse can also happen. I saw no problem in replacing phosphorylation with ion-coupling as the driving force because I had long believed that any given reaction, among those of equivalent energy yield, found to occur in metabolism must be viewed as a mechanistic, not a thermodynamic requirement. Indeed, this was the crux of the issue with regard to the mechanism of glyceraldehyde-3-phosphate dehydrogenase which I had solved nearly 10 years earlier. Also, it obviously greatly helped that, coming into membrane transport from the outside, no one had ever told me how active transport was supposed to work.

The first model, convective-coupling

By July, I was seriously trying out ideas on how to couple an influx of Na^+ at the brush border membrane to the transport of sugar and diagrammed for Clark Read the one which is described in Crane [50] (see Fig. 3). Precious though the story seems in the telling, this diagram of Na^+-coupling between the Na^+-pump and glucose transport was actually drawn for the first time in what little sand there was on Stony Beach at Woods Hole.

My reasoning in developing this convective-coupling mechanism is best explained by quoting myself [50]:

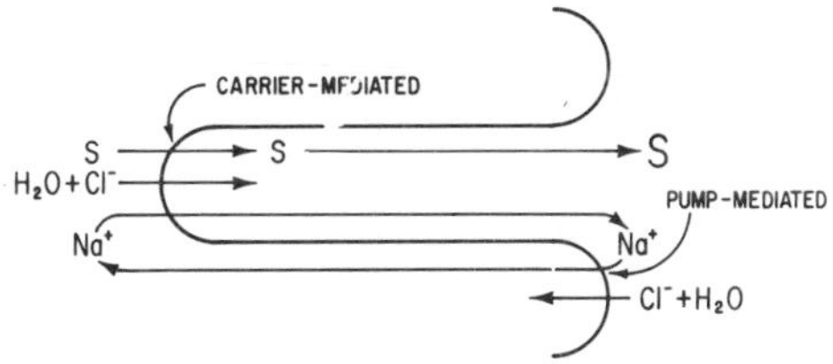

Fig. 3. Model of possible convective-coupling of glucose transport to an Na^+pump by an Na^+ circuit. From Crane [69].

"All animal cells that have been examined seem to possess in their membrane structure-specific sites through which sugars may enter and other sites out of which Na^+ is 'pumped'. In cells such as the erythrocyte which have relatively smooth limiting membranes these sites, whatever their location on the surface, are probably about equidistant from the main intracellular compartment. These cells do not actively transport sugar. In the intestinal and tubular epithelial cells, on the other hand, there is a typically rough, or brush-like border on the side facing their respective lumina which is composed of finger-like, presumably cytoplasm-filled, projections called microvilli. These cells carry out active transport of sugar. Could this be the result, one may ask, of the distribution along the brush border in a specific way of the same kind of structure-specific sugar entry sites and sites of Na^+ pumping present in other cells."

In reviewing this idea, it seems to me that it was not such a bad first attempt. It obeys the principle called Occam's Razor. Also it provides a means of transducing energy to glucose without a covalent reaction. Up to this time, I had done no experiments with Na^+, it was all thought. We dispensed with the convective-coupling idea very quickly when we found that penetrating ions like Cl^- were not required for active transport and that the presence of Na^+ was necessary for the entry of glucose. It was left in the review [50] because I did not have a better idea to substitute for it.

On returning to St. Louis after Labor Day, I greeted Ivan Bihler who was newly arrived from Rothstein's laboratory and put him to work to test my idea. I busied myself with the writing of the review of the field of Intestinal Absorption of Sugars which was due at the end of the year [50]. I also went in to say "hello" to Carl Cori and to tell him about my idea for Na^+ as the driving force for transport. He disconcerted me by sitting back in his chair and saying, with a note of surprise, "Riklis and Quastel".

I instantly knew what he meant because I had reviewed three papers from Quastel's laboratory for *Archives of Biochemistry and Biophysics* and sent my evaluation on 30 September 1957. The papers were actually published in *Can. J. Biochem. Physiol.* Nonetheless, in the first of the series [63] it was shown that when Na^+ was completely replaced by K^+, absorption of glucose by isolated surviving guinea pig intestine was abolished and the statement was made, "sodium ions seem to be essential for the process of active transport". I had completely forgotten.

Why I had forgotten I cannot say. Perhaps, it was because the focus of their work was on the effects of K^+ concentrations on transport, effects with which I disagreed [50] and the effect of zero Na^+ was only observed incidentally to the use of the highest K^+ concentration. Riklis and Quastel [63] made no speculation on the meaning of the Na^+ effect so there was no question of any priority on my concept of coupling and it is actually too bad that I forgot. Had I remembered, I could have saved myself some valuable months of time and gone to Woods Hole that summer with experiments done and not just ideas. We also would have been well ahead of Tim Csaky rather than slightly behind.

I found out about Csaky's work [64] when I told Dan Tosteson about my coupling concept. Dan knew that Csaky had taken a sabbatical leave in Ussing's Institute in Copenhagen and had somehow learned of the work. I seem to recall mention of an abstract but, if there was one, it is not in my collection. The final net result of being made to remember about Riklis and Quastel and of being told about Csaky was, of course, an unshakeable conviction that, indeed, we had chosen the right track. I remarked to Bihler that "Csaky may be ahead experimentally but we are ahead conceptually". Csaky was thinking of direct interaction of glucose with the Na^+-pump. I was sure that the driving force came indirectly from the pump by way of the Na^+ circuit.

The writing of the review [50] had come at a most propitious moment. We had begun the process of leaving the past behind and entering the future of membrane transport and no better time could have been chosen for an exhaustive reading of all that had been published before, along with what I could gather together of manuscripts in press including that of Csaky and Thale [64]. I scanned or read close to 1000 papers, I referenced 300 and I knew the field, I thought, well enough.

Coupling by cotransport

The experiments which Bihler carried out in the fall of 1959 and the spring of 1960 were done against my personal deadline. I had been invited to present a paper at the International Conference on Membrane Transport and Metabolism to be held in Prague in August of 1960. I wanted very much to have the answer to the mechanism in time. The experiments which were finally to yield the answer were all completed well before time but I left for Prague without a final commitment to mechanism. Part of the reason may have been that I had left a copy of my review article with Fred Snell and he sent me a letter of comment. In this letter, dated 11 February 1960, Fred took strong issue with my thoughts that glucose transport might be a reversible process. He came down hard on the point that active transport must be a thermodynamically irreversible process. Inasmuch as I would naturally have granted a superior knowledge of membrane phenomena to Fred and since there was no model I could make using the

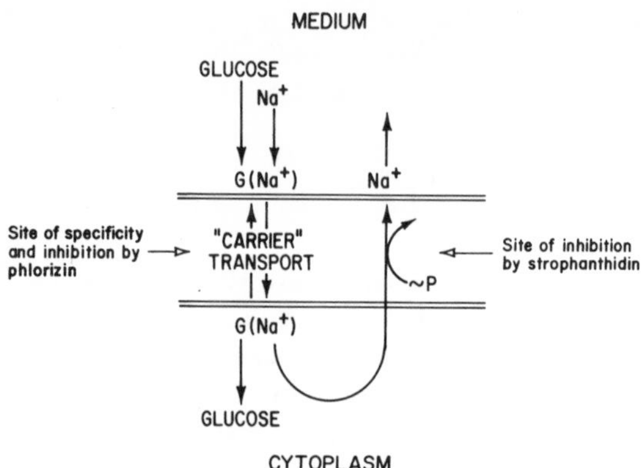

Fig. 4. Model of cotransport coupling of glucose transport to an Na^+pump by an Na^+ circuit. Redrawn from Crane et al. [66].

Na^+ circuit coupled to a glucose carrier that was thermodynamically irreversible, I would have been stuck. However, the undoubtedly largest part of the reason for delay is that I had also to fit in the work with David Miller on the brush border membrane enzymes [65]. I am certain that if we had not got into this area at the same time we were trying to work out the transport mechanism, the latter would have been conceptually easier. In any event, at Prague I finally decided that our experiments did actually define the mechanism and went ahead with the idea of Na^+-glucose cotransport in a mobile carrier model. What I presented at Prague [66] is redrawn (Fig. 4) but solely to leave out the confusion of the hydrolytic enzymes so as to permit seeing that the concept was clearly and explicitly presented. On returning to St. Louis, I showed my drawing to Bihler and Miller and asked them if they liked it. They did.

At that moment in time, that is, at the Prague meeting, we knew or had good reason to believe the following:

(1) Carrier mediated entry and active transport of glucose both occur in the brush border membrane [59] making it likely that the two events, entry and accumulation, occur at the same time.

(2) There is no chemical reaction of glucose [53]. Therefore, given that the convective-coupling mechanism is incorrect (see 5 and 6 below) energy for active transport is not transduced to glucose itself, it must be transduced to the carrier.

(3) Uncouplers of aerobic phosphorylation suppress active glucose

transport [58]. This probably means that ATP is used.

(4) Energy transduction to the carrier is not made directly by ATP in a cycle of phosphorylation-dephosphorylation (inference explained above). Therefore, ATP must be used indirectly as by an ion pump.

(5) Among cations, Na^+ is uniquely required at the mucosal surface for active transport to occur ([63, 64], and Bihler, experiments in September, 1959). Replacement of chloride with slowly or non-penetrating anions has no appreciable effect on active transport. Whatever happens it is question only of Na^+ (Bihler, various experiments from September, 1959 to June, 1960).

(6) Na^+ is also required at the mucosal surface for glucose entry to occur when active transport is suppressed by nitrophenols and replacement of oxygen with nitrogen (Bihler, experiments in January, 1960).

(7) Since phlorizin prevents entry of glucose (see text above, 1957) and Na^+ is required for entry of glucose it is likely that phlorizin and Na^+ act at the same step.

(8) The Na^+-pump inhibitor, strophanthidin, inhibits active transport of glucose but only active transport. It has no effect on the anaerobic Na^+-dependent entry of glucose (Bihler, experiments in June, 1960). Thus, the pump is necessary for active transport but it is not a part of the entry mechanism.

(9) The level of accumulated substrate varies with the external concentration of Na^+ in the manner expected of a steady-state (Bihler, experiments in February, 1960). This means that the force available to establish a glucose gradient is proportional to the Na^+ gradient.

(10) Under these apparent steady-state conditions, there is exchange of added labelled substrate with the accumulated intracellular substrate. The rate of exchange depends upon the external Na^+ concentration (Bihler, experiments in April, 1960).

Most of the data, available on the dates indicated above, were published in final form a good deal later [67, 68] and only the most critical few were presented at Prague because of the restrictions of time imposed. However, all of it was available to me and I saw that the only model which fitted all the criteria was the gradient-coupled cotransport model as I offered at Prague (Fig. 4) and as it has come finally to be proved beyond reasonable doubt [69].

Aftermath

At a symposium held in Zurich in 1976 [70] the feeling was rather widespread among the participants that the modern era of the biochemistry of membrane transport was born at Prague in August, 1960 and that the current meeting celebrated its coming of age. It was certainly true that the concept of ion-coupled membrane processes, having survived hundreds of attempts to do away with it, was finally believed without serious reservation in Zurich. Two years later the chief advocate of the most startling example of ion-coupled membrane processes, Peter Mitchell, was awarded the Nobel Prize.

Beyond surviving, the concept of ion-coupled membrane processes has, I understand, come to be viewed by some historians as a paradigm and there is considerable interest in its origins. I cannot tell them everything about that but I believe I can tell them the date of birth. The reason is that I seem to have played at Prague the unique role of midwife to this paradigm.

The diagram in Fig. 4 was presented at Prague. Search as I will, and have done, in the proceedings of the Prague meeting [71], in all of the papers read for my review and in others of that period come to hand since, I cannot find another which clearly expresses the concept of ion-coupling. It is, moreover, right. There are plenty of descriptions of membrane processes involving ions. There are ATP-driven and electron transport-driven transmembrane pumps for Na^+, K^+ and H^+. The fact that ATP-using kinase reactions were reversible [72, 73], and that the equilibrium position depended upon the concentration of H^+ [74] was well known to biochemists, at least to those involved in carbohydrate metabolism. There is even mention by Mitchell [75] of "the possibility that the movement (or translocation) of ions can actually take place through a chemical reaction, which is being catalyzed by an enzyme, such as an ATPase". The effect of K^+ on amino acid transport was extensively discussed. Only with bacterial transport was there yet no hint of ion involvement. However, with all of this the concept of coupling was represented only by Fig. 4.

From what is published one finds the first approach to proton-coupling in a lecture given by Mitchell on 17 September 1960 [76] to be followed in a few months by a description of his views with respect to proton-coupling in ATP formation [77]. Proton-coupling in bacterial transport came later. The published record does not, of course, tell us when an idea was formed. Nor is an idea such as Na^+-coupling which I invented for intestinal glucose transport a year before Prague or H^+-coupling which Mitchell invented, I know not when, really a paradigm until it is shown to be general. When did the generalization occur? The answer is in what follows.

I gave my talk. At the end of it, Peter Mitchell cried out, "You've got it". He then hurried to the table I shared with Aharon Katchalsky and I re-explained the model to them both. I had proposed Na^+-coupling. Together with what Mitchell surely had in mind, H^+-coupling, the generalization was, on the instant, made. The paradigm, gestated simply as a mechanism for the active transport of sugar by the intestine for a full 12 months since that day on the beach at Woods Hole, was born on 24 August 1960.

Epilogue

In 1962, I left Carl Cori's department. Carl said to me, "Bob, it was very difficult living with you, but immensely worthwhile." I cannot phrase better what I feel about my years with Carl. Some months after my arrival in Chicago, a new friend, wishing to make a point, sent me a copy of one of my granduncle Stephen's poems:

I saw a man pursuing the horizon;
Round and round they sped.
I was disturbed at this;
I accosted the man.
"It is futile," I said,
"You can never – "
"You lie," he cried
And ran on.

Stephen Crane

REFERENCES

1 M. Kunitz and M.R. McDonald, J. Gen. Physiol., 29 (1946) 393–412.
2 R.K. Crane, Biol. Bull., 93 (1947) 192–193.
3 B. Libet, Biol. Bull., 93 (1947) 219–220.
4 W.F. Loomis and F. Lipmann, J. Biol. Chem., 173 (1948) 807–808.
5 R.K. Crane and E.G. Ball, J. Biol. Chem., 188 (1951) 819–832.
6 R.K. Crane and E.G. Ball, J. Biol. Chem., 189 (1951) 269–276.
7 H. Niemeyer, R.K. Crane, E.P. Kennedy and F. Lipmann, Bol. Soc. Biol., Santiago, 10 (1953) 54–58.
8 R.K. Crane and F. Lipmann, J. Biol. Chem., 201 (1953) 235–243.
9 F. Lipmann, Enzymol., 1 (1941) 99–162.
10 J. Harting and S.F. Velick, J. Biol. Chem., 207 (1954) 857–866.
11 J. Harting and S.F. Velick, J. Biol. Chem., 207 (1954) 867–878.
12 R.K. Crane and F. Lipmann, J. Biol. Chem., 201 (1953) 245–246.
13 H. Weil-Malherbe and A.D. Bone, Biochem. J., 49 (1951) 339–347.
14 R. Levine, M. Goldstein, S. Klein and B. Huddlestun, J. Biol. Chem., 179 (1949) 985–986.
15 C.F. Cori, in Proceedings of the First International Congress of Biochemistry (1949) 1309–1318.
16 R.K. Crane and A. Sols, J. Biol. Chem., 203 (1953) 273–292.
17 A. Sols and R.K. Crane, J. Biol. Chem., 206 (1954) 925–936.
18 A. Sols and R.K. Crane, J. Biol. Chem., 210 (1954) 581–595.
19 R.K. Crane and A. Sols, J. Biol. Chem., 210 (1954) 597–606.
20 E.E. Machado de Domenech and A. Sols, FEBS Lett., 119 (1980) 174–176.
21 E.E. Smith, P.M. Taylor and W.J. Whelan, in F. Dickens, P.J. Randle and W.J. Whelan (Eds.), Carbohydrate Metabolism and its Disorders, Academic Press, London, 1968 pp. 89–138.
22 C.F. Cori, in D.E. Green (Ed.), Currents in Biochemical Research, Interscience Publishers, New York, 1956 pp. 198–214.
23 T. Bücher, Angew. Chemie, 71 (1959) 744.
24 J.V. Passonneau and O.H. Lowry, Biochem. Biophys. Res. Commun., 7 (1962) 10–15.
25 O.H. Lowry and J.V. Passonneau, J. Biol. Chem., 241 (1966) 2268–2279.
26 I.A. Rose and J.V.B. Warms, Biochem. Biophys. Res. Commun., 92 (1980) 1030.
27 H.E. Umbarger, Science, 123 (1956) 848.
28 R.A. Yates and A.B. Pardee, J. Biol. Chem., 221 (1956) 757–770.
29 Z. Dische, Bull. Soc. Chim. Biol., 23 (1941) 1140–1148.
30 J. Monod, J.-P. Changeux and F. Jacob, J. Mol. Biol., 6 (1963) 306–329.
31 H.J. Fromm and V. Zewe, J. Biol. Chem., 237 (1962) 1661–1667.
32 P.A. Lazo, A. Sols and J.E. Wilson, J. Biol. Chem., 255 (1980) 7548–7551.
33 C.F. Cori, J. Biol. Chem., 66 (1925) 691–715.
34 J.B. Wittenberg, Biol. Bull., 117 (1959) 402–403.
35 P.F. Scholander, Science, 131 (1960) 585–590.
36 R.K. Crane, R.A. Field and C.F. Cori, J. Biol. Chem., 222 (1957) 649–662.
37 J.F. Kachmar and P.D. Boyer, J. Biol. Chem., 200 (1953) 669–682.
38 J.R. Imler and G.A. Vidaver, Biochim. Biophys. Acta, 288 (1972) 153–165.
39 A. Sols, Biochim. Biophys. Acta, 19 (1956) 144–152.
40 G.T. Cori, Harvey Lectures, Ser. 48 (1952–1953) 145–171.

41 W. Wilbrandt, Arch. Exp. Pathol. Pharmakol., 212 (1950) 9–31.
42 W. Wilbrandt, Soc. Exp. Biol. Symp., 8 (1954) 136–164.
43 G.N. Cohen and J. Monod, Bacteriol. Rev., 21 (1957) 169–194.
44 T.H. Wilson and G. Wiseman, J. Physiol., 123 (1954) 116–125.
45 R.K. Crane and S.M. Krane, Biochim. Biophys. Acta, 20 (1956) 568–569.
46 A.S. Keston, Science, 120 (1954) 355–356.
47 S.M. Krane and R.K. Crane, J. Biol. Chem., 234 (1959) 211–216.
48 B.R. Landau and T.H. Wilson, J. Biol. Chem., 234 (1959) 749–753.
49 R.K. Crane, Biochim. Biophys. Acta, 45 (1960) 477–482.
50 R.K. Crane, Physiol. Rev., 40 (1960) 789–825.
51 T.H. Wilson and R.K. Crane, Biochim. Biophys. Acta, 29 (1958) 30.
52 M. Cohn, Biochim. Biophys. Acta, 20 (1956) 92–99.
53 R.K. Crane and S.M. Krane, Biochim. Biophys. Acta, 31 (1959) 397–401.
54 P. Mitchell, Nature, 180 (1957) 134–136.
55 P. Mitchell, Biochem. Soc. Symp., 16 (1959) 73–94.
56 P. Mitchell and J. Moyle, Nature, 182 (1958) 372–373.
57 J.P. Kaltenbach, M.H. Kaltenbach and W.B. Lyons, Exp. Cell Res., 15 (1958) 112–117.
58 R.K. Crane and P. Mandelstam, Biochim. Biophys. Acta, 45 (1960) 460–476.
59 D.B. McDougal Jr., K.D. Little and R.K. Crane, Biochim. Biophys. Acta, 45 (1960) 483–489.
60 J.C. Skou, Biochim. Biophys. Acta, 23 (1957) 394–401.
61 T.I. Shaw, Ph. D. Thesis, Cambridge University. Quoted by I.M. Glynn, J. Physiol., 134 (1956) 278–310.
62 C.F. Cori, in Atti del 1 Symposium sull'estere di Cori e glucidi fosforilati, 1960.
63 E. Riklis and J.H. Quastel, Can. J. Biochem., 36 (1958) 347–362.
64 T.Z. Csaky and M. Thale, J. Physiol., 151 (1960) 59–65.
65 D. Miller and R.K. Crane, Biochim. Biophys. Acta, 52 (1961) 293–298.
66 R.K. Crane, D. Miller and I. Bihler, in A. Kleinzeller and A. Kotyk (Eds.), Membrane Transport and Metabolism, Academic Press, New York, 1961 pp. 439–449.
67 I. Bihler and R.K. Crane, Biochim. Biophys. Acta, 59 (1962) 78–93.
68 I. Bihler, K. Hawkins and R.K. Crane, Biochim. Biophys. Acta, 59 (1962) 94–102.
69 R.K. Crane, Rev. Physiol. Biochem. Pharmacol., 78 (1977) 99–159.
70 G. Semenza and E. Carafoli (Eds.), Biochemistry of Membrane Transport, Springer Verlag, Berlin, 1977.
71 A. Kleinzeller and A. Kotyk (Eds.), Membrane Transport and Metabolism, Academic Press, New York, 1961.
72 H.A. Lardy and J.A. Ziegler, J. Biol. Chem., 159 (1945) 343–351.
73 J.L. Gamble Jr. and V.A. Najjar, Science, 120 (1954) 1023–1024.
74 E.A. Robbins and P.D. Boyer, J. Biol. Chem., 224 (1957) 121–135.
75 P. Mitchell, in A. Kleinzeller and A. Kotyk (Eds.), Membrane Transport and Metabolism, Academic Press, New York, 1961 p. 319.
76 P. Mitchell, in T.W. Goodwin and O. Lindberg (Eds.), Biological Structure and Function, Vol. II, Academic Press, London, 1961 pp. 581–603.
77 P. Mitchell, Nature, 191 (1961) 144–148.

G. Semenza (Ed.) Selected Topics in the History of Biochemistry: Personal Recollections (Comprehensive Biochemistry Vol. 35)

Chapter 4

From Fructose to Fructose-2,6-bisphosphate with a Detour through Lysosomes and Glycogen

HENRI-GÉRY HERS

Laboratoire de Chimie Physiologique, Université Catholique de Louvain and International Institute of Cellular and Molecular Pathology, UCL 75.39, 75, avenue Hippocrate, B-1200 Brussels (Belgium)

From the title of this chapter, it is apparent that my scientific interest during the last 34 years has been confined to a rather restricted area of biochemistry. Taking glucose-6-phosphate as a crossroad, my metabolic investigation extended from fructose and glucose to glycogen and pyruvate, the liver being my favourite field of investigation. Only the discovery of the lysosomal diseases introduced me to a much broader biological and chemical domain but this was a parenthesis which I closed as soon as the message that I wanted to convey had been heard.

My investigation of carbohydrate metabolism was performed at a time when the major advances in that field had already been made by Carl and Gerty Cori, by Luis Leloir, Earl Sutherland and others. The dominant interest in biochemistry was shifting to molecular and cellular biology and, even in my immediate surroundings, Christian de Duve, who had introduced me to carbohydrate metabolism, gave up the field to study the subcellular structures and to eventually discover lysosomes. This often gave me the impression of being behind the times, in the rear guard of biochemistry. It is indeed remarkable that fructose-2,6-bisphosphate was discovered 71 years after that Young [1] had isolated the fructose bisphosphate which was later identified as fructose-1,6-bisphosphate by Levene and Raymond [2]. Obviously, the pioneers had left behind for the enjoyment of the newcomers quite a few things to be discovered. The purpose of this chapter is to explain how some of these discoveries, in which I had the privilege to participate, were made.

Plate 4. Henri-Géry Hers.

As a medical student, I participated in the rediscovery of glucagon

A medical student has the tedious obligation to memorize a long series of badly understood pathological conditions. But it is known that "the awareness of the present unsatisfactory state of knowledge is the stimulus to the production of ideas" [3]. This awareness oriented me to medical research.

The "Université Catholique de Louvain", my Alma Mater, has always encouraged students to participate in research during their free time. I myself spent many hours in the Laboratory of Physiology directed by J.P. Bouckaert and, as early as in 1943, I had the good fortune to be associated with the research of Christian de Duve, who had got his M.D. degree a few years before. The major preoccupation of our group was to clarify the mode of action of insulin. De Duve had reached the conclusion that the glycogenolytic action of commercial insulins on the liver, described by several American investigators, was due to an impurity; he himself had found no such effect with the Danish (Novo) insulin that we were currently using during war time. The impurity was likely to be glucagon, a polypeptidic contaminant of the amorphous preparations of insulin [4], which was generally believed to have been eliminated from the crystalline preparations of the hormone with the mother-liquor. This was actually true for the initial Abel procedure of crystallisation but not for the more currently used Scott insulin-zinc method (for a review see [5]).

In September 1944, the American and British armies liberated our country and the most highly purified Lilly insulin became available for our research. We could rapidly show that this insulin, when injected intravenously into rabbits, caused an initial hyperglycaemia, quickly followed by hypoglycaemia.

Since the hyperglycaemic effect was transient, I proposed to administer the Novo or the Lilly insulins by constant infusion, following a saturating dose of Novo insulin; glucose was simultaneously infused to maintain a normal level of glycaemia. Most remarkably, the glucose requirement, which was 23 mg/min/kg body weight upon infusion of the glucagon-free insulin, decreased to 3.2 mg when the glucagon-contaminated insulin was used. The difference (20 mg/min/kg) had obviously been mobilised from endogenous sources. It is noteworthy that a 2-kg rabbit received as many as 90 units of insulin with little effect on the level of glycaemia, when given in this way [6]. Our conclusions were confirmed and extended one year later by Sutherland and Cori [7], who reported that the glycogenolytic action of the Lilly insulin was due to a "hyperglycaemic glycogenolytic factor" (HGF; now recognised to be glucagon) not detectable in the Novo insulin. This was actually the first publication of Earl Sutherland in a field of research which eventually led him to the discovery of cyclic AMP. It took many more years until the presence of glucagon in the circulating blood was demonstrated by Unger [8], establishing its hormonal nature.

This early work, performed during war time, had been done with very poor technical means but with much thought given to the problem beforehand, essentially by de Duve. Its success is at the origin of my deep belief that reflection is much more important in research than heavy equipment, a conviction that was obviously reinforced by my natural lack of interest for elaborate instrumentation. It is at that time that I became aware that doing research was my natural way of life, and also, that the elucidation of the mode of action of insulin became for me a permanent, although inaccessible goal of research.

How the agglutination of microsomes at pH 5 opened the way to several discoveries

In the summer of 1948 I obtained my M.D. degree and a fellowship from the Fonds National de la Recherche Scientifique to work as a research associate with Christian de Duve. With the hope to discover the mode of action of insulin, we were studying, in collaboration with Jacques Berthet, the mechanisms by which glucose is formed in the liver from a mixture of hexose phosphates, mostly glucose-6-phosphate and glucose-1-phosphate, maintained in equilibrium by phosphoglucomutase. In an attempt to purify the phosphatase by isoelectric precipitation, I found that this enzyme was not only insoluble at pH 5 but that this insolubility was quite irreversible. Doing so, I had rediscovered the pH 5 agglutination of subcellular structures previously described by Albert Claude [9]. The separation of the pellet from the clear supernatant (which is equivalent to the high-speed supernatant) could be accomplished in a few minutes by low-speed centrifugation. When performed in the cold, this procedure allowed to separate glucose-6-phosphatase from all soluble enzymes, including phosphoglucomutase, and to define its specificity [10]. In a next step, the microsomal localisation of glucose-6-phosphatase was established by differential centrifugation [11].

This was the start of a long series of investigations by Christian de Duve, Jacques Berthet, Henri Beaufay and others on the intracellular localisation of enzymes, which led them to the discovery of lysosomes. This story has been beautifully told by C. de Duve [12]. As explained below, the pH 5 agglutination of microsomes gave also a clue to the problem of the metabolism of fructose-1-phosphate in the liver.

The metabolism of fructose in the liver

A sustained interest in the mode of action of insulin led me to an investigation of liver glucokinase. This work was not very successful and it was not until 10 years later that the high K_m glucokinase was recognised in the laboratory of Weinhouse [12a] and of Sols [13]. However, control experiments made with fructose introduced me to fructokinase (see also next section) and to fructose metabolism. The main problem to be solved in this field was how fructose-1-phosphate, the product of fructokinase, is converted to a glycolytic intermediate. Shortly before, Cori et al. [14] had concluded that a purified fraction from rat liver catalysed, in the presence of Mg^{2+}, the following conversions:

$$\text{Fru-1-P} \rightarrow \text{Fru-6-P} \rightarrow \text{Glc-6-P} \rightarrow \text{Glucose} + P_i$$

in which the formation of glucose-6-phosphate had been measured specifically by reaction with the corresponding dehydrogenase but glucose, assumed to be formed by glucose-6-phosphatase, had been identified only by its reducing power; the authors had tentatively concluded that the conversion of fructose-1-phosphate into fructose-6-phosphate occurred through the action of a Mg-dependent "phosphofructomutase".

In collaboration with T. Kusaka [15], I was soon able to confirm these data, using a pH 5 extract as a source of enzyme. Since, as explained in the preceding section, the pH 5 supernatant is devoid of glucose-6-phosphatase, we knew that the reducing sugar formed was not glucose and that the production of P_i was not due to the hydrolysis of glucose-6-phosphate.

These negative conclusions forced us to consider another pathway for the metabolism of fructose-1-phosphate. Fifteen years earlier, Meyerhof et al. [16] had observed the synthesis of fructose-1-phosphate from D-glyceraldehyde and phosphodihydroxyacetone, catalysed by muscle aldolase. Assuming that this reaction would be sufficiently reversible, fructose-1-phosphate could then be converted to fructose-6-phosphate by the series of reactions shown in Fig. 5. In this mechanism glyceraldehyde accumulates (at least in the absence of ATP) and the triose phosphates are

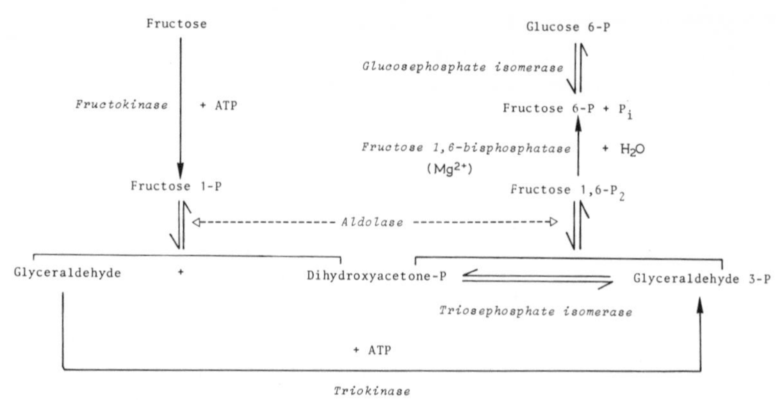

Fig. 5. The metabolism of fructose in the liver [15].

removed by their condensation into fructose-1,6-bisphosphate and hydrolysis of the latter to fructose-6-phosphate and P_i. Accordingly, we could identify the reducing sugar as D-glyceraldehyde thanks to its property of reducing copper at a relatively low temperature and we could attribute the magnesium-dependent formation of P_i to the action of fructose-1,6-bisphosphatase. For 2 mol of fructose-1-phosphate that had disappeared, 2 mol of D-glyceraldehyde, 1 mol of hexose-6-phosphate and 1 mol of P_i were formed. Furthermore, and in strong contradiction with the phosphofructomutase hypothesis, this conversion was quite irreversible, because of the phosphatase step. We also found that the liver contains an enzyme which phosphorylates D-glyceraldehyde into the corresponding triose phosphate; we called it "triokinase".

In the same publication [15] we also described some remarkable properties of liver aldolase and of fructose-1,6-bisphosphatase. We found that the aldolase present in a liver extract is characterised by its ability to split both fructose-1-phosphate and fructose-1,6-bisphosphate at the same maximal velocity whereas the muscle enzyme is very poorly active on fructose-1-phosphate (see Fig. 1 in [15]). Later on it was found [17] that liver and muscle aldolases are very similar proteins, the typical substrate specificities of which can be modified by a slight chemical alteration. All these properties found an interesting application when, as a result of my new interest in inborn errors of metabolism, I investigated the biochemical basis of hereditary fructose intolerance. This was a newly discovered inborn disturbance of fructose metabolism, extensively studied by Rudi

Froesch and others [18], in which the patients become deeply hypoglycaemic when they receive fructose. With Guy Joassin [19], I could show that in these patients liver aldolase is abnormal in that it has lost a great deal of its activity on fructose-1-phosphate and now resembles muscle aldolase.

The hydrolysis of fructose-1,6-bisphosphate is an essential step in the conversion of fructose to glucose in the liver. It explains its magnesium dependency, its irreversibility and the formation of equimolar amounts of P_i and hexose-6-phosphate. However, the only fructose-1,6-bisphosphatase known at that time was the alkaline enzyme described by Gomori [20], which was completely inactive in the pH zone in which we were working. We could show that a fresh liver extract contains a neutral fructose-1,6-bisphosphatase which had all the properties required to explain our results [15]. Because this neutral enzyme catalyses one of the irreversible steps of gluconeogenesis, it attracted much interest during the following 30 years (for a review see [21]). In 1965, Taketa and Pogell [22] described its inhibition by AMP. In my laboratory we kept an intermittent [23] but more often dormant interest in that enzyme. It is only in November 1980 that after this long hibernation, it was discovered by Emile Van Schaftingen that fructose-2,6-bisphosphate, at micromolar concentration, inhibits the enzyme synergistically with AMP, and changes the substrate saturation curve from hyperbolic to sigmoidal [24].

The metabolism of fructose-1-phosphate in a liver extract was also investigated by Leuthardt et al. [25] who could rapidly confirm our conclusions. The cleavage of the hexose into two 3-carbon fragments in vivo needed to be established by an isotopic study. In 1953, the use of radioactivity in metabolic studies was still limited and the proper technology was not available in our laboratory. With the help of the Rockefeller Foundation, I went to Berkeley, where I spent a year in the Department of Plant Biochemistry with Zev Hassid who was well known for his pioneer work with radioactive sugars in plants [25a]. Zev was very hospitable to me, but I could never get him interested in liver metabolism. I spent a very enjoyable year (1953–1954) in his laboratory in the company of Victor Ginsburg and Jack Edelman. Together, we synthesised [1-^{14}C]glucose and converted it to a mixture of glucose, fructose and mannose [26]. It is also at that time that I became acquainted with Elizabeth Neufeld, who was working next door; much later we found a common interest in the lysosomal diseases, a field to which she greatly contributed [27].

I injected [1-^{14}C]fructose into rats (which were considered very incongruous in the plant biochemistry department) and could show that, as expected from the participation of triose phosphate isomerase in the scheme shown in Fig. 5, the label found in liver glycogen had been randomised between C-1 and C-6 of the hexose molecule [28]. To do this work, it was necessary to degrade glucose into its 6 constitutive carbons and to measure the radioactivity of each of them. It quickly appeared, however, that the published procedures were far from satisfactory. I was, therefore, in great despair until the February issue of the *Journal of Biological Chemistry* came with a paper [29] from Harland Wood's laboratory describing the complete degradation of glucose by heterolactic fermentation with *Leuconostoc mesenteroides*. In this type of fermentation, as described by Gunsalus and Gibbs [30], glucose is converted into CO_2 (carbon 1), ethanol (carbons 2+3) and lactic acid (carbons 4–6), the latter 2 constituents being easily degraded chemically. I could immediately obtain the correct strain of *Leuconostoc* from the Microbiology Department, and the method worked rapidly without problems.

I met Carl and Gerty Cori for the first time in September 1954 at a scientific meeting held in Cape Cod. They were very kind to me and had no objection to my views on fructose metabolism. Hassid introduced me by saying that I had had a lot of luck in my work in Berkeley.

"Luck does not exist, replied Carl Cori at once, because it is always the same people who are lucky".

I often remembered this comment when, becoming older, I had to work on committees that distribute money to research projects; in evaluating the chance of success of applicants, I always paid much more attention to what they had accomplished before than to their actual programme. In doing so, I was applying a rule expressed by Beveridge [3] in his book on *The Art of Scientific Investigation:*

"In applied research, it is the project which is given support, whereas in pure research, it is the man."

Unfortunately, few foundations apply this policy but, particularly in the U.S., they demand very detailed applications, which are tiresome for the applicant to write and almost impossible for the reviewer to read.

The ATP-magnesium complex

I still remember the day and even the time of the day when, looking at a graph (Fig. 1 in [31]) which showed the effect of magnesium concentration on the activity of fructokinase measured at two different ATP concentrations, I got the idea that the role of magnesium was to form a complex with ATP and that this ATP-Mg complex was the true substrate of the reaction. During the following days, I had a quick look at the literature and could verify that practically all known reactions which involved ATP also required magnesium and that the optimum rate was obtained at a 1:1 ratio of the two components.

I had been prepared for the idea of the ATP-magnesium complex by what could be considered as an irritating and trivial observation made at the bench. When I was preparing the reaction mixture for the determination of fructokinase, which included ATP, magnesium and fluoride I often observed a cloudy precipitate. I also found that the abundance of this precipitate as well as the rate of fructokinase reaction varied according to the order in which the reagents were added. The maximal rate was obtained when ATP could meet magnesium before fluoride. The precipitate usually dissolved upon slight dilution.

I think none of my findings has been so rapidly and easily accepted as the ATP-Mg complex. Indeed, well before my paper [31] on the role of magnesium was published (this paper was received for publication on May 9, 1951), Christian de Duve met Henry Lardy and informed him of my data. Lardy gave a full account [32] of them in the Symposium on Phosphorus Metabolism (held on June 18, 1951 at the Johns Hopkins University). Judging from the comments of Bentley Glass who summarised the Symposium [33], the idea was immediately accepted. He concluded that the actual substrate of fructokinase was ATP-Mg and kindly qualified my work as an "illuminating study".

The publication of the sorbitol pathway was delayed by thermodynamical considerations

My interest in foetal and seminal fructose came from the fact that, having completed my work on the metabolism of fructose in the liver, I found it natural to investigate the mechanism of fructose formation before writing a monograph entitled *Le Métabolisme du Fructose* [34].

The presence of fructose in the amniotic fluid of the cow had already been recognised by Claude Bernard in 1855 [35] thanks to its laevorotary properties and had been confirmed with more precise analytic methods by Bacon and Bell [36] in 1948. Huggett and his coworkers [37] made an extensive investigation of the mechanism of biosynthesis of this fructose, which was present not only in the foetal blood of the cow but also of the sheep, other ungulates and the whale. They showed that fructose is formed from glucose and that the placenta plays a role in this conversion. They had no hypothesis for the biochemical mechanism by which this fructose was formed.

The presence of fructose as the sole sugar in the semen was firmly established by Mann [38] in 1946. This fructose originates from the seminal vesicles and the prostate and its formation is controlled by male hormones. It is formed from glucose and, according to Mann and Lutwak-Mann [39], this conversion would occur by a phosphorylation-dephosphorylation mechanism: a glucokinase would allow the phosphorylation of glucose but not of fructose whereas alkaline phosphatase would hydrolyse the hexose-6-phosphates formed in a mixture of glucose and fructose (about twice as much glucose as fructose). The selective reutilisation of glucose would explain fructose accumulation. Experimental facts in favour of this hypothesis were the more rapid glycolytic utilisation of glucose than fructose by seminal vesicles and the existence of an alkaline phosphatase in this tissue. The major weakness of the hypothesis was that the mechanisms proposed involved enzymes which were present in most tissues and, therefore, could not explain the specific localisation of fructose synthesis in particular organs.

It was in 1954 that Williams-Ashman and Banks [40] observed that tissues in which seminal fructose is formed are rich in sorbitol dehydrogenase and suggested that fructose could originate by dehydrogenation of sorbitol. They, however, made no suggestions as to the mechanism by which sorbitol could be formed from glucose. I could rapidly confirm their observation and also show that radioactive sorbitol and fructose are formed from labelled glucose by slices of seminal vesicles whereas fructose only is formed from sorbitol [34]. To convert glucose into sorbitol the most obvious mechanism was a reduction; accordingly, NADPH, but not NADH, was oxidised by an extract of seminal vesicles when glucose or other aldehydic sugars were present, and this oxidation was stoicheiometric with the disappearance of the reducing group of the sugar. I called the enzyme responsible for this reaction the "aldose reductase" and the overall mechanism of glucose to fructose conversion could now be written [41]

$$\text{Glucose} + \text{NADPH} + \text{H}^+ \underset{\textit{reductase}}{\overset{\textit{Aldose}}{\leftrightarrows}} \text{Sorbitol} + \text{NADP}^+$$

$$\text{Sorbitol} + \text{NAD}^+ \underset{\textit{dehydrogenase}}{\overset{\textit{Sorbitol}}{\leftrightarrows}} \text{Fructose} + \text{NADH} + \text{H}^+$$

$$\text{Glucose} + \text{NADPH} + \text{NAD}^+ \leftrightarrows \text{Fructose} + \text{NADP}^+ + \text{NADH}$$

A high K_m of aldose reductase for glucose explained why the amount of fructose formed was proportional to glucose concentration. The lack of specificity of the enzyme was not an objection against its physiological role because glucose is the only aldehydic sugar present in large amounts in blood under normal conditions.

This mechanism was discovered in the fall of 1955. It provided, however, no explanation for the fact that fructose accumulates from glucose. At that time a similar role was attibuted to NAD and NADP-dependent dehydrogenases and it was also known that the redox potentials of the two pyridine nucleotides were practically identical. The proposed mechanism contained, therefore, no source of free energy and could only bring glucose and fructose to their thermodynamic equilibrium, i.e. about twice as much glucose as fructose without accumulation of the latter sugar.

It was only at the end of 1955 that Glock and McLean [42] published their remarkable investigation on the relative proportion of the reduced and oxidised forms of the two coenzymes. They showed that NADP is essentially present in its reduced form in all tissues whereas the reverse is true to some extent for NAD. The reason for this difference is that NADPH is continuously generated by the pentose phosphate pathway and does not transfer its electrons into the mitochondria, whereas NADH does. The glucose–fructose conversion, in which NAD^+ is reduced by NADPH, is therefore coupled with the pentose phosphate pathway and the mitochondrial reoxidation of NADH. This provides the energy required for the accumulation of fructose.

The same mechanism also explained the formation of foetal fructose since aldose reductase was present in the placenta of the ungulates and sorbitol dehydrogenase in the foetal liver [34]. The accumulation of fructose in foetal blood was also a consequence of the absence of fructokinase before birth [43].

The sorbitol pathway was published in the fall of 1956 and was enthusiastically received by Williams-Ashman and by Huggett. The mechanism was also easily accepted by the biochemists who had not been previously involved in other theories. Thaddeus Mann remained reluctant. In 1962 he himself obtained evidence in favour of what he called the "Hers's views", but still concluded:

"It is difficult to assess the relative importance of the two pathways of fructose formation, one involving sorbitol and the other phosphorylated sugar esters." [44]

In the 1964 [45] and still in the 1981 [45 a] edition of his classical book devoted to the biochemistry of semen, he was still faced with the same "unsolved dilemma". Huggett organized in November 1957 a Ciba Symposium devoted to the Glucose–Fructose Conversion. This symposium was very instructive for me. I was confident in the sorbitol pathway and considered it rather futile to discuss the previous hypotheses. I learned that it takes time for scientists to give up an idea that they have cherished for many years and that it is hard, for each of us, to see the facts to contradict our views.

Several years later aldose reductase was also discovered in the lens and nervous tissue and was assumed to play an important role in the pathogenesis of cataract and of other disturbances in congenital galactosaemia [46] and in diabetes [47].

A call from a paediatrician introduced me into glycogen storage disease

In 1957, glycogen storage disease had been subdivided by Gerty Cori [48] into 5 subgroups among which type I was the deficiency in glucose-6-phosphatase and type III deficiency in amylo-1,6-glucosidase (debranching enzyme). Type II, also called Pompe's disease, from the name of the Dutch pathologist who described it in 1932 [49], was the most mysterious one because all enzymes known to act in glycogen degradation were active in the tissues of the patients. Furthermore, the affected children were apparently able to mobilise their glycogen, since they responded normally to the hyperglycaemic action of glucagon. Nevertheless, they accumulated large amounts of the polysaccharide in all their tissues and they usually died in the first year of life.

In January 1958, Professor Pierre Denys, who was chief of the Department of Paediatrics in the Louvain University, asked me to analyse a biopsy from one of his patients affected with glycogen storage disease. I accepted this proposal because I thought that it would take me only a few days to demonstrate the deficiency of glucose-6-phosphatase, an enzyme in the discovery of which I had participated and which I knew fairly well. It appeared that glucose-6-phosphatase was normal. It was therefore necessary to investigate the activity of amylo-1,6-glucosidase which was much more difficult to assay. I remembered then that one year before, in a first attempt to investigate the control of glycogen synthesis in the liver, I had observed that [^{14}C]glucose could be incorporated into glycogen by a crude liver homogenate; it had soon appeared, however, that this incorporation was due to a slight reversibility of amylo-1,6-glucosidase and I had given up the work. Now I could use this procedure for a simple assay of the enzyme and found that the incorporation of [^{14}C]glucose was easily measurable in normal tissues but was completely deficient in the liver of my patient [50]. Because of its simplicity, this method became very popular and has since allowed the diagnosis of several hundreds of cases of type III glycogenosis. Soon afterwards, I had the opportunity to present my results in the Geigy symposium on Carbohydrate Metabolism in Children, held in Bern on June 2–7, 1958 [51]. Not only Carl Cori was present, but also most of the paediatricians who in Europe were interested in glycogen storage disease and in other congenital disorders of carbohydrate metabolism. It was at that time that I started with them a collaboration which has now lasted over 24 years and which has brought to my laboratory biopsies from thousands of patients with various types of congenital disorders. I initiated this collaboration with the intention of helping these clinicians in taking care of their patients; it may, however, be more honest to admit that my main motivation was my intense desire to obtain biopsies of type II glycogenosis which had for me the fascinating appeal of the unknown.

Among the liver biopsies sent to my laboratory, I found some with a deficiency of phosphorylase and I published three such cases in a short review article [50] written at the request of the editors of the *Revue Internationale d'Hépatologie.* Having been informed that Hans Stetten was preparing an important review on the subject, I sent him a copy of my manuscript. Stetten [52] called the newly recognised liver phosphorylase deficiency type VI glycogenosis or Hers' disease. This is how a rather small piece of work made my name famous since I now find it in the subject index of most textbooks of biochemistry and of paediatrics and in medical dictionaries. I often noticed a note of admiration in the voice of people who introduced me to their friends, saying: "He's got a disease named after him". Occasionally there were some jokes, too, about "his" and "hers" disease.

The first lysosomal disease was discovered because of the complexity of the glycogen debranching system

In 1963, I published [53] that type II glycogenosis (Pompe's disease) was due to the deficiency of a lysosomal acid α-glucosidase (acid maltase). As explained in the next section, this observation was important because it introduced for the first time the concept of inborn lysosomal disease. The question that one could ask, retrospectively, is how a sensible enzymologist could think of assaying a maltase in an investigation of glycogen metabolism in patients with glycogenosis. Indeed, glycogenolysis was known to occur by phosphorolysis followed by the hydrolysis of the hexose phosphates in the liver and by glycolysis in all tissues. There was no role for a maltase in this pathway.

My observation was pure serendipity. Until 1960, it was generally believed that the limit dextrin which is formed by the exhaustive action of phosphorylase on glycogen was an amylo-1,6-glucoside, in which the outer chains were unequal; the side chains were assumed to be made of single glucose units whereas the outer portions of the main chains would be six to seven units long. This was because amylo-1,6-glucosidase liberates glucose from the limit dextrin which becomes then susceptible to further phosphorolysis. In 1960, Walker and Whelan [54] elegantly demonstrated that the side chains and the outer portions of the main chains of the limit dextrin were four units long. The conversion of this structure into an amylo-1,6-glucoside was believed to occur by an intramolecular transfer of a maltotriose from the side chains to the main chains of the limit dextrin but could also occur by the repetitive transfer of a single glucosyl unit. This discovery indicated that one more enzyme, an oligoglucan transferase or a glucosyl transferase was involved in glycogenolysis; a deficiency of that enzyme could be responsible for a new type of glycogen storage disease.

My problem was then to design an assay for the transferase. A simple one was to incubate a tissue extract with a labelled glucoside, [^{14}C] maltose, and glycogen and to measure the radioactivity incorporated in the polysaccharide. This reaction indeed occurred at a measurable rate. I then performed this assay with a series of pathological livers and muscles originating from patients with various types of glycogenosis. To my surprise, all types III catalysed the reaction normally, whereas all types II were inactive.

It then took me some time to realise that my radiochemical assay was measuring the transferase activity of a maltase and had nothing to do with amylo-1,6-glucosidase, which actually operates as an oligoglucan transferase. The optimal pH of the maltase was remarkably low (pH 4) and routine investigation revealed that the enzyme was sedimentable in a fresh liver homogenate but was recovered in the supernatant when the homogenate had been frozen and thawed. Working in the laboratory of Christian de Duve, I was well aware that these properties are characteristic of a lysosomal enzyme. The lysosomal appurtenance of acid α-glucosidase was then clearly established by its property of structure-linked latency and of sedimentability [55].

The concept of inborn lysosomal disease is a logical deduction made from my observations in type II glycogenosis

Since type II glycogenosis was clearly associated with a deficiency of a lysosomal acid α-glucosidase, it was now necessary *to explain all its manifestations by this single enzymatic defect.* The role of a lysosomal α-glucosidase in glycogen metabolism was far from being clear at that time. The concept of lysosomes had been defined by de Duve and coworkers in 1955 [56] and in the early speculations on the role of these "lytic bodies" it was postulated that some kind of injury to the lysosomal membrane was a prerequisite to the action of the hydrolases on cytosolic structures. It is almost exactly when the deficiency in α-glucosidase was discovered that the "suicide-bag" hypothesis [57] was progressively replaced by the concept of cellular autophagy. Whereas lysosomes had been identified by Alex Novikoff with the pericanalicular dense bodies [58, 59], Ashford and Porter [60] extended the concept to the digestive vacuoles in which mitochondria in various stages of breakdown could be recognised. In the autophagic concept [61], a small area of cytoplasm, containing various cellular constituents, including glycogen, becomes walled off the rest of the cytoplasm by a membrane. An autophagic vacuole is formed and fuses with lysosomes. All the cytoplasmic constituents are then digested by the lysosomal hydrolases. This is a normal process which continuously occurs in living cells and ensures the catabolic part of cellular turnover. The role of my α-glucosidase was then obviously to destroy the glycogen segregated in the vacuoles.

This view has never been seriously challenged and has been illuminat-

ing in various respects:
(1) It indicated that there are two pathways of glycogenolysis in cells; one of them is extralysosomal, phosphorolytic and hormonally controlled. This mechanism is normal in type II glycogenosis and its operation explains why the patients never become hypoglycaemic and respond normally to the hyperglycaemic stimulation by glucagon. The second mechanism is autophagic and lysosomal. Its role is not to supply glucose to the cell but to destroy the glycogen that has been engulfed in the autophagic vacuoles. This glycogen is not accessible to phosphorylase from which it is separated by the lysosomal membrane. In the absence of acid α-glucosidase, it cannot be degraded and it accumulates in the vacuoles; it corresponds to the polysaccharide present in excess in type II glycogenosis.
(2) If the above interpretation were correct, *the glycogen-loaded vacuoles would be visible with the electron microscope.* I was unable to use an electron microscope myself but could convince my friend Pierre Baudhuin, an expert cytologist, of the importance of this investigation. Soon afterwards Dr. H. Loeb from Brussels University offered me the

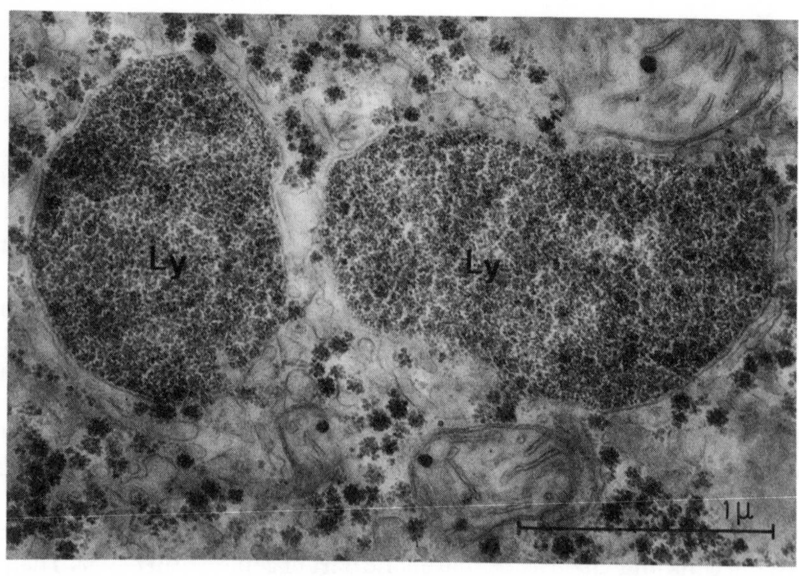

Fig. 6. Ultrastructure of the liver of a child with type II glycogenosis. Two lysosomes overloaded with glycogen are visible. The glycogen particles are also clearly apparent in the cytoplasm (× 36 000) [62].

opportunity to investigate a new case of type II glycogenosis and the ultrastructural study of the liver of this patient was performed. It was in October 1963 that Baudhuin obtained the first micrographs which showed exactly what had been predicted by logical deduction from the biochemical data [53]: both the parenchymatous and the Kupffer cells contained large (upto 8 μm in diameter) vacuoles delineated by a single membrane and filled with easily recognisable particulate glycogen (Fig. 6). Except for the absence of pericanalicular dense bodies (the normal lysosomal structure), the rest of the cytoplasm was normal and also contained a normal amount of particulate glycogen [62].

Because of its internal logic, publication [62] describing the ultrastructure of type II glycogenosis is one of those things which gave me the greatest satisfaction in my entire scientific career. I have therefore no shame to mention that two editors of the *Journal of Cell Biology* agreed to reject the manuscript on the basis that it did not provide new and substantial information concerning lysosomal function.

(3) My previous experience with inborn errors of metabolism and particularly glycogen storage disease had told me that practically *any enzyme can be made inactive by a mutation of the corresponding gene.* It was therefore almost certain that the accident that had happened to acid α-glucosidase also happened to other lysosomal hydrolases involved in the degradation of other cellular structures [53]. *The deficiency of one of these hydrolases* in man or in animals *would be responsible for a specific storage disease characterised by the same morphological hallmark: hypertrophy and overloading of lysosomes.* I proposed that several well known storage diseases like Gaucher's disease, Tay-Sachs disease, metachromatic leukodystrophy and the mucopolysaccharidoses would be explained on that basis. In 1965, at the invitation of Hans Popper, I wrote an editorial for *Gastroenterology* in which the concept of "inborn lysosomal disease" was fully expressed for the first time [63].

(4) Inborn errors of metabolism are experiments of Nature. The pictures of lysosomal overloading obtained by Pierre Baudhuin in type II glycogenosis and later on by François Van Hoof in mucopolysaccharidosis [64] could leave no doubt on the role of lysosomes in cellular autophagy. The discovery of the inborn lysosomal diseases played a major role in the general acceptance of the lysosomal concept. I still remember meeting Efraim Racker in 1962 and telling him the story; his comment was: "So, lysosomes do exist".

This story is a good illustration of the advantage of serendipity over oriented research. It was with reluctancy that de Duve had allowed me to continue my investigation of carbohydrate metabolism rather than to follow him in the lysosomal field. It was a quite independent and unforeseeable detour that brought my personal contribution to lysosomes. However, if serendipity played a major role in the discovery of the deficiency of acid α-glucosidase in Pompe's disease, the rest of my work on inborn lysosomal diseases was entirely based on deductive reasoning derived from my faith in the several propositions which are italicised in the preceding paragraphs. This is in agreement with the general belief [3] that logical reasoning is of little value in making new discoveries but plays a dominant part in extending knowledge.

It took 10 years to have the concept of inborn lysosomal diseases accepted

To establish that a well-defined storage disease was of lysosomal origin, it was necessary to associate it with the deficiency of a single lysosomal enzyme. On this basis, metachromatic leukodystrophy, in which a deficiency of lysosomal sulfatase had been demonstrated by Austin et al. [65] could rapidly be classified in the group. In the following years, a deficiency of β-glucosidase in Gaucher's disease [66, 67], of sphingomyelinase in Niemann-Pick's disease [68], of β-*N*-acetyl-hexosaminidase A in Tay-Sachs disease [69], and of α-galactosidase in Fabry's disease [70] was demonstrated. Similar deficiencies were also found in several less known disorders. The second criterion was ultrastructural: in each of these diseases, the overloading of lysosomes in various tissues was easily observed (see [71] for more complete information).

The main challenge for the lysosomal theory was the relatively ill-defined group of storage diseases known as gargoylism, mucopolysaccharidosis or Hurler's syndrome. This pathological condition had been initially classified among the sphingolipidoses, but was later on called mucopolysaccharidosis on the basis of an excessive excretion of mucopolysaccharides in the urine. In fact, it could be called a lipomucopolysaccharidosis because of the excess of gangliosides in the central nervous system and of mucopolysaccharides in most other tissues. The cause of the disease was believed to be an overproduction of mucopolysaccharides by the fibroblasts [72] or a defective binding of mucopolysaccharides to proteins [73]. None of these theories could explain the heterogeneity of the depot material which, in contrast, had been predicted by the lysosomal theory [63]. This is because digestive enzymes are not specific for one molecule but for one linkage, which may be the same in a glycolipid and a mucopolysaccharide. Nothing was known at that time of the enzymic breakdown of mucopolysaccharides in mammalian tissues.

For testing the lysosomal origin of the mucopolysaccharidoses, the ultrastructural approach was, therefore, the only one available. I was fortunate to receive then the skilful collaboration of François Van Hoof who could rapidly master both the biochemical and ultrastructural technology. We undertook the investigation of the ultrastructure of the liver of a child with Hurler's disease and observed the huge vacuoles shown in Fig. 7 [64]. The lysosomal nature of these vacuoles was again suggested by the complete disappearance of the dense bodies from this liver. For several years, this micrograph remained the main argument in favour of the hypothesis that the various forms of mucopolysaccharidosis were inborn lysosomal diseases. It was not until 1968, with the discovery of fucosidosis, that the first convincing proof was presented that the lysosomal theory could adequately explain the Hurler syndrome. Fucosidosis was a new form of lipomucopolysaccharidosis described by Durand et

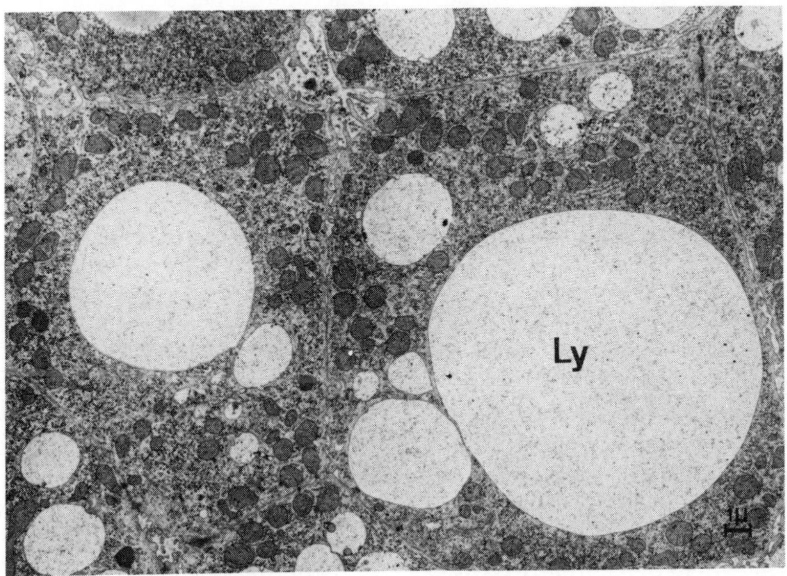

Fig. 7. The structure of the liver of a child with mucopolysaccharidosis (× 3900) (Photo F. Van Hoof).

al. [74], in which François Van Hoof observed a complete deficiency of a lysosomal acid α-fucosidase [75, 76]. This deficiency was discovered because the synthetic substrate, p-nitrophenyl-α-L-fucoside became available from Koch-Light and because Van Hoof assayed systematically the activity of this enzyme in all biopsies obtained from patients with any kind of storage disease. We could rapidly demonstrate that fucosidosis had all the biochemical and morphological characteristics of an inborn lysosomal disease.

It still took several years to demonstrate the enzymatic defect in Hurler's disease, which is the most common type of mucopolysaccharidosis. One likely candidate for the deficiency was L-iduronidase, an enzyme that had not been identified yet and that would hydrolyse the iduronic linkage present in several mucopolysaccharides. The synthetic substrate was not available and we had asked a well recognised carbohydrate chemist to synthesise it for us. Unfortunately, he was not successful and it was Weissmann and his coworkers who, in 1972, synthesised the p-nitrophenyl-α-L-iduronide [77] and described L-iduronidase [78]. They gave the substrate to our friends Albert Dorfman and Liz Neufeld who could rapidly demonstrate the deficiency in the pathological samples [79, 80]. In the same year, 1972, several other deficiencies characteristic of mucopolysaccharidoses were also described, leaving no doubt about the lysosomal origin of the disorder. When, in 1973, I edited with François Van Hoof the book entitled *Lysosomes and Storage Disorders* [71] we could account for more than 20 congenital disorders that were explained by the same general concept.

The hepatic threshold to glucose

As early as 1940, Soskin [81] proposed that glucose itself is the trigger which converts the liver from an organ of glucose output into an organ of glucose uptake. This control was compared to a thermostat-furnace arrangement; in this view, the liver was a glucostat, able to sense glucose concentration and to react differently according to the concentration being above or below a threshold value. The biochemical mechanism of this control by glucose had not been elucidated and it was generally believed that glucose stimulates glycogen synthesis in the liver by a "push" given on the whole biosynthetic pathway. An increased level of glycaemia was assumed to increase the rate of glucose phosphorylation and the concentration of glucose-6-phosphate with a secondary activation of glycogen synthase by the latter ester. Glycogen synthase had been identified in 1957 by Leloir and his coworkers [82] who also discovered the stimulation of that enzyme by glucose-6-phosphate [83]. The inactivation of the enzyme by phosphorylation and its reactivation by dephosphorylation had been demonstrated shortly afterwards by Joe Larner and his coworkers [84].

At the end of 1966, Henri De Wulf in my laboratory made three fundamental observations [85]: first, the intravenous administration of glucose to mice produced an increase in the rate of glycogen synthesis which was many-fold greater than the change in the level of glycaemia; second, this effect of glucose was only observed after a few minutes latency and reached its maximum when the level of glycaemia was already decreasing; third, when the rate of glycogen synthesis reached its maximum, the intrahepatic concentration of glucose-6-phosphate was not increased but decreased. In agreement with this latter observation, we could show that glucose-6-phosphate was without effect on glycogen synthase in the presence of a physiological concentration of P_i [86] and that the effect of glucose was to convert glycogen synthase from its inactive (synthase *b*) into its active form (synthase *a*). The rate of glycogen synthesis was always proportional to the concentration of synthase *a* in the liver [87].

The precise mechanism, in which – in agreement with Soskin's views – a change in glucose concentration favours the activation of glycogen synthesis, could be elucidated when the glucose effect, with its latency period, could be observed in a cell-free system. Henri De Wulf and Willy Stalmans were mostly responsible for this work which led to the following conclusions:

(1) the primary effect of glucose is to bind phosphorylase *a*, and, by changing its conformation [88], to favour its conversion into *b* by phosphorylase phosphatase [89];

(2) phosphorylase *a* is a strong inhibitor of synthase phosphatase [90].

These properties explain why the activation of glycogen synthase by its phosphatase occurs only after a latency period, the duration of which is considerably shortened by glucose, and is the time required for the inactivation of phosphorylase. Our interpretation of the glucose effect was later elegantly supported by the crystallographic investigations of Fletterick and Madsen [91] who demonstrated that the effect of glucose on phosphorylase *a* was to expose the serine phosphate residue, rendering it easily accessible to the phosphatase.

Although this mechanism of control was clearly understood as early as

1970 [92], it took almost 10 years before it could be considered generally accepted and was described in textbooks. This delay originated from a very trivial detail: the presence of a small amount of caffeine in the assay of phosphorylase *a* (a procedure introduced by W. Stalmans) was essential to avoid the interference of phosphorylase *b*. This detail was overlooked by several groups of investigators who were therefore unable to confirm that the conversion of phosphorylase *a* into *b* is a prerequisite to the activation of glycogen synthase.

The control of synthase phosphatase by phosphorylase *a* explains not only the effect of glucose to stimulate glycogen synthesis in the liver but also that of insulin. Gérald van de Werve and Willy Stalmans [93] observed that insulin favoured the conversion of phosphorylase *a* into *b* and that glycogen synthase became activated only after this conversion had been completed. This conclusion was confirmed by Witters and Avruch working with isolated hepatocytes [94]. The same sequence of events was also observed by Francisco Sobrino [95] in adipose tissue of rats following the administration of glucose and insulin.

The stimulation of liver glycogen synthesis by glucose is a pull mechanism (for reviews, see [96–98]) in which the control is on the last enzyme of the sequence, with a subsequent decrease in the concentration of intermediary metabolites, UDPG and glucose-6-phosphate. Because the hydrolysis of glucose-6-phosphate by glucose-6-phosphatase is a first-order reaction, a decrease in its concentration causes a proportional decrease in the rate of glucose formation, allowing then a net uptake of sugar by the liver [99].

The last publication of Samuel Soskin was dated 1952 and, for a long time, I thought that he had died. In 1975, however, I learned that he was alive and well and practicing medicine in California. I sent him several of our reprints in which the Soskin effect was explained in molecular terms. He answered with a kind letter in which he said:

> "I have been out of the research laboratory and in the practice of medicine for the past 25 years. But, like the Old War Horse, the sound of the bugle (in your reprints) still thrills me."

Fructose revisited with a glance at primary gout

When, in 1957, I finished writing my book on fructose metabolism [34], I naively thought that I had exhausted the subject and that little more could be done on this sugar; I then oriented my research to glycogen. The fructose field remained very quiet for about 10 years but, in 1967, it was spectacularly reopened by a group of Finnish paediatricians and biochemists who described the remarkable hyperuricaemia which follows the oral or intravenous administration of high doses of fructose (0.5 g/kg body weight) to normal children as well as to those with hereditary fructose intolerance [100]. This fortuitous observation was rapidly followed by an extensive study in the rat made by the same group of investigators [101]. The hyperuricaemia was shown to be related to a dramatic fall in hepatic ATP and the subsequent loss of a major fraction of the adenine nucleotide pool. In the presence of a high concentration of fructose, the liver phosphorylates this sugar faster than it can convert fructose-1-phosphate to glucose and lactate. Consequently, fructose-1-phosphate accumulates (up to 10 mM) and, acting as a phosphate trap, exhausts the ATP (2.5 mM) and P_i (4 mM) of the cell. The loss of the adenine nucleotides was attributed to a decrease in the inhibitory action of these metabolites on the two enzymes that were thought to catalyse the initial degradation of AMP, namely 5′-nucleotidase and AMP deaminase [102].

The control of purine catabolism had been little studied and contained many obscure points. In my laboratory, Georges Van den Berghe made a series of fundamental studies on that subject. He discovered that the cytosolic 5′-nucleotidase had regulatory properties that rendered it inactive on AMP although not on IMP under physiological conditions as well as after a load of fructose [103]. Furthermore, he made clearly apparent that the AMP deaminase constitutes the limiting step in the catabolism of the adenine nucleotides, leading to the formation of allantoin in the rat. Under normal circumstances this enzyme displays only about 5% of its full potential activity, mostly because it is inhibited by P_i and GTP [104]. The hyperuricaemic effect of fructose could be explained by a decrease of both effectors. It was on a Sunday afternoon, when I was walking along the Meuse river, talking with my wife, Suzanne, about a very different subject that the idea struck me that an abnormality of AMP deaminase was a possible explanation for the overproduction of uric acid in primary gout. This idea was based on the same kind of logic that I had developed in the investigation of lysosomal disease. Indeed, it is very probable that there exists in the human population a series of mutants with an abnormal deaminase that might be less sensitive to its natural inhibitors. If the enzyme were only 90% inhibited instead of 95%, this would significantly increase the catabolism of AMP in the liver and cause hyperuricaemia [105]. The verification of this hypothesis was, however, difficult because of the great difficulty in obtaining liver samples from gouty patients. In 1980, we did obtain such a liver and found that its AMP deaminase was less sensitive to GTP inhibition [106]. It is, however, too early to know if this will be a general finding and more cases need to be investigated to evaluate the hypothesis.

The futile cycles and the discovery of fructose-2,6-bisphosphate

In 1975, the arrival in my laboratory of a young Spanish biochemist, Juan Felíu, who wanted to study the properties of pyruvate kinase, gave me the opportunity to start an investigation of gluconeogenesis. This was just at the time when Engström et al. [107] had shown that liver pyruvate kinase can be inactivated through phosphorylation by cyclic AMP-dependent protein kinase. In collaboration with Louis Hue, Juan made a beautiful demonstration of the role of this inactivation in the hormonal control of gluconeogenesis [108].

A great attractiveness of gluconeogenesis lies in its futile cycles. A futile cycle is a metabolic interconversion, the net balance of which is the hydrolysis of ATP to ADP and P_i (for reviews see [109, 110]). There are three such cycles in gluconeogenesis, one at the level of the glucose/glucose-6-phosphate interconversion, the second between fructose-6-phosphate and fructose-1,6-bisphosphate and the third at the level of pyruvate and phosphoenolpyruvate. Following a discussion with Hans Krebs, who pointed out that Nature is never futile and that these cycles should have some kind of usefulness, I wrote with Louis Hue a paper [99] entitled *Utile and futile cycles* in which the role of the first of these cycles in the control of glucose uptake and output by the liver was explained. The third cycle, which operates between pyruvate and phosphoenolpyruvate, could also make some sense in relation to cyclic AMP-dependent phosphorylation and inactivation of pyruvate kinase. The role of the fructose-6-phosphate/fructose-1,6-bisphosphate cycle was much more obscure and even its operation was controversial. An investigation of this cycle involves a very complex utilisation of isotopes the interpretation of which is difficult.

Nevertheless, this was the subject of research that I proposed in September 1978 to Emile Van Schaftingen when he started his scientific career in my laboratory as a fellow of the Fonds National de la Recherche Scientifique. Making a long story short, the evidence that Emile got, in collaboration with Louis Hue, was that the cycle was operating only in the fed condition and was suppressed by glucagon [111, 112]. In the meantime, several reports [113–115] were published indicating that the kinetic properties of liver phosphofructokinase (PFK) could be modified by treatment of hepatocytes with glucagon. These kinetic modifications were thought to be stable and related to a phosphorylation of the enzyme by cyclic AMP-dependent protein kinase. It was finally in March 1980 that Emile obtained a clear evidence that the effects of glucagon on PFK were not stable but could be mimicked by gel filtration of a liver extract. Furthermore, these effects could be reversed by the addition of a low-M_r fraction obtained by ultrafiltration of a liver extract. I could therefore announce at the symposium on Biological Cycles held in Dallas (March 17–19) in honour of Sir Hans Krebs' 80th birthday that glucagon causes the removal of a low-M_r stimulator of phosphofructokinase [116].

It took only a few weeks for Emile Van Schaftingen and Louis Hue to identify the stimulator as fructose-2,6-bisphosphate. If I could be of some help in this investigation it is because I had known, in the 50s, a period when biochemists had to prepare their own substrates like ATP or hexose phosphates before starting any enzymatic investigation. It rapidly appeared that the stimulator was a non-nucleotidic phosphoric ester because it was destroyed by alkaline phosphatase and was not adsorbed on charcoal. It was also extremely acid-labile and could be isolated at neutral pH by the standard methods used for isolation of phosphoric esters some 20 or 30 years before. The identification as fructose-2,6-bisphosphate came from the fact that upon a mild acid hydrolysis, fructose-6-phosphate and inorganic phosphate were formed together with a reducing power in parallel with the disappearance of the stimulatory power [117].

The discovery of fructose-2,6-bisphosphate calls for further comments. A first question is: how did fructose-2,6-bisphosphate escape discovery until 1980? One reason for this is its extreme acid lability which caused its destruction in the trichloroacetic extracts used in systematic surveys of phosphoric esters in tissues. This, however, cannot be the only reason, since other phosphoric esters, like ribose-1-phosphate are equally acid-labile and had been discovered at least 25 years earlier. The main reason is probably that fructose-2,6-bisphosphate is not an intermediary in any important metabolic interconversion but, like cyclic AMP, a pure regulator; it is a signal formed or destroyed only in relation to the control of metabolism. Thanks to this property it was eventually discovered.

One can also wonder: how important is fructose-2,6-bisphosphate in biology? It is obviously too early to answer this question. The fact that fructose-2,6-bisphosphate has been found in nearly all mammalian cells, in yeast, other fungi and higher plants [118] suggests that it may have a very general biological role in the whole living world.

The other face of the story

As the story has been told, it might seem that everything went easily. I made no mention of the numerous attractive hypotheses which have been disproven and were as soon forgotten; the death of each of them has usually been a relief since, after a short moment of disappointment, it liberated my mind and that of my coworkers for the formation of other ideas. I have not said either that many experiments were performed and never published because I thought that they did not make enough sense to be considered progress in science. Finally, the reader has presumably noticed that I have not made an important contribution to our understanding of the mode of action of insulin.

More important could be the fact that the participation of my coworkers both in the emergence of ideas and in their experimental verification has not been made sufficiently apparent. During the last 15 years my immediate professional surroundings have been made up of a dozen young scientists. Their moral and intellectual qualities have brought some of my greatest satisfactions in life. It is therefore my pleasure to gratefully acknowledge the contribution not only of those whose name has been mentioned in the text but also of Ephrem Eggermont, Walter Verhue, Nicole Lejeune, Claci Zancan, Wilfried Den Tandt, Pierre Devos, Béatrice Lederer, Thierry de Barsy, Monique Laloux, Françoise Bontemps, Mary Frances Jett, Françoise Vincent, Dewi Davies, Ramon Bartrons and Jean François.

The emergence of ideas is a complex phenomenon to which many books have been devoted [3]. The scientific community is made up of a great variety of individuals, the most praised ones being the brightest and most erudite scientists. Some of them have made great discoveries. Their role is also to maintain knowledge, to write books, to construct mathematical models and complex theories and to deliver brilliant lectures. I do not belong to this group. In contrast, it has often been my feeling that my ignorance was the price I had to pay for being a discoverer; I knew by instinct that "a mass of information makes it more difficult for the mind to conjure up original ideas ... particularly because some of that information may be actually false" [3]. Often I made use of the method known as "day-dreaming" which is currently recognised as an efficient method of originating new ideas [3]. It can happen that Suzanne asks me: "are you dreaming?" and that I will answer: "no, I am working". For a long time, I believed that my mistrust of too bright scientists was only an excuse for my own dullness until I discovered the following comment (quoted in [3]) made by Claude Bernard:

"It is that which we do know which is the great hindrance to our learning that which we do not know",

a dilemma which apparently all creative workers are facing. The same idea seems to apply to the non-scientific domain since a well-known Belgian poet, Henri Michaux, born as I in Namur, also said:

"Il ne faut pas trop apprendre, car il faut toute une vie pour désapprendre." [119]

At 59, I still have some time ahead of me to learn and unlearn a few things. I hope not to have to unlearn too many of the things described in this chapter.

REFERENCES

1 W.J. Young, Roy. Soc. Proc. B, 81 (1909) 528–544.
2 P.A. Levene and A.L. Raymond, J. Biol. Chem., 80 (1928) 633–638.
3 W.T.B. Beveridge, The Art of Scientific Investigation, Heinemann, London, 1951, 172 p.
4 M. Bürger and W. Brandt, Z. Ges. Exp. Med., 96 (1935) 375–397.
5 C. de Duve, Lancet, 265 (1953) 99–104.
6 C. de Duve, H.G. Hers and J.P. Bouckaert, Arch. Int. Pharmacodyn. 22 (1946) 45–61.
7 E.W. Sutherland and C.F. Cori, J. Biol. Chem., 171 (1947) 737–750.
8 V.H. Unger, A.M. Eisentraut, M.S. McCall and L.L. Madison, J. Clin. Invest., 41 (1962) 682–689.
9 A. Claude, J. Exp. Med., 84 (1946) 61–89.
10 C. de Duve, J. Berthet, H.G. Hers and L. Dupret, Bull. Soc. Chim. Biol., 31 (1949) 1242–1253.
11 H.G. Hers, J. Berthet, L. Berthet and C. de Duve, Bull. Soc. Chim. Biol., 33 (1951) 21–41.
12 C. de Duve, in J.T. Dingle and H.B. Fell (Eds.), Lysosomes in Biology and Pathology, Vol. 1, North Holland, Amsterdam, 1969, 3–40.
12aD.L. Dipietro, C. Sharma and S. Weinhouse, Biochemistry, 1 (1962) 455–462.
13 E. Viñuela, M. Salas and A. Sols, J. Biol. Chem., 238 (1963) PC 1175–1177.
14 G.T. Cori, S. Ochoa, M.W. Slein and C.F. Cori, Biochim. Biophys. Acta, 7 (1951) 304–317.
15 H.G. Hers and T. Kusaka, Biochim. Biophys. Acta, 11 (1953) 427–437.
16 O. Meyerhof, K. Lohmann and P. Shuster, Biochem. Z., 286 (1936) 319–335.
17 J.W. Rutter, O.C. Richards and B.M. Woodfin, J. Biol. Chem., 236 (1961) 3193–3197.
18 E.R. Froesch, A. Prader, A. Labhart, H.W. Stubert and H.P. Wolf, Schweiz. Med. Wschr., 87 (1957) 1168–1171.
19 H.G. Hers and G. Joassin, Enzymol. Biol. Clin., 1 (1961) 4–14.
20 G. Gomori, J. Biol., 148 (1943) 139–149.
21 B.L. Horecker, E. Melloni and S. Pontremoli, Adv. Enzymol., 42 (1975) 193–226.
22 K. Taketa and B.M. Pogell, J. Biol. Chem., 240 (1965) 651–662.
23 H.G. Hers and E. Eggermont, in R.W. McGilvery and B.M. Pogell (Eds.), Fructose-1,6-diphosphatase and its Role in Gluconeogenesis, Port City Press, Baltimore, MD, 1964, pp. 14–19.
24 E. Van Schaftingen and H.G. Hers, Proc. Natl. Acad. Sci. USA, 78 (1981) 2861–2863.
25 F. Leuthardt, E. Testa and H.P. Wolf, Helv. Chim. Acta, 36 (1953) 227–251.
25a E.C. Ballou and H.A. Barker, Adv. Carbohydr. Chem. Biochem., 32 (1976) 1–14.
26 H.G. Hers, J. Edelman and V. Ginsburg, J. Am. Chem. Soc., 76 (1954) 5160.
27 E.F. Neufeld, W. Timple and L.J. Shapiro, Annu. Rev. Biochem., 44 (1975) 357–376.
28 H.G. Hers, J. Biol. Chem., 214 (1955) 373–381.
29 P. Schambye, H.G. Wood and G. Popjak, J. Biol. Chem., 206 (1954) 875–882.
30 I.C. Gunsalus and M. Gibbs, J. Biol. Chem., 194 (1952) 871–875.
31 H.G. Hers, Biochim. Biophys. Acta, 8 (1952) 424–430.
32 H.A. Lardy, in W.D. McElroy and B. Glass (Eds.), Phosphorus Metabolism, Vol. 1, 1951, pp. 477–499.
33 B. Glass, in W.D. McElroy and B. Glass (Eds.), Phosphorus Metabolism, Vol. 1, 1951, pp. 658–741.
34 H.G. Hers, Le Métabolisme du fructose, Arscia, Brussels, 1957, 200 p.
35 C. Bernard, Leçons de Physiologie Expérimentale, Baillère, Paris, Vol. 1, 1855, 405 p.

36 J.S.D. Bacon and D.J. Bell, Biochem. J., 42 (1948) 37–405.
37 A. St. G. Huggett, F.L. Warren and N.W. Warren, J. Physiol., 113 (1951) 258–275.
38 T. Mann, Biochem. J., 40 (1946) 481–491.
39 T. Mann and C. Lutwak-Mann, Biochem. J., 48 (1951) 16–17.
40 H.G. Williams-Ashman and J. Banks, Arch. Biochem. Biophys., 50 (1954) 513–515.
41 H.G. Hers, Biochim. Biophys. Acta, 22 (1956) 202–203.
42 G.E. Glock and P. McLean, Biochem. J., 61 (1955) 388–390.
43 D.G. Walker, Biochem. J., 84 (1962) 118p–119p.
44 L.T. Samuels, B.W. Harding and T. Mann, Biochem. J., 84 (1962) 39–45.
45 T. Mann, The Biochemistry of Semen and of the Male Reproductive Tract, Methuen, London, 1964, 240 p.
45a T. Mann and C. Lutwak-Mann, Male Reproductive Function and Semen, Springer-Verlag, Berlin, 1981, 495 p.
46 R. van Heyningen, Nature, 184 (1959) 194–195.
47 K.H. Gabbay, L.O. Merola and R.A. Field, Science, 151 (1966) 209–210.
48 G.T. Cori, Mod. Probl. Paediat., 3 (1957) 344–358.
49 J.C. Pompe, Ned. Tijdschr. Geneesk., 76 (1932) 304–311.
50 H.G. Hers, Rev. Int. d'Hépatol., 9 (1959) 35–55.
51 H.G. Hers and H. Malbrain, Probl. Act. Péd., 4 (1959) 203–209.
52 D. Stetten Jr. and M.R. Stetten, Physiol. Rev., 40 (1960) 505–537.
53 H.G. Hers, Biochem. J., 86 (1963) 11–16.
54 G.J. Walker and W.J. Whelan, Biochem. J., 76 (1960) 264–268.
55 N. Lejeune, D. Thines-Sempoux and H.G. Hers, Biochem. J., 86 (1963) 16–21.
56 C. de Duve, B.C. Pressman, R. Gianetteo, R. Wattiaux and F. Appelmans, Biochem. J., 64 (1955) 604–617.
57 C. de Duve, in T. Hayashi (Ed.), Subcellular Particles, Ronald Press, New York, 1959, pp. 128–159.
58 A.B. Novikoff, H. Beaufay and C. de Duve, J. Biophys. Biochem. Cytol., 2 (1956) 635–637.
59 E. Essner and A.B. Novikoff, J. Ultrastruct. Res., 3 (1960) 374–391.
60 Th.P. Ashford and K.R. Porter, J. Cell Biol., 12 (1962) 198–202.
61 C. de Duve and R. Wattiaux, Annu. Rev. Physiol., 28 (1966) 435–492.
62 P. Baudhuin, H.G. Hers and H. Loeb, Lab. Invest. 13 (1964) 1140–1152.
63 H.G. Hers, Gastroenterology, 48 (1965) 625–633.
64 F. Van Hoof and H.G. Hers, Compt. Rend., 259 (1964) 1281–1283.
65 J. Austin, D. McAffee, D. Armstrong, M. O'Rourke, L. Shearer and B. Bachhawat, Biochem. J., 93 (1964) 15C–17C.
66 R.O. Brady, J.N. Kanfer and D. Shapiro, Biochem. Biophys. Res. Commun., 18 (1965) 221–225.
67 A.D. Patrick, Biochem. J., 97 (1965) 17–18.
68 R.O. Brady, J.N. Kanfer, M.B. Mock and D.S. Fredrickson, Proc. Natl. Acad. Sci. USA, 55 (1966) 366–369.
69 S. Okada and J.S. O'Brien, Science, 165 (1969) 698–700.
70 J.A. Kint, Science, 167 (1970) 1268–1269.
71 H.G. Hers and F. Van Hoof (Eds.), Lysosomes and Storage Diseases, Academic Press, New York, 1973, 666 p.
72 K. Meyer and Ph. Hoffman, Arthr. Rheum., 4 (1961) 552–560.
73 A. Dorfman, in J.B. Stanbury, J.B. Wijngaarden and D.S. Fredrickson (Eds.), The

Metabolic Basis of Inherited Disease, 2nd ed., McGraw Hill, New York, 1966 pp. 963–994.
74 P. Durand, C. Borrone and G. Della Cella, Lancet, ii (1966) 1313.
75 F. Van Hoof and H.G. Hers, Lancet, i (1968) 1198.
76 F. Van Hoof and H.G. Hers, Eur. J. Biochem., 7 (1968) 34–44.
77 R.B. Friedman and B. Weissmann, Carbohydrate Res., 24 (1972) 123–131.
78 B. Weissmann and R. Santiago, Biochem. Biophys. Res. Commun., 46 (1972) 1430–1433.
79 G. Bach, R. Friedman, B. Weissmann and E.F. Neufeld, Proc. Natl. Acad. Sci. USA, 69 (1972) 2048–2051.
80 R. Matalon and A. Dorfman, Biochem. Biophys. Res. Commun., 47 (1972) 959–964.
81 S. Soskin, Endocrinology, 26 (1940) 297–308.
82 L.F. Leloir and C.E. Cardini, J. Am. Chem. Soc., 79 (1957) 6340–6341.
83 L.F. Leloir, J.M. Olavarria, S.H. Goldenberg and H. Carminatti, Arch. Biochem. Biophys., 81 (1959) 508–520.
84 D.L. Friedman and J. Larner, Biochemistry, 2 (1963) 669–675.
85 H. De Wulf and H.G. Hers, Eur. J. Biochem., 2 (1967) 50–56.
86 H. De Wulf, W. Stalmans and H.G. Hers, Eur. J. Biochem., 6 (1968) 545–551.
87 H. De Wulf and H.G. Hers, Eur. J. Biochem., 6 (1968) 558–564.
88 W. Stalmans, M. Laloux and H.G. Hers, Eur. J. Biochem., 49 (1974) 415–427.
89 W. Stalmans, H. De Wulf, B. Lederer and H.G. Hers, Eur. J. Biochem., 15 (1970) 9–12.
90 W. Stalmans, H. De Wulf and H.G. Hers, Eur. J. Biochem., 18 (1971) 582–587.
91 R.J. Fletterick and N.B. Madsen, Annu. Rev. Biochem., 49 (1980) 31–61.
92 H.G. Hers, H. De Wulf and W. Stalmans, FEBS Lett., 12 (1970) 75–84.
93 G. van de Werve, W. Stalmans and H.G. Hers, Biochem. J., 162 (1977) 143–146.
94 L.A. Witters and J. Avruch, Biochemistry, 17 (1978) 406–410.
95 F. Sobrino and H.G. Hers, Eur. J. Biochem., 109 (1980) 239–246.
96 H.G. Hers, Annu. Rev. Biochem., 45 (1976) 167–189.
97 W. Stalmans, Curr. Top. Cell. Regul., 11 (1976) 51–97.
98 H.G. Hers, in L. Hue and G. van de Werve (Eds.), Short-term Regulation of Liver Metabolism, Elsevier North-Holland, Amsterdam, 1981, pp. 105–117.
99 L. Hue and H.G. Hers, Biochem. Biophys. Res. Commun., 58 (1974) 540–548.
100 J. Perheentupa and K. Raivio, Lancet, ii (1967) 528–531.
101 P.H. Mäenpää, K.O. Raivio and M.P. Kekomäki, Science, 161 (1968) 1253–1254.
102 H.F. Woods, L.V. Eggleston and H.A. Krebs, Biochem. J., 119 (1970) 501–510.
103 G. Van den Berghe, Ch. van Pottelsberghe and H.G. Hers, Biochem. J., 162 (1977) 611–616.
104 G. Van den Berghe, M. Bronfman, R. Vanneste and H.G. Hers, Biochem. J., 162 (1977) 601–609.
105 H.G. Hers and G. Van den Berghe, Lancet, i (1979) 585–586.
106 G. Van den Berghe and H.G. Hers, Lancet, ii (1980) 1090.
107 O. Ljungström, G. Hjelmquist and L. Engström Biochim. Biophys. Acta, 358 (1974) 289–298.
108 J.E. Feliu, L. Hue and H.G. Hers, Proc. Natl. Acad. Sci. USA, 73 (1976) 2762–2766.
109 J. Katz and R. Rognstadt, Curr. Top. Cell. Regul., 10 (1976) 237–289.
110 L. Hue, Adv. Enzymol., 52 (1981) 247–331.
111 E. Van Schaftingen, L. Hue and H.G. Hers, Biochem. J., 192 (1980) 263–271.

112 E. Van Schaftingen, L. Hue and H.G. Hers, Biochem. J. 192 (1980) 887–895.
113 J.G. Castano, A. Nieto and J.E. Feliu, J. Biol. Chem., 254 (1979) 5576–5579.
114 T. Kagimoto and K. Uyeda, J. Biol. Chem., 254 (1979) 5584–5587.
115 S. Pilkis, J. Schlumpf, J. Pilkis and T.H. Claus, Biochem. Biophys. Res. Commun., 88 (1979) 960–967.
116 H.G. Hers, L. Hue and E. Van Schaftingen, Curr. Top. Cell. Regul., 18 (1981) 199–210.
117 E. Van Schaftingen, L. Hue and H.G. Hers, Biochem. J., 192 (1980) 897–901.
118 H.G. Hers and E. Van Schaftingen, Biochem. J., 206 (1982) 1–12.
119 H. Michaux, Poteaux d'Angle, Herne, Paris, 1971, 48 p.

G. Semenza (Ed.) Selected Topics in the History of Biochemistry: Personal Recollections (Comprehensive Biochemistry Vol. 35)

Chapter 5

Sir Frederick Gowland Hopkins (1861–1947)

N.WILLIAM PIRIE

Rothamsted Experimental Station, Harpenden, Herts. AL5 2JQ (U.K.)

Character and outlook

Even irreverent students noticed that there was something special about Hopkins. He occasionally went round among the students in the elementary practical Biochemistry class, and more regularly among those in the advanced class. No one could have been less aloof or awe-inspiring, and yet we treated him with a respect and affection that I did not see accorded to any other member of Cambridge University staff. When I first met him, in 1927, he was already 66 and so, under rules widely operating today, would have been retired. He had earlier rejected the option of retiring on a pension when the University structure was reorganised. When I told him that research in his laboratory was my ambition, he said that that would be a matter for his successor. Nevertheless, he remained a stimulating and effective professor until 1943. The students' judgement was shared by others. Marjory Stephenson wrote of "... tenacity of purpose concealed from the superficial observer by his gentle, slightly hesitating, courteous manner"; Sir Rudolph Peters of "... a critical mind with immense human sympathy"; and Sir Edward Mellanby remarked that the atmosphere of his laboratory "doomed anyone to research".

Concern and human sympathy were not restricted to the needs of individuals. Sir Charles Sherrington [1] commented:

"I fancy that, after Biochemistry, his greatest interest lay in 'Socialism'. His views were quite far to the 'Left'."

Plate 5. Sir Frederick Gowland Hopkins (1861–1947).
(Reproduced with permission of the Fitzwilliam Museum, Cambridge.)

Sherrington was probably thinking of early days when he knew Hopkins well. During the 1930s he was not overtly active in politics, but he needed little persuasion to sign protests against preparations for war and the growth of fascism. Hopkins had supported the campaign to get Bertrand Russell reinstated in the lectureship from which he had been dismissed by Trinity College Council because of opposition to aspects of legislation during the 1914–18 war [2]. In 1925 he staunchly supported J.B.S. Haldane in his successful appeal against dismissal by Cambridge because of his involvement in a divorce. He accepted presidency of the then left wing Association of Scientific Workers (later amalgamated into ASTMS) and, in his presidential address in 1938, stressed the need for better press coverage of science. He told me that he insisted on a reallocation of seating when he found that he had been put beside General Dyer (of the Amritsar massacre in 1919) at a public dinner.

During the 1914–18 war, Hopkins was a member of the Royal Society Food (War) Committee, chairman of a vitamin committee of the Medical Research Committee (later Council) and the Lister Institute, and author of, or contributor to, seven or eight reports on the food supply. As a result, the average state of nutrition at the end of that war was probably better than it had been at the beginning. However, in the inter-war years, it was far from satisfactory. Malnutrition in Britain was the main theme of one of his presidential addresses to the Royal Society [3]. In a presidential address to the Children's Nutrition Council* he said

"Few I think will doubt that it would be logical enough to remove essential foodstuffs from the crude operation of the law of supply and demand, but we have to realise that to do this effectively would call for considerable readjustments in our social structure."

He often spoke and wrote favourably of the publicity the Committee Against Malnutrition was giving to the desirability of that objective. Members of the Committee Against Malnutrition were prime movers in starting the Nutrition Society in 1941. It therefore had a strong bias towards the practical aspects of human nutrition. Hopkins, now 80, was its first president. His address (1944) lacks the sparkle of some earlier addresses on similar occasions. A passage from an earlier lecture [5] gives a sharper impression of his theme:

---

*Quoted from [4].

"If we prove justified in ascribing to nutritional errors more — even a little more — of failure of health and development in children, the circumstance will surely be a happy one. A little more of evil becomes referrable to nurture instead of nature: to environment instead of inheritance and therefore to the remediable instead of the irremediable."

In his prime, Hopkins read the biochemical literature thoroughly. He wrote the sections on Biochemistry for the Annual Reports on the Progress of Chemistry in 1914 to 1918. These Reports show a sustained concern with fact rather than theory. They contain many shrewd comments on the validity of the experiments described and suggestions for other experiments that should be done. Nevertheless, their tone is kindly. In 1915 he commented:

"I find, on the whole, very few papers which it is a pleasure to ignore."

Acerbity is confined to comments on those who, like Rubner with his "isodynamic equivalence" and "physiological law" tried to trammel Biology: examples are quoted later in this article. In 1917, rebutting Arrhenius' criticism that biologists did not present their results in proper physicochemical form, he commented that, unlike physical-chemists, biologists understand the nature of the problem to be solved.

By the 1930s he tended to rely on others to tell him what was happening in the biochemical literature. He once said to me that ignorance could be useful: it kept you from knowing the seemingly valid reasons why what you had just observed could not happen! By that time he no longer lectured to elementary classes, and his occasional lectures to the advanced class were fascinating accounts of the early history of Biochemistry and Physiology, and of Biochemistry during the period in which he had watched it develop. Only one of these lectures, on the centenary of Wöhler's synthesis of urea, was published [6]. In it he made the important point that:

"...we may be apt to overestimate the effect of this or that discovery upon the thought of its own period. In our backward perspective we see it stand out in all its importance, but it does not follow that this importance was immediately felt. I think, indeed, that our textbook writers in their familiar references to the event which we now have specially in mind have tended to exaggerate its effect upon the thought of the time."

Hopkins went on to quote from Berzelius' ribald letter of congratulation, and to comment on the small impression the synthesis made on the vitalistic outlook of Wöhler's contemporaries.

Hopkins presided with subtle sagacity at our weekly colloquia, usually called, "Tea Club Meetings". He had a remarkable faculty for seeing significance in what at first sight seemed rather dull and inadequate material. He was also expert at mollifying those who might otherwise have resented the frank, or even hostile, comments of an unusually critical audience.

Hopkins' public lectures during the 20s and 30s dealt mainly with the philosophy of Biochemistry and the various notions and illusions that had obstructed its development. He did not write a textbook. He and Plimmer edited the series *Monographs on Biochemistry* published by Longmans, Green and Company. Some of the early volumes in the series list "The development and present position of Biological Chemistry", not as "forthcoming", but in among those that actually existed. This must have been confusing for librarians and others searching the early literature of the subject. I do not think he even started to write that book.

Hopkins was short and slender. He usually had a quick and rather jerky walk, but it became a slow shuffle when he was worried. He smoked cigarettes incessantly and the edge of his bench bore innumerable traces of those he had put down when doing something that needed his full attention. This was not slovenliness; he was very neat and dextrous (he never wore a lab jacket), it was simply preoccupation with what really mattered at the time. In his late 70s he suffered from arthritis and failing vision, till then he had excellent health and said he never had a cold until he was 50. Although he always walked the 2 km between home and laboratory, he did not, at any stage of his life, take deliberate exercise; he said he never sweated. One hot summer afternoon he met me, streaming with sweat after putting several kg of leaves through a hand mincing machine: he said he wondered whether he had been missing anything!

Education and early employment

Two passages in the brief *Autobiography* that Hopkins wrote in 1937 shed much light on his education and development. When eight or nine he was given a microscope that had belonged to his father and realised that the things

"...thus revealed to me were something very *important* – the most important thing I had yet come up against; so much more significant than anything I was being taught at school."

When 14 his truancy from school was undetected for several weeks. He spent the time wandering round docks, and in libraries and museums. Of this episode he wrote

"If by digging deep in memory I endeavour to recall what was the state of mind which first led to this lapse from right conduct, so soon regretted, I come honestly to the conclusion that it was sheer boredom with my life at school."

Nevertheless, he did impressively well in a chemistry examination and also won a science prize offered by the College of Preceptors. He went to four schools but writes that

"...of education in any adequate sense of the word I received very little."

Families nowadays often put themselves to considerable inconvenience to avoid "disturbing a child's education". Hopkins' success in spite of, in his opinion, negligible school influence, suggests that schooling may not be all-important. During his infancy, his father, who had been president of an amateur "Scientific and Literary Society", died; his mother was a forceful character (her mother had eloped to get married) and was well-read but not scientific. Hopkins comments that "she knew nothing of such things". The pattern will be familiar to those who have read "Hereditary Genius" by Galton [7]. Galton asks the readers to

"...note how irregularly many of the men and women have been educated whose names appear in my appendix..."

and also comments:

"It therefore appears to be very important to success in science that a man should have an able mother. I believe the reason to be, that a child so circumstanced has the good fortune to be delivered from the ordinary narrowing, partisan influences of home education."

In essence, the pattern was apparent to Hopkins who wrote, at the end of his *Autobiography:*

"Looking back I wonder whether, had my training been more orthodox – had I come to Cambridge, for instance, in 1878 instead of 1898 – I should not then have followed some more conventional career in which I might well have found it difficult to reach distinction."

On the advice of an uncle, Hopkins went into an insurance office, left it after six months and became an articled pupil in a commercial analytical laboratory. He worked there for three years during which he learnt little "... save perhaps some manipulative skill ...". Aware that he needed a more orthodox scientific education, he then enrolled in the Royal School of Mines in South Kensington, did well in the end of term examination, and was given a job in a private analytical laboratory. After six months he went to University College and did so well in the examination that Stevenson (later Sir Thomas), the Medical Jurist at Guy's Hospital, made him his assistant. That job lasted from 1883 till 1888 and gave him a vast experience of both human nature and the detection of poisons. During this time he studied for an external degree from London University; there was no formal teaching except for a practical course in Biology at Birkbeck College. It is clear from the *Autobiography* that he does not recommend this method of getting an education and ends this section with the words

"... lack of early education in the basal aspects of science has made me, I feel, an amateur intellectually."

While working on Medical Jurisprudence he met Martin (later Sir Charles) who remained a lifelong friend. Martin was a medical student and urged Hopkins to take a medical degree as a prelude to studying Physiological Chemistry. The idea harmonised with his early interests. In 1935 in an address to the London Natural History Society [8] he said that, as a result of an attempt at the age of seventeen to study the discharge of the bombardier beetle (*Brachinus crepitans*):

"Though the designation was not yet invented, I became there and then a Biochemist at heart."

A small legacy, and a research studentship from the hospital, made enrollment as a medical student possible in 1888. With his usual ability to shine in examinations he got a gold medal for Chemistry in the Intermediate M.B., and honours in Materia Medica as a result of two days intensive study. At the unusually late age of 33 he qualified in 1894. As well as working for his medical degree, he was able during this period to study the wing pigments of butterflies, and to find a week in which to work out a method for measuring the concentration of uric acid in urine.

As soon as (or perhaps just before) he qualified, Hopkins was appointed to the teaching staff of Guy's Hospital with responsibility for Physiology, Toxicology and elementary Chemistry and Physics. He also edited the *Guy's Gazette* which included a section in which members of the staff commented on pathological specimens that had been sent in by medical practitioners. Hopkins enlarged this system into a service that sent replies by post and then, with two colleagues, set up the Clinical Research Association as a commercial venture. The financial arrangements are not clear from the *Autobiography*, but it contains the comment:

"I was paid a decent salary, but the work was hard, and usually done after 5 pm."

Besides all this, Hopkins managed to find time for research on porphyrins, bile pigments, the halogenation of proteins and protein crystallisation.

Establishment of biochemistry in Cambridge

In 1898, Hopkins accepted a lectureship in the department of Physiology in Cambridge where little emphasis had hitherto been given to what was still called Physiological Chemistry. To supplement a lecturer's salary of £ 200 a year he also supervised medical students. His financial position gradually improved; he became a Reader in 1902, then Tutor in Emmanuel College and Praelector in Biochemistry in Trinity College, with Fellowships in both colleges. He was runner up for the Quick Professorship of Biology and, in 1914, was made Professor of Biochemistry. But that appointment was unpaid [9]. Earning a living in these circumstances involved a very heavy teaching load. In his *Autobiography* he laments that every hour of supervision in Anatomy called for several hours of preparation because, not unexpectedly, he had forgotten what he had learnt as a student. He also had to teach Embryology and Comparative Anatomy. Overwork, coupled with the effects of hitting his head on a spiral staircase in the laboratory in 1910 caused

"... months of extreme mental and bodily discomfort. Only those I think who have suffered from neurasthenia at its worst can properly appraise the misery involved. After six months, however, I completely recovered...."

Although he now had a chair, Hopkins still had no adequate laboratory facilities. At first he had a small room and a cellar in the old Physiology laboratory in Corn Exchange Street. When Physiology moved to its present building, he got more space and then moved into a building in Downing Place, which had been built in 1790 as a Nonconformist Chapel, with an annexe attached by a bridge. The original plan had been to put Biochemistry into a wing of the new Physiology laboratory, but that wing went to Psychology instead.

Before and during the 1914–18 war Hopkins was acutely aware that the demand for biochemists greatly exceeded the supply. He could find neither suitable candidates for jobs, nor space for many would-be visitors. His problems, both spatial and financial, worried his old friends Fletcher (later Sir Walter) and Hardy (later Sir William). Fortunately, the former was secretary of the Medical Research Committee, and the latter was a secretary of the Royal Society and was influential in several other organisations, e.g. the Department of Scientific and Industrial Research. They were therefore well-placed to be helpful. Largely as a result of Fletcher's persistent advocacy, the trustees of the Dunn Estate agreed at the end of 1919 to build and endow a Biochemistry laboratory in Cambridge. They had at the time 500 other applications for support. Hitherto, Dunn benefactions had gone to organisations with a direct connection with the relief of suffering. Most of Fletcher's effort was directed towards persuading the trustees that the prevention of disease was as fit an object for charity as the treatment of disease. All formalities were settled early in 1920 and the building was completed in 1924. Hopkins had at last managed to get an institute for general Biochemistry, not firmly tied to a hospital or medical school, and had established Biochemistry as an independent discipline in Cambridge. The story is told by Kohler [9] in interesting detail, with many quotations from published documents, papers in the MRC archives, and personal letters from the various participants in the argument. At Hopkins' instigation, the Dunn trustees later endowed a laboratory in Cambridge exclusively for research on Nutrition.

Research

Purines and pterins

As already mentioned, the explosive emission of the bombardier beetle directed Hopkins' adolescent interest towards Biochemistry. He retained an interest in insects, especially butterflies, throughout life. Their pigments were the theme of his D.Sc. thesis, and the word *lepidoporphyrin* (the name he gave to a purple substance formed when the wing pigment of the brimstone butterfly is heated with sulfuric acid in the presence of air) and its spectrum appear, presumably at his suggestion, on the sheet of paper in front of him in the portrait that hangs in the Royal Society. He concluded [10] that the material on the wings of the cabbage white butterfly was uric acid and that the pigments on several other butterflies were related to uric acid. That conclusion depended on some physical properties and on the murexide reaction. He commented that the substance did not crystallise in quite the same manner as uric acid from urine – but attributed that to the presence of impurities that differed from those in urine. Wieland and his colleagues, in a series of papers between 1925 and 1940, extended these studies on wing pigments. They gradually came to the conclusion that the parent compound is not uric acid but pterin – a novel structure containing an extra carbon atom. Pterins have more recently been identified as components of several substances involved in other biochemical processes. These observations slightly spoil Hopkins' jest [8]:

> "I had thus come across a case of the use of excretory substances in ornament, a phenomenon which may shock or please the aesthetic sense according to the point of view..."

He went on to say that his error

> "... had I been younger might have seemed to me a tragedy, but now I can bear it with equanimity."

It is interesting to note that the analysis which he published in 1895 shows the presence of a little more carbon, and less nitrogen, than is correct for uric acid and is more in accord with a pterin – a tribute to his analytical skill. He returned to these problems towards the end of his life and added some more sections to the complex story of wing pigments and the products that can be made from them and from uric acid. Furthermore, he was not altogether wrong: there is a little uric acid on the wings of cabbage white butterflies.

Anomalies in uric acid metabolism were, at the end of the last century, more often invoked as the basic cause of various clinical signs than they are today. While in Guy's Hospital, Hopkins devised an analytical method that depends on precipitating ammonium urate from urine saturated with ammonium chloride: it remained the standard technique for about 40 years. At that time he wrote the chapter on "The Chemistry of Urine" for the *Textbook of Physiology* edited by Schafer; this deals comprehensively with all the urinary components about which there was any information. Detailed consideration of the composition of urine caused him to doubt the facile generalisation that nucleic acids in food were the primary source of the uric acid in urine, and he made thymus extracts that contained little nucleic acid but enhanced the excretion of uric acid.

Although Ure, early in the 19th century, had found hippuric acid in the urine of people who had eaten benzoic acid, and several similar biological syntheses were discovered soon after, it was widely believed at the end of the century that animals, unlike plants, had only a limited capacity for synthesis. They were thought of as essentially catabolic: they built their bodies from molecules that were made by plants and which then underwent little modification. Hopkins' studies on uric acid mark the beginning of his lifelong opposition to this outlook. Much of one of his addresses to the British Association [11] is devoted to descriptions of the synthetic powers of animals and to a discussion of the source of the glycine with which benzoic acid is conjugated when hippuric acid is made. He pointed out that much more glycine may be used in that conjugation than is present in the protein eaten.

Aldehyde oxidase was found in milk at the beginning of the century. More or less by accident, a fraction separated from yeast extracts during the preparation of glutathione was found to resemble an aldehyde in its manner of oxidation in the presence of milk. As Pasteur emphasised, lucky accidents happen to the prepared mind! The active substance turned out to be hypoxanthine [12]. This was the start of a prolonged series of studies by Hopkins and his colleagues on the ability of various tissues to oxidise xanthine and hypoxanthine to uric acid and on the relationship between xanthine oxidase and aldehyde oxidase. As the closing paragraphs of his lecture on the centenary of Wöhler's synthesis of urea show, Hopkins was fascinated by the teleologically rational use of insoluble uric acid in place of soluble urea as the main vehicle for nitrogen excretion by those animals which have a manner of life that necessitates the economical use of water. He attached particular importance to the need for end-product segregation in the cleidoic egg, and often discussed Morgan's [13] curious observation that xanthine oxidase appears suddenly in chick liver immediately before hatching.

Proteins

Crystals of, or containing, protein had often been observed in vivo or made in vitro during the second half of the 19th century. At that time, before pH was either understood or controlled, crystallisation was erratic. Hopkins noticed the smell of ammonia when he mixed egg white with ammonium sulfate; he added acetic acid and, using the appearance of turbidity as an internal indicator, got crystallisation infallibly without even going to the trouble of starting with genuinely fresh eggs. He was well aware that crystallinity was in itself no guarantee of purity, however, working with proteins as varied as egg albumin, serum proteins, lactalbumin and Bence-Jones protein, he presented evidence that each was a definite chemical substance. He claimed that in analytical composition, optical rotation etc., many proteins had as much individuality as most other substances of biological origin.

As a result of this evidence for the uniformity and individuality of proteins, Hopkins became a vigorous opponent of the whole set of vague and mystical notions that had accreted round the concepts of "protoplasm" and "giant molecules". He thought, as Emil Fischer thought at about the same time, of proteins with 20 or 30 amino acids in them and was somewhat disquieted when, 30 years later, evidence accumulated that proteins were very much larger than that. He feared a return to the old defeatist outlook on proteins and expressed his fear repeatedly in lectures given at the time.

In spite of their individuality, proteins share the ability to give several colour reactions. By 1900 the particular amino acid(s) responsible for some of these colours had been discovered. No known amino acid seemed to be implicated in the violet colour produced when strong sulfuric acid is added to some proteins in the presence of acetic acid. One of the bottles of acetic acid used in class-room work failed to give this reaction because it did not contain the glyoxylic acid which contaminated most samples. Hitherto, it had been thought that the colour depended on the formation of furfural as a result of the action of strong acids on carbohydrate components in those proteins which gave the reaction. Having realised that a completely different type of reaction was involved, Hopkins and Cole used the colour reaction as a guide while fractionating tryptic digests and, ultimately, isolating tryptophan – the amino acid responsible for the colour. This isolation depended on the use of a novel precipitant – mercuric sulfate in sulfuric acid – which is more specific than those which had been used by others. A later paper [14] shows an interesting facet of his character. He was as anxious to give proper credit to others as to claim what was his due. A footnote says that this reagent should not be credited to Hopkins and Cole but to Denigès because, unknown to them, it had been used by Denigès two years earlier for a different purpose.

Experiments on dogs and rats had shown that some proteins, e.g. gelatine and zein, were nutritionally inadequate; analysis showed that certain amino acids were missing from them. Hopkins found that the survival and well-being of mice were increased when tryptophan was added to a diet in which zein was the dominant protein. But the mice did not grow. Although he knew that both lysine and tryptophan were absent from zein, he seems not to have tried the effect of adding both amino acids. A further 40 years were needed before the accepted list of 8 or 9 "essential amino acids", e.g. those that mammals and birds (and presumably other vertebrates) cannot synthesise at a rate adequate for growth, could be drawn up. He did not think of amino acids solely as the parts out of which proteins were built, but thought of them also as the precursors for other forms of synthesis. This was one theme of a review [15] in which he suggested that amino acids were transported in the blood, and that protein was not stored in a reserve as fats and carbohydrates are. The review ends with the words:

> "... the demands of the body are complex, and the constituents of protein may have uses which come under neither of these heads. What is the optimum supply for such purposes we cannot even yet be said to know."

During the 1914–18 war the question of protein requirement exercised some of the committees on which he sat. He quickly became aware of the diversity of human likes and, perhaps, requirements. For example, he was amused to find that Chittenden who, mainly for ideological reasons, advocated a very low-protein diet, reacted to that diet worse than other scientists from USA when confronted with it in British war-time rations. In a review [16] in which he quoted experiments that seemed to show that nitrogen equilibrium could be maintained on 19 g of potato protein per day, and that some Greenland Eskimos ate 500 g of protein per day, he commented:

"... physiological optimum of protein can only be defined in relation to personal and racial peculiarities, as well as habits and environment."

Though he did no personal research on amino acid requirement after 1916, he retained an active interest in the subject. He often spoke of Corry Mann's observations on the improved growth and vitality of those boys in an institution who were given extra milk daily, and he attributed the poor physique of army recruits about 1930 primarily to protein deficiency.

While studying various enzyme reactions, Hopkins observed a few phenomena connected with protein denaturation. He did not examine the process in detail until, in 1930, he returned to work he had done 30 years earlier on the effects of urea on egg albumin and serum proteins. By comparing the structures of 10 substances that denatured proteins with those of 15 related substances that did not, he suggested the essential qualities of what are now often called "chaotropic" substances [17]. These dissociate the cross-links that hold proteins in their native form. One consequence is the appearance of –SH groups to which, because of his interest in glutathione, he attached particular importance. An unexpected observation was that denaturation is faster at 0° than at 23°C.

Vitamins

One of the more surprising features of the history of the growth of our knowledge of nutrition is the slow incorporation of practical experience, even of repeated experience, into the canon of orthodox medicine. The beneficial effects of supplementing restricted diets, especially those used on long voyages in sailing ships, with small amounts of fresh foodstuffs had often been stressed before experience was systematised by Cook and Lind, and established as a matter for disciplinary regulation by Blane in

1796. For example: Hawkins used lemon juice in 1593 and Woodhall in 1612. Nevertheless, Hopkins told me that, when he was a medical student, some sailors with scurvy came into Guy's Hospital, one of them suddenly got better and he had *wondered* whether this was because that man's wife had brought him a bag of oranges!

Although his teachers had not mentioned what he was to call "accessory food factors", he became well aware of their importance by the time he started the feeding experiments on proteins and amino acids described in the last section. His Nobel Prize in 1929 was awarded for the discovery of vitamins: his lecture on getting the award deals mainly with his forerunners. He often expressed disappointment that that facet of his work was the one for which he was being honoured – he considered other things he had done more important. Futhermore, he did not regard vitamins as a properly definable category. In 1916 he suggested that they should be called exogenous growth hormones, and reasserted a disbelief in any clear distinction between vitamins and hormones in his Linacre Lecture in 1938 [18]. Those who strive to fit all natural entities into defined categories have recently taken a step in the same direction and suggest that the active material that the kidney makes from vitamin D should be recategorised as a hormone.

It is obvious that the diets which induce vitamin deficiency in both seamen and laboratory rats are monotonous and uninteresting. Hopkins considered the possibility that the cause of growth failure and the other signs was simply loss of appetite, and that appetite was stimulated when a more interesting food became available. His elaborate records of the amount of food eaten by rats on deficient diets disproved that: growth failure preceded loss of appetite.

When lecturing to the Society of Public Analysts in 1906, Hopkins made the confident statement:

> "Scurvy and rickets are conditions so severe that they force themselves upon our attention; but many other nutritive errors affect the health of individuals to a degree most important to themselves, and some of them depend upon unsuspected dietetic factors.
>
> I can do no more than hint at these matters, but I can assert that later developments of the science of dietetics will deal with factors highly complex and at present unknown."

That lecture was given about the middle of the 12-year period during which he studied "accessory food factors" and attempted to isolate the substances postulated. Though the experimental results he published in

1912 convinced most of his contemporaries, they were soon found to be irreproducible in other laboratories. This mattered little in the long run because the concept of vitamins was by then well established. Many years later his own work and that of others suggested that the use in the rats' diets of raw potato starch, which is relatively indigestible, was largely responsible for the effect he had observed. The minute amount of milk he had used probably enabled the gut microflora to proliferate on undigested starch and synthesise the necessary vitamins; the rats then got the necessary supply as a result of their habitual coprophagy. This interesting historical accident is discussed by Kon [19].

Awareness of widespread malnutrition in Britain, and of the importance of the school milk program, focussed Hopkins' attention on the amount of vitamin C lost when bottled milk sits on doorsteps in the sun. He found that lactoflavin sensitised vitamin C to photocatalytic destruction [18]. During 1937, whenever a suitable period of sunshine was probable, he took his equipment, mounted on a laboratory stool, out onto the lawn beside the laboratory. Some of the more imaginative occupants of rooms overlooking the lawn were surprised at the sight of a distinguished professor on his knees in the sun at what seemed to be a small altar – they wondered whether he had adopted a new form of Zoroastrianism!

In spite of general acceptance by biochemists and by those directly involved in research on nutrition and in its application, the vitamin concept was hotly contested for ideological reasons by much of the medical profession. Some critics seem to have had philosophical difficulties in attributing misfortune to the absence of something, some thought vitamins acted by curing an infection or a poisoning of some sort, some thought they might be necessary in youth but could not possibly be needed by adults, and some referred to the whole issue as quackery and a stunt. Hopkins discussed these semantic or philosophical matters in several articles – perhaps most cogently in the Huxley Lecture [5] in which he wrote:

"I put it to you (as counsel would say) that, considering the nature of the evidence, such endeavours can only be due to the obsession that some positive agent—some intruder so to speak—must always dominate aetiology. Several recent writers, concerning themselves with the definition of deficiency disease, have criticised the use of the term 'negative cause' and doubtless to formal logic such a term would be abhorrent. But it is better to use a somewhat illogical term than to miss some solid meaning which is conveniently conveyed by it. We must either speak of the cause of scurvy being the absence of something from the diet, or, because of its casuistic accretions, avoid the word cause altogether. Objection to a term is a small matter, but apparently with some the objection involves the practical standpoint that the production of phenomena like the symptoms of disease can only be due to the influence of some positive factor. If not to a poison or to infection, then to excess of something. Certain authors find it logical to refer causation to excess, but illogical to refer it to defect. They feel perhaps that the latter course takes liberties with the old adage 'Ex nihilo nihil fit'. But do not such views carry us back to the popular or metaphysical conception of cause as enforcement, and away from the scientific conception of cause as the routine of experience? Causation, in the words of John Stuart Mill, is recognised by uniform antecedence. Antecedent to normal nutritional phenomena is the consumption of a particular set of essential nutrients. If in the absence of one of them the sequence proves to be consistently different, such absence is then the reason (for we may avoid the word cause) for that difference. So we may say the reason for scurvy is the absence of the unstable antiscorbutic vitamine, and it is a sufficient reason."

As that passage shows, Hopkins' real contribution to the study of vitamins is that, for the vital first 30 years, he had a perfectly clear, and as we now think correct, picture of the complexity of our nutritional needs.

Today, when some trophologists write as if protein and energy were in some sense alternatives rather than complements, it is salutary to read Hopkins' [5] extension of this thesis. Having written:

"It is always a delight and relief to the mind when by the generalisation of knowledge it is enabled to neglect details. This relief is given in many branches of science by the application of the principles of energetics or thermodynamics."

He went on to write:

"Contrast Rubner's claim that the animal body makes no demand for specific material but only for energy, with the fact that when deprived of materials which contribute at most some few hundredths of its total food, the body fails altogether to make proper use of other of the energy supply. Rubner and those who think with him would speak of energy supply as the one essential 'limiting factor' in nutrition. What we have actually to recognise is that each of several factors may become that which limits efficiency, and that no one of these is in any strict sense more important than any other. Normal nutrition calls for a certain minimum of each one and every one. If a diet is harmoniously balanced in a chemical sense, then indeed energy does become the sole limiting factor. Nutrition then fails, of course, only when too little of the diet is eaten to yield the essential minimum of energy. But the supply of fat may become the limiting factor, and no less that of carbohydrate. Or, again, when the supply of energy consumed is ample, with fat and carbohydrate duly adjusted, the circumstance that a single essential amino-acid in one case, or a vitamine in another, is present in amount below the necessary minimum converts each of these in turn into the factor which limits utilisation. Small as the necessary minimum in either case may be, unless it is reached the proper use of the rest of the diet is reduced to a degree which is proportional to the deficiency.

If the deficiency be complete normal utilisation is altogether impossible.

This is a statement of the working of the Law of the Minimum, a much more significant principle in nutritional phenomena than the Law of Isodynamic Equivalence.

It is of course true that in civilised and prosperous communities the large variety of foods upon the market and the natural instincts of those who can afford to purchase them, secure for the most part the consumption of well-balanced dietaries, so that when, say, the requirements of a particular calling in such a community are to be estimated, energy measurements yield a justifiable criterion. It is nevertheless important to remember that even in the midst of plenty individuals, and even communities, may consume chemically deficient diets, and a proof that their energy consumption is normal may then lead to very wrong conclusions."

Muscular contraction

While working on vitamins, Hopkins collaborated with Fletcher in a study of the chemical processes in muscular contraction. The old idea, advocated by Liebig and others, that protein supplied the energy, had been disproved by Lawes. It was replaced by what Fletcher and Hopkins [20] called "the mystical complexes of irritable protoplasm". One of these was "inogen"; it was advocated "with great wealth of rhetoric, but without significant change and without fresh experimental support" by Pflüger and others. "Inogen", a giant molecule that was thought to absorb oxygen and then explode during the contraction, was the type of notional biochemical to which Hopkins was rootedly opposed. It was already known that muscle could evolve carbon dioxide, and that chopped, irritated or fatigued muscle, but not muscle dropped suddenly into boiling water, contained lactic acid. Fletcher and Hopkins improved the technique by grinding frog muscle in ice-cold alcohol and so were able to follow the appearance of lactic acid during anaerobic stimulation, and its disappearance with evolution of carbon dioxide during recovery in air. They suggested that most of the lactic acid was reconstituted into the original fuel during this recovery. Their suggestions about the role of lactic acid in the actual process of contraction have, obviously, been completely superseded by more recent work but they established the separation in time of the processes of contraction and oxygen uptake and they made what was probably the first attempt to fractionate a tissue after minimal manipulative damage. In an earlier paper they had said:

"... no treatment for the extraction from muscle can be accepted which, acting itself as a stimulus, has among its effects an increase of the acid to be estimated... in no recorded observations has a genuinely resting muscle been available for examination."

Glutathione

With characteristic candour Hopkins remarked [21] that, while studying the possible role of keto acids in muscular contraction, he did not realise that a mercaptan could be responsible for the nitroprusside reaction he observed. When he found that many tissues gave the reaction, he methodically set about isolating the substance(s) responsible because a widely distributed reactive substance able to "play a real part in cell dynamics" was just what his outlook on Biochemistry demanded.

Many years earlier, de Rey-Pailhade had published a series of papers on the reduction of elementary sulfur to H_2S by tissue extracts. He called the hypothetical substance responsible for this phenomenon "philothion", and attributed to it varied and often improbable properties. Hopkins' choice of glutathione (GSH) as the name for the substance he isolated was influenced by the earlier name although it was the reaction with sulfur that justified the – ϑειον root in the first name, and its presence in the molecule that justified it in the second. There is no justification for the -one, and it would not be accepted by a modern editor. Hopkins said he was influenced by the analogy with peptone: he may also have been influenced by the fact that he had originally been looking for a ketone. Some casualness about the application of rules of nomenclature is not uncommon in Biochemistry. Vitamin B_2, for example, is still called ribo- instead of ribi-flavine.

GSH yielded glutamic acid and cysteine on hydrolysis: no other amino acid was found. Hopkins can perhaps be criticised for not searching more assiduously and for not determining which of the three possible structures the supposed dipeptide had. But those were points with which, rightly, he was not greatly concerned. He had a substance and knew enough about its structure and properties to be able to think of it in relation to metabolic processes; for him biochemistry was not the study of the structure of substances but of their activities in tissues. Nearly half of the first paper [21] is devoted to discussion about, and experiments on, oxidations, reductions and the part that an –SH to –SS-reaction may be playing in them, both in vivo and in vitro.

Such interest as Hopkins might have had in studying personally the precise structure of GSH was dissipated when, in his own laboratory, Stewart and Tunnicliffe claimed that they had synthesised glutamyl cysteine and that it had the properties of GSH. Papers from other laboratories cast doubt on the structure, and the existence of scepticism, or even ribaldry, about the "synthesis" gradually filtered up to professorial attention. He therefore reinvestigated GSH and wrote [14] of "my junior colleagues'... supposed accomplishment". His critical observation that GSH, when boiled in water at its own pH, yielded glycine-cysteine diketopiperazine and pyrrolidone carboxylic acid, convinced him that GSH was a tripeptide. As an example of the complete absence of secretiveness in his character I should add that, although he was distressed by his original mistaken conclusion and naturally anxious to correct it himself, he spoke freely about the diketopiperazine for some years before publishing the reinvestigation.

The original method of preparation involved precipitation with copper. Hopkins realised that what was precipitated was a cuprous compound and that cupric copper was being reduced to cuprous by the oxidation of half the GSH to GSSG. That GSSG was not recovered. Cuprous precipitation is more specific as a means for separating mercaptans than precipitation with the other commonly used metals, and the silky sheen of suspensions of cuprous glutathione shows that it is microcrystalline and so not as likely as an amorphous precipitate to adsorb other components of tissue extracts. Hopkins therefore prepared GSH from yeast, liver and some other animal and plant tissues (e.g. peas after germination [22]) by partly purifying the extracts, acidifying them with sulfuric acid and stirring in a suspension of cuprous oxide. Considering the insolubility of cuprous oxide, this is a remarkable reaction.

Many papers by Hopkins and others, in the Cambridge Biochemical Department and elsewhere, dealt with the role of GSH. Sometimes so many functions were being assigned to it that it seemed to be primarily involved in no more than the maintenance of a reducing environment in the cell. This idea was sometimes called the "euphoristic theory" and GSH was flippantly called a "happiness factor". It is now given several more specific functions. Early work is discussed elsewhere [23]; Meister [24] discusses recent work. The situation is still fluid: 13 papers and demonstrations at a meeting of the Biochemical Society in December 1981 were concerned with GSH. It is much the most abundant mercaptan in tissue extracts, several analogues are known, and other γ-glutamyl peptides are widely distributed in small amounts. Hopkins studied GSH during 30 years; it is unlikely that his instinct that it is a fundamentally important substance was mistaken.

Intellectual and physical management of his Department

Although Hopkins had a thorough chemical training and initially became well-known in medical circles because of his analyses, he was primarily a biologist. Unlike most biologists at the end of the last century he was not, as his work on muscle and many comments in his published addresses show, content to take refuge in notional chemical agents and hypothetical processes. He stigmatised these as mere words coined to cover up ignorance and assumed experimental intractability. His attitude is summed up in his description of Biochemistry as the study of

"...simple substances undergoing comprehensible reactions."

But this quest for chemical rigour was tinctured by biological enthusiasm. It was that enthusiasm which differentiated him from his predecessors such as Liebig who, he said [25]

"...though so brilliant a chemist, lacked a biologist's instincts."

The distinction between the approaches to phenomena of chemists and biochemists, i.e. between the study of the structure of a substance and of its function or metabolism, was made in many of Hopkins' articles. As an example, a passage from his lecture on the centenary of the synthesis of urea [6] may be quoted:

"...there is all the difference between the methods of the living body and the classical methods of the laboratory. That the agencies which control biochemical events work on lines of their own is the basic circumstance which alone justifies the claim of biochemistry to rank as a self-standing scientific discipline. It is those special agencies and their impact upon molecular structures which constitute the peculiar field of biochemical study, and should be its final concern. The information which artificial syntheses afford in proof of constitution is of course absolutely essential to the progress of biochemistry, but for the genuine biochemist its attainment is a means to an end and not an end in itself. We can never be content with what Berzelius would have called "Probabilitätsbiochemie."

When Hopkins started research the term Biochemistry had not superseded Physiological Chemistry in English-speaking countries. The subject consisted of little more than a continuation of the structural studies that had given rise to Organic Chemistry, the study in vitro of the enzymic activity of some biological fluids and tissue extracts, and the measurement of urinary constituents. He wanted to study processes more intimately and often quoted appreciatively Mulder's old disparagement of the methods of Liebig and Dumas, that it was like studying what went on in a household by examining what went in at the door and out through the chimney and drain. Having acute biological instincts, he realised that this would not be easy because, to quote another of his aphorisms:

"The life of the cell is the expression of a particular dynamic equilibrium which obtains in a polyphasic system."

During the period when his Department was most productive (about 1925 to 1939) this attitude influenced almost all our work. Apart from GSH, new substances of biological importance were now isolated and characterised there. He subtly directed our interests towards processes rather than substances. The word *subtly* should be stressed. He did not direct research but influenced it by his obvious interest in some aspects rather than others. To take a personal example: when he suggested that I should collaborate with Miles (who worked in the adjacent Pathology Department) on the antigens of *Brucella*, he explained the curious immunological relationship between *B. abortus* and *B. melitensis* and the possibility that we might uncover a family of antigens in which serological specificity was controlled by a set of ascertainable chemical steps. He thought that clarification of the situation in one group of organisms might shed light on the general nature of immunological specificity. He always showed great interest in our immunological results – and much less interest in my excursions into marginally relevant sugar chemistry.

Hopkins would have been puzzled by the present-day tendency for teams to work on a problem and publish together. He seldom published in collaboration with anyone except James Morgan, who was his assistant for 20 years, and he discouraged collaboration between others unless they had dissimilar skills. Dale [4] records him saying that he could have worked happily on his own in a private laboratory. It seemed to him that someone's time was being wasted once collaborators had got beyond the teacher and pupil stage. There were therefore many research workers in his laboratory who were nominally independent but still getting much help and advice from a senior worker. Individual merit was consequently sometimes hard to assess. Lest this article should seem to be too much a eulogy, I should add that Hopkins put little effort into trying to assess merit. He saw good aspects in the character and work of everyone. When two of his students applied for the same grant or job, he was apt to praise both in almost identical terms. This engendered disbelief in both testimonials.

Not only did Hopkins not direct our research, he exercised little control over our expenditure on it. This was probably a consequence of his own financial struggles when he first came to Cambridge. His efforts to ensure that we had what we wanted involved constant argument with the financial authorities of the University. Fortunately for us, he had had ample experience of dealing with obstructive officialdom during the 1914–18 war. A taste of his attitude is given by a letter to Fletcher in April 1918 [9]:

"Things have been muddled by the official class, and, among the young men at any rate, there is impatience and intended rebellion. Who, after all, are these people who have taken our destinies into their inefficient hands? A producer of privilege, a narrow type of education, and a rotten examination system: A la lanterne! Voilà!"

Twenty years later, circumstances seemed to him still much the same. In the address to the Children's Nutrition Council from which I have already quoted, he said

"Unfortunately, the type of education which the upper and governing classes have usually received (at least until quite lately) leaves them ignorant of the most elementary facts of physiology."

Financial matters did not, however, get out of hand. Hopkins himself set an admirable example of economical working. He made great use of simple colour reactions (biuret, glyoxylic, Millon, nitroprusside etc.) and did them in tiny test tubes: it is as easy to see a colour in 0.5 ml as in 5 ml. His economical outlook spread throughout the laboratory, and there was strong social pressure on anyone who seemed to be extravagant.

With the techniques of the first half of this century, it was not easy to make progress along the lines that Hopkins had in mind. In his work on muscle he suggested the possibility that the formation of an acid would cause a change in protein conformation that could result in contraction. His work on xanthine oxidase suggested that part of the enzyme was attached to the cell membrane. He devoted much time to studying the relationship between GSH, GSSG, and the fixed –SH groups of proteins. He was fully aware of the possible complexity of these processes and the need for extra components in them. He used the phrase *tertium quid* regularly in conversation and lectures: it is odd that it appears infrequently in his publications. Towards the end of his life he was deeply interested in the evidence that several enzymes are associated on the finely subdivided material in tissue pulps in such a manner that their activities are integrated. On all such issues he had, and gave us, the vision, all that was lacking was the technique.

After the 1939–45 war the techniques became available – but too late for Hopkins to exploit them. With the new array of ultracentrifuges, electron microscopes, isotopic techniques etc., biochemists could move into the territory he had glimpsed. Much of what is being done in this territory is now called Molecular Biology and there has been some acrimonious discussion about the nature of that subject, the qualifications of its exponents, and the line of demarcation between it and Biochemistry. The transition resembles the transition from Physiological Chemistry to Biochemistry. Each transition depends partly on the exploitation of new techniques, but there are also differences in psychology and motivation. Biochemists had a more biological outlook than physiological chemists; molecular biologists have a more physical outlook than biochemists. The development of ideas about protein structure shows this clearly. Throughout the 1930s many protein structures were proposed which depended on quasi-rational geometrical or numerical considerations. Biochemists ridiculed all of them. X-ray analysis had for many years been trusted by biochemists as a reliable source of information about muscle, steroids, viruses etc. By 1958, the technique had advanced so far that an objectively based 3-dimensional structure could be drawn for myoglobin. To biochemists the structure seemed eminently reasonable – it was just what they had expected a protein molecule to look like. By contrast, the molecular biologists [26] were dismayed by the apparent absence of a rational plan.

In spite of the newness of its name, Biochemistry is an ancient science. It systematises, and springs directly from, the arts of the brewer, dyer, pharmacologist and poisoner. From time to time, offshoots from the main stem get such financial and technical endowment that they can become independent sciences. Organic Chemistry did this in the second half of the last century; Molecular Biology in the second half of this century. From time to time in future it will probably be convenient to separate other facets of the subject from the main stem and give them new names. But Biochemistry will remain, as Hopkins defined it, the chemical study of biological processes.

Some biographical points

Married Jessie Anne Stevens in 1898. She had been a Lady Probationer in the Royal Free Hospital (London) and was very active as a V.A.D. during the 1914–18 war.
Children: F.E.G. Hopkins, a medical practitioner; B.E. Holmes, a biochemist; J.J. Hawkes, an archaeologist.
Fellow of the Royal Society 1905, Royal Medal 1918, Copley Medal 1926, President 1930–1935.
Knighted 1925; Order of Merit 1935.
Nobel Prize 1929.

REFERENCES

1 C.S. Sherrington, Lancet, i (1947) 728.
2 G.H. Hardy, Bertrand Russell and Trinity, Cambridge University Press, 1942.
3 F.G. Hopkins, Proc. Roy. Soc. (B), 119 (1935) 88.
4 J. Needham and E. Baldwin (Eds.), Hopkins and Biochemistry, Heffer, Cambridge, 1949, contains a full list of Hopkins' publications, and his Autobiography. It also reprints 15 of his addresses in full and excerpts from some other papers by him. There is more information in the obituary written by H.H. Dale (Obituary Notices of Fellows of the Royal Society 17, 115, 1948). In selecting passages to quote in this article, I have tended to select those that have not already been reprinted in "Hopkins and Biochemistry".
5 F.G. Hopkins, Lancet, i (1921) 1.
6 F.G. Hopkins, Biochem. J., 22 (1928) 1341.
7 F. Galton, Hereditary Genius, 1869.
8 F.G. Hopkins, London Naturalist, 40 (1936).
9 R.E. Kohler, Isis, 69 (1978) 331.
10 F.G. Hopkins, Phil. Trans. Roy. Soc. (B), 186 (1895) 661.
11 F.G. Hopkins, Brit. Ass. Rep., 652 (1913).
12 E.J. Morgan, C.P. Stewart and F.G. Hopkins, Proc. Roy. Soc. (B), 94 (1923) 109.
13 E.J. Morgan, Biochem. J., 24 (1930) 410.
14 F.G. Hopkins, J. Biol. Chem., 84 (1929) 269.
15 F.G. Hopkins, Sci. Prog., 1 (1906) 159.
16 F.G. Hopkins, Ann. Rep. Prog. Chem., Chem. Soc., 11 (1914) 188.
17 F.G. Hopkins, Nature, 126 (1930) 328 and 383.
18 F.G. Hopkins, C.R. Lab. Carlsberg Ser. Chim., 22 (1938) 226.
19 S.K. Kon, Proc. Roy. Soc. (B), 156 (1962) 351.
20 W.M. Fletcher and F.G. Hopkins, Proc. Roy. Soc. (B), 89 (1917) 444.
21 F.G. Hopkins, Biochem. J., 15 (1921) 286.
22 F.G. Hopkins and E.J. Morgan, Nature, 152 (1943) 288.
23 N.W. Pirie, Proc. Roy. Soc. (B), 156 (1962) 306.
24 A. Meister, Trends in Biochemical Science, 6 (1981) 231.
25 F.G. Hopkins, Science, 84 (1936) 258.
26 J.C. Kendrew, G. Bodo, H.M. Dintzis, R.G. Parrish and H. Wyckoff, Nature, 181 (1958) 662.

G. Semenza (Ed.) Selected Topics in the History of Biochemistry: Personal Recollections (Comprehensive Biochemistry Vol. 35)
© 1983 Elsevier Science Publishers

Chapter 6

*A Short Autobiography**

JUDA HIRSCH QUASTEL

TRIUMF, The University of British Columbia,
4004 Wesbrook Mall, Vancouver, BC (VGT 2A3 Canada)

I am now over 84, having been born in Sheffield, England, on October 2, 1899 and not yet decrepit though I am, I dare say, well started. I think I have a little time left, but not much, and if I am to write an autobiography I had better do it now whilst I am still able to write fairly legibly. Ideas still come to me, though not as frequently as in earlier days, and whilst a scientific man has new ideas there is still useful life in him. I recall talking at dinner in Trinity College, U.K., some years ago, to J.E. Littlewood, one of the world's greatest mathematicians, and his saying to me "I must be getting old; for the first time in my life, as I remember, I have not had a new idea in the last 48 hours!" He died not long after that, in his early nineties.

I was born in a room over a sweet shop (candy store) rented by my father and I think I remember becoming interested in my surroundings for the first time when my mother lifted me up, when I was about a couple of years old, and allowed me to look through the opened window. In the manner of Mr. Pickwick, I saw Ecclesall Road stretching to my right and Ecclesall Road stretching to my left whilst opposite I saw shops of the butcher, the furniture man, the grocer, the hair-dresser and at the corner, where the road turned sharply to the left, was the pawnbroker. The corner was also distinguished by the presence of two public-houses or taverns, one on each side of the road and these formed a social center for this particular location of Sheffield, that erupted into song, frolic and occasional violence every Saturday night. This was my environment until I was inducted into the British Army in October 1917.

---

* Based upon an article by the author, reprinted with permission from the *Bulletin of the Canadian Biochemical Society,* 18 (1981) 13–34.

Plate 6. Juda Hirsch Quastel.
(Photo by M. Saffran, 1979)

My parents were very orthodox Jews but nevertheless liberal in outlook. When I was about 5 years old, my father walked with me to the elementary school in Pomona Street about a mile away – and often I sat astride his shoulders. I entered the infants' class, which always began its labours for the day with prayers. I stood up, like all the other children and folded my hands to pray, when I happened to glance through the classroom window. There I saw my father in his most irate mood, indicating that I must sit down and not fold my hands, that in doing so I was committing some awful sacrilege. When prayers were over, my father entered the classroom, and spoke vigorously to the teacher who apparently understood the situation at once. Thereafter, I was excused from morning prayers. I do not think I have ever folded my hands in prayer since that one occasion. But this does not mean I have failed to say my prayers either then or during the rest of my life. On the contrary, I said my morning prayers, under the watchful eye of my father who accompanied me in this observance, every day every year from the age of 5 until I entered the army. This incident in the infant school, insignificant as it was, had a marked effect on my life, for it, at once, estranged me from those around me. I entered the class alone after prayers and the other children saw in me something strange and alone. From that time onwards, there seemed to be a gulf between me and the majority of my companions. I accepted this as part of my life. I did not rebel or wish to bridge the gulf. My father, religious, hard working and a scrupulously honest man, taught me my birthright and I never had the slightest desire to escape from it. I longed for companionship however, and a little later became friendly with a boy a couple of years older than I was, Alex Scott, whose friendship I valued and enjoyed for over 60 years. He died only a few years ago. There was no racism in him, only intelligent understanding, sympathy and consideration. He was a great Englishman. The years I spent at the elementary school were not only important in my education and in my ripening years but were marked by my first interest in chemistry. I think I first heard this word when I was 7 or 8 years old. There were no laboratories in the school, but there was a weekly visit to which I always looked forward. This visit was by an itinerant science teacher who came along with his black box. This contained a Bunsen burner with a rubber tube which he attached to a gas jet which was part of a ring of such jets that hung, as a chandelier, from the ceiling. The box contained, apart from some chemicals, a wash-bottle and water, lots of test tubes, various small jars, glass cylinders, spatulas and even a retort and condenser apparatus. He was a remarkably good teacher and told us, for the first time, about mixtures and elements, solids and gases, atoms and molecules. I will never forget, one day, when he mixed two perfectly clear solutions, one in each test tube, and there appeared a white solid. I learned later that this was silver chloride precipitated by adding a solution of salt to one of silver nitrate. But that two clear solutions when mixed should produce a solid – this seemed magic! At or before that time I loved to read about all things magical; fairy tales I revelled in. I saw before me a world that I never dreamed existed. The teacher explained everything as well as he could with his limited means. But his greatest triumph was when he prepared hydrogen using a sort of Kipp apparatus, led the gas into a cylinder by water displacement, allowed it to mix with air in a particular proportion, and let it explode on applying a light. The sound of that explosion still lingers in my ears. I knew from then on I wanted to learn all I could about this strange world where things could undergo changes, that what we saw around us was the result of changes that must have happened before. This burning curiosity of mine was heightened by my passing each day, on my way to school, a steel foundry at whose open doors I would stand fascinated as I watched men pour molten steel from the Bessemer furnaces into the floor moulds. I visited later other factories and steelplating plants and knew by the age of nine that of all things I wanted to learn about, chemistry was the subject I had most at heart. It was so then and it is so now.

A year later I wanted to go to a secondary school, but my parents who were devoted to their children and their education, had no money for school fees – so I had to get a scholarship, if only to obtain free tuition. I happened to see a brochure of the Sheffield Central Secondary School and there I saw, to my delight, pictures of laboratories. I must get to that school. I took the necessary entrance examinations and was lucky enough to be placed in the top form and I had no tuition fees to pay. This, I think, was in the autumn of 1910. I had a first-rate education at that school. French was compulsory and I did sufficiently well in the language to earn a certificate of merit. I was better at German, which I took in preference to Latin, because I knew I would need German if I had, what I already wanted, a scientific career. I had excellent teachers of chemistry, physics and mathematics. I also had excellent tuition in history and, to my astonishment, at the year's final examination became head of the class, earning a prize in history, of all subjects! But one subject was not taught – not to me at any rate. This was biology. I believe it was taught in the girls' section of the school, and I had been brought up to believe it was an effeminate subject – a matter of learning the names of flowers and trees, etc. But I soon learned that this was not so, that it was a remarkably interesting subject – as wonderful in some ways as chemistry. But the manner in which I came to know this was also truly remarkable.

I was in my early teens when my father, a shopkeeper as I have said, in a rather small way, but with a dislike for business, decided I should enter the Civil Service, earning what seemed to him was a luxurious living, a steady and sure salary with automatic promotion and pension. For this I must take the Civil Service Examinations; for these, I was to be trained in précis-writing, book-keeping, copperplate writing, mathematics (commercial arithmetic), English composition and analysis. The final examinations were to be taken eventually in 1917. But in the previous year, 1916, I heard that all Civil Service Examinations were cancelled. My destiny suddenly became nebulous. I asked my father what I should do, and in a gesture of irritation, he said "Do what you want". Then came one of those moments which seem, in retrospect, to determine the course of one's life. I was walking up the Sheffield Moor – a busy thoroughfare – on the way to school when I met a good friend, Lawrence Horton, a year or two older than I. I chatted with him about the war (he was due to go into the army very soon) and about my perplexities as to my future. He suggested that I might try for a scholarship, which would enable me to go to the Imperial College of Science in London, but only one or two such scholarships were awarded each year, the examinations being held at Sheffield University. The exams were stiff, the competition severe and the subjects were chemistry, physics, pure and applied mathematics. I decided that this I would do. I had already taken various examinations, the results of which would enable me to enter a number of universities but they conferred no scholarships. And I must have sufficient funds to provide me with a livelihood. The scholarship examination, for which I entered, was for me, at that time, at a high standard, considerably higher than that in the top form of my high school. So in 1916, I began to cram all I could of chemistry and physics and mathematics, but though the work was hard, it was a labour of love. I had the habit, on leaving school in the afternoon, to visit a second-hand book shop. There I often picked up a used textbook, or some old work of interest, for a few pence. Among these I purchased cheaply a paperbacked book entitled *Riddle of the Universe* by E. Haeckel who I discovered later to be an eminent German biologist. It was published by the Rationalist Press Association. I was immersed in this fascinating book at home when my father caught me reading it. I have seldom seen him in such a fury. He tore the book from my hands, ripped it to pieces and for good measure belted me. Reading such a book would turn me into the worst of all creatures – a Free Thinker! All the knowledge I needed to know was in the Bible. Never again was I to be found reading, in his house, such abominable literature as that published by the Rationalist Press Association! This prohibition immediately resulted in my going, on leaving school, to the Sheffield Public Reference Library where I became one of its most frequent visitors. There I read, at my leisure, all the forbidden literature. I studied the elements of anatomy, physiology, zoology, botany, but more than this I learned much that was known, at that time, about growth and reproduction, about sex and genetics. Biology became open to me in a most unexpected manner. I also found advanced textbooks of chemistry, physics and mathematics which later I found to be of the greatest help for my scholarship examination. I began to appreciate the wonders of the arts, to learn about the great creative artists, Rembrandt, Leonardo, Michelangelo, about the scores of the great writers. I dimly realized how much I had missed in my formal education at school. I owe the Sheffield Public Library a great debt which I know I cannot repay and all I can do is to acknowledge it with gratitude.

I took the scholarship examination for which I had worked hard and to my great delight I was successful. The scholarship was tenable at the Imperial College in London. This covered all tuition fees together with the magnificent sum of 50 pounds a year which might have been sufficient to live on in some comfort in 1914, but not early in 1919 when I was demobilized from the army. But, luckily for me, Lloyd George was then Prime Minister and, through him, ex-servicemen obtained grants to enable them to pursue a higher education.

On Surrey Street, that I walked along almost daily from my school to the Public Library, was a house with a shining plate on its portals inscribed "Sheffield Public Analyst Laboratory. Mr. J. Evans, F.I.C.". Early in 1917, when I knew already what my destiny was likely to be, if I survived the war, for by then I had obtained my scholarship, I found time hanging heavily on my hands. I plucked up my courage, and knocked on the door of 67 Surrey Street. Mr. Evans, himself, let me in, took me to his office and asked me what I wanted. Mr. Evans was a short stocky grey-haired man, terse in speech, and undoubtedly, as I later discovered, of considerable analytical-chemical ability. I told him that I would like to be of help in his laboratory, that I wanted no payment but I wanted experience. I was willing to do anything he wished. He questioned me closely, was evidently impressed by my having gained a scholarship to the Imperial College, knew that I would enter the army towards the end of the year and decided to take a chance with me. I was first put on to the analyses of milks, following a routine procedure which I soon mastered. There were literally hundreds of milks to be analysed every week, for total solids, fats and possible preservatives such as boric acid. Many farmers thought to swell their meagre profits by adding water to a high-fat milk, but this was illegal if the contents of solids (proteins etc.) were over-diluted. Soon my results tallied with those of others in the laboratory and Mr. Evans himself tested those milks which had been illegally tampered with and gave rise to court cases. He usually confirmed our results. For the first time I learned how to use a balance correctly – those were the days long before the modern torsion – and other – balances which a babe-in-arms can use today with fair accuracy. I learned to analyse alloys and oils, sewage and drinking waters, beverages for their alcohol content, etc. I learned how to estimate CO_2 in the atmosphere using the Haldane apparatus and entered movie-houses and other places of public gathering to estimate the CO_2 of the air. This became quite a problem when young children gathered round me in dense clusters, as I manipulated my apparatus, and so unwittingly sent up the atmospheric CO_2 content.

My best friend in the laboratory was Stanley Dixon, older than I and very experienced; he later became Public Analyst to Cardiff and S. Wales. He was never averse to coaching me, giving me tips of great use in analytical practice. But above all he taught me how to use a microscope. I gazed at a drop of sewage water under the microscope with, I feel sure, the same interest and wonder as Leeuwenhoek himself experienced almost three centuries before. I never tired of using the microscope myself, looking at all sorts of sections, metals, ores, rocks, milks, etc. Dixon took me to his lodgings from time to time and we, together, examined a multitude of objects under a microscope of his own. I mention this rather trifling matter, of the microscope, as it was to have a profound effect on the course of my life as I will now show.

The War 1917–1919

Soon after my 18th birthday, I was called up and I became a private in the British Army. As a private, I did what I was told to do and went where I was sent. One raw December day, in 1917, I stood at ease with my unit on a barrack square after a long march when the sergeant, in charge of my company, suddenly yelled out "Anybody 'ere who can use a microscope? if so, one step forward, march!". The whole company looked at the sergeant as though he had suddenly taken leave of his senses. I decided that whatever happened could not be for the worse and I stepped forward. To me this step turned out to be every bit as important as the first step of the famous astronaut on the moon's surface. I was sent from one office to another until I was finally told to report next morning to the pathological laboratory of the local military hospital. It seemed that the pathologist in charge was in dire need of help and that there was apparently little hope of obtaining such help except from the ranks. There were no, or few, reservations in those days.

The pathologist was Dr. R. Donaldson, later Professor of Pathology in the University of London and Director of the Pathology Department of St. George's Hospital, London. We became firm friends and I owe very much to him. As soon as he saw me, he realized I could be of little use to him without preliminary coaching but he was a good teacher and I was a ready learner. I was at work each day from 7 a.m. until 7 p.m., learning to make and sterilize bacteriological media, to recognize and culture all manner of organisms, to make the various sugar-media in order to differentiate between organisms, to make vaccines, to carry out Wassermann reactions, to kill guinea pigs and prepare complement, to help run a bacteriological and pathological laboratory and to help Donaldson in his many post-mortem examinations. I saw the dead and the wounded of the war for the first time and I learned how easy it was to make false diagnoses on too little data. I was kept in the hospital, under Donaldson, who seemed to have considerable influence, through 1918, owing to the difficulty of finding a suitable replacement for the pathological laboratory. One of the workers in the laboratory was Ann Barbara Clark, who taught me much of microbiological techniques. We became good friends, discussing many things including her plans to go to Cambridge eventually. This was the first time that I had heard of Cambridge as a place to go for research in the field of microbiological research. I knew it was beyond my means ever to go there, but Miss Clark urged me to think about it. The great influenza epidemic came towards the end of 1918 and Donaldson, with myself as helper, carried out innumerable post-mortems. So many died suddenly, that they had to be put in a large marquee in the hospital grounds as there was no room for them elsewhere. It is strange that, although neither Donaldson nor I wore face masks or took no more than elementary precautions and we were in constant contact with victims of the disease, we never developed the slightest symptoms of influenza. I had to help Donaldson open up the corpses and extract fluids which he thought must house the offending organism. But although we worked till after midnight, on many days, isolating all sorts of organisms and injecting many animals, Donaldson never detected the responsible factor. Nowadays we know much about these virus diseases but in 1918 little was known. Donaldson died of influenza in 1933.

One major effect on me of this war experience was to arouse an intense curiosity as to the nature of the chemical factors operating in microorganisms that made it possible to differentiate between them by the use of different culture media. I felt that one day I would like to find the reason for this; the textbooks at the time gave little indication.

I was demobilized early in 1919, enabling me to proceed to the Imperial College in London, and the income I derived from the scholarship I had won in Sheffield in 1917 was supplemented by the welcome ex-servise grant raising my income to 180 pounds a year. I lived in rented lodgings in London with barely sufficient funds (board and lodging for 95 ponds a year, the rest for books, travel, lunches, clothes, amusement, etc.), but I realized how fortunate I was to be taught physics and chemistry by such men as Lord Rayleigh, H.L. Callendar, A.O. Rankine, H.B. Baker, J.C. Philip, J.F. Thorpe, and C.K. Ingold; mathematics by A.R. Forsyth and A.N. Whitehead; botany by J.B. Farmer, and zoology by E. MacBride. These men were giants in those days and I do not think that there was any place in the world where I could have received a more thorough and stimulating instruction in the basic sciences. Chemistry, of course, was my main subject, but, curiously, I did well enough in botany to prompt Sir John Farmer to offer me a job on a tea plantation in Ceylon. This I did not accept. I happened also to do well in physical chemistry and the Imperial College was good enough to offer me a scholarship to stay on and do research in that field. But my mind was by now set on biochemistry, a science that seemed new in the U.K. as few people had ever heard of it. Although some plant biochemistry was being carried out in the Imperial College I was not attracted to it. I had heard of Hopkins and wanted to go to his laboratory but I had no financial means whatever to enable me to do so.

Cambridge

One day I saw, on the screens of the Imperial College, a notice concerning the possibility of obtaining a studentship at Trinity College, Cambridge. I knew that Trinity, as a seat of learning, surpassed any other College anywhere and I was determined to go there somehow. Once again I took a major step forward. I called Hopkins by phone from Sheffield, told him I wanted to work under him and would he see me? I was both astonished and delighted when he agreed at once and arranged to meet me in Cambridge in a few days' time. In the meantime I had obtained letters of reference from Baker and Farmer and MacBride and armed with these I made my first trip to Cambridge about the summer of 1921. I went to the old Balfour Laboratories, where I was later informed that Girton and Newnham science students used to be taught before they were admitted to University classes, and which now housed the Biochemical Department. In some fear and trembling, I called on Hopkins. As soon as I saw him all my fears disappeared. He put me at my ease immediately. I could not have met a more charming and courteous and understanding man. He supported my application for admission to Trinity and in September 1921 my scientific life began. I entered Trinity as a graduate student and candidate for Ph.D. No words of mine can adequately express my sense of thankfulness, or my happiness, at my becoming a member of that great College and on my good fortune in becoming a participant in the active community of men and women who worked in Hopkins' laboratories.

Perhaps I should mention at this point that, earlier on in the year, only a month or two before the final examinations in the Imperial College, I was carrying out some little research work in the organic chemistry laboratory when there was a minor explosion and a stream of hot nitric and sulphuric acids poured over the left side of my face and into my left eye. Fortunately, assistance was close at hand and much of the acid was washed away before I was taken for treatment at St. George's Hospital near Hyde Park. There my eyes were bandaged and washed frequently and the fear that I would lose my eyesight was ultimately dispelled. Nevertheless the cornea of my left eye was scarred, so that, since that time, I have never been able to see clearly with that eye. My entire observational work, during my scientific life, has been carried out with the use of my right eye only. I mentioned this to Hopkins when we first met and I asked if this accident might interfere with research work which would involve accurate measurements. He laughed at my fears and he turned out to be, as he usually was, correct.

Hopkins gave me no problem but told me to go ahead with whatever I liked. I had to do something quickly, for it was necessary for me to apply for a scholarship or exhibition. I was at that time entirely without funds and had to borrow a little from my father, who could ill afford it, to keep going. Obviously in my application for funds I would have to say something of what I was doing. I recall thinking that I would like to see whether bacteria, during their growth, would discriminate between *cis* and *trans* unsaturated acids and how they would handle a substance such as glutaconic acid. I made cultures of various organisms and inoculated them into media containing the sodium salts of fumaric acid and maleic acid, and having devised a method of estimating these substances, found to my surprise that, with certain cultures, fumarate, in contrast to maleate, disappeared very quickly from the culture media. I then turned my attention to succinic acid and found it to disappear only slowly in the presence of growing organisms that disposed of fumarate rapidly. These results introduced me at once to the problems of succinate metabolism – a line of work that I followed for a long time. Little did I realise how important these problems were to become in later years. At any rate, what few results I had, at the time, helped me to obtain an 1851 Senior Exhibition and once again I had something to live on and to repay my debt. This came almost miraculously at a time when I seriously thought of quitting research work. I had been invited to the laboratory Christmas party and everybody was in evening dress, except me for I possessed no such luxury. I was very embarrassed. But Hopkins saw me and came to me where I sat at the end of the dinner table trying to look as inconspicuous as possible and in his usual kindly way he put me at my ease. I enjoyed the party and Hopkins' witty after-dinner speech, full of friendly personal remarks about each one of us, that was a joy to listen to. Hopkins radiated kindness and friendliness to all around him and the tone of the laboratory was set by him, at any rate in those early days. I encountered nothing but generous friendship from all my colleagues. However, I was in despair, for I thought my work extremely dull and unpromising. The little money I had borrowed was running out and I thought that maybe I should quit and try to get a real job somewhere soon. But as fate would have it, I not only succeeded in obtaining the scholarship, but just as important I soon discovered why fumarate disappeared so quickly with certain growing organisms; it was being converted rapidly to pyruvate. It was about this time I realised how much I owed to Hopkins. When things looked most hopeless, he would be so encouraging and sympathetic that I felt that I just dare not abandon the work. He suggested no new ideas or techniques or plans, but engendered the feeling that with patience, persistence and some originality of thought I would get over this difficult period.

At about that time, the idea that pyruvate was produced as an intermediate stage in the course of sugar fermentation had already been suggested through the early work of the German pioneers. Neuberg's scheme, according to which glucose was converted to methylglyoxal and this into a mixture of glycerol and pyruvic acid, the latter being decomposed into acetaldehyde which then formed alcohol, was current at the time and this, and later schemes of a similar sort, involving fixation procedures, were very valuable in Germany during the first World War, where there was a great demand for glycerol for production of explosives, especially dynamite. But nobody, so far as I knew at that time, had demonstrated that succinate or fumarate was a biological precursor of pyruvate. Nowadays, it would be only a matter of hours to detect the products of bacterial breakdown of fumarate, but in 1923 I had not only to devise a colour test for the unknown product so that I could detect the most suitable time during the fermentation procedure for its isolation but to make a guess at its chemical nature, to form a crystalline derivative of it and to analyse it. This took me some months of painstaking work but finally I was satisfied that the major product of oxidative metabolism of fumarate (by *B. pyocyaneus*) was pyruvate. Publication of this work in 1924 put me at once in touch with scientists such as Neuberg, Aubel, and others and I formed many friendships with noted investigators in the fermentation field, particularly in Paris, at the Pasteur Institute, which I visited often. We exchanged letters and greetings and reprints. Life for the young investigator was, I think, much more pleasant in those days than today, when competition is so fierce and ruthless, when shedding reprints is a pure formality dependent on whether "Current Contents" has been read or not and when scientific courtesies so delightful in my young days are now on the whole conspicuous by their absence.

To say that discovering pyruvate as a metabolite of fumarate and therefore of succinate oxidation in microorganisms was a milestone in my career is to put things mildly. It set me on a course of events which I think I have followed most of my life. As soon as I saw that fumarate was rapidly converted to pyruvate by *B. pyocyaneus* I realized from a study of the rates of oxygen consumption and carbon dioxide evolution and from the construction of a balance sheet that the organism must be using pyruvate for growth.

In the words of my article [1] the ratio $\frac{\text{carbon dioxide output}}{\text{oxygen uptake}}$ was 2.86 and in another 2.72:

A third experiment gave the ratio $\frac{CO_2 \text{ output}}{O_2 \text{ uptake}} = 2.61$ and we have as the average of these three results $\frac{CO_2 \text{ output}}{O_2 \text{ uptake}} = 2.73$.

If we assume the early part of the fermentation of fumaric acid to follow the equation

$$\begin{array}{l} CH.COOH \\ \| \\ CH.COOH \end{array} + O = \begin{array}{l} CH_3 \\ | \\ CO.COOH \end{array} + CO_2$$

we have the ratio

$$\frac{CO_2 \text{ output}}{O_2 \text{ uptake}} = \frac{44}{16} = 2.75.$$

Moreover, since 116 g of fumaric acid will require an uptake of 16 g of oxygen for this reaction, the amount subjected to fermentation, i.e. 0.5 g, will require 0.069 g of oxygen. Experiment shows that a little more than this amount of oxygen is utilised, and this is probably used in the oxidation of pyruvic acid to acetic acid:

$$\begin{array}{l} CH_3 \\ | \\ CO.COOH \end{array} + O = \begin{array}{l} CH_3 \\ | \\ COOH \end{array} + CO_2$$

where the ratio CO_2/O_2 is the same as in the previous reaction. If this is the case, then the difference in oxygen uptake between the experimental and the theoretical should be equivalent to the amount of acetic acid produced. This may be shown to be approximately the case. In the second example quoted above, the amount of fatty acids present calculated as acetic acid was 0.077 g. The theoretical quantity was 0.071 g.

All the oxygen taken up could be accounted for by the oxidation of the fumarate. Was not oxygen needed for the mechanism of proliferation itself? Why should only the pyruvate molecule be needed to make the bacterial substance? Why not fumarate itself or perhaps some other metabolite? These questions perplexed me at that time and Hopkins gave me no advice.

The results in fact posed the question: why was not oxygen needed for the bacterial proliferation process? All of it seemed to be required to convert fumarate to pyruvate and a little acetate. Why was pyruvate so important for providing organic nutrient to the microorganisms? In what was one of the first, modern-type, metabolic maps [2], it was suggested that organic compounds, supplying bacterial substance, must be broken down to pyruvate before they could act as bacterial nutrients.

Only after the lapse of about 20 years, was it realised that not oxygen, but ATP was required to sustain proliferation and that not pyruvate, but a product of pyruvate oxidation, acetyl-CoA, was needed to provide the organic basis for bacterial development.

So I decided, in 1924, to see what would happen if the bacteria were examined in a non-proliferant condition. I abandoned *B. pyocyaneus* as an organism for investigation; it tended to form sticky, mucilaginous growths and I eventually concentrated on *B. coli* (or *E. coli*) as a very suitable organism, easily grown, and saline suspensions of it could be washed and handled with considerable accuracy. At this time I was joined by Margaret Whetham who had previously worked with Miss Stephenson and together we began systematic studies of bacterial suspensions which were termed "resting cells" to discriminate them from proliferating bacteria. We used them under conditions (absence of added nitrogenous material, short time exposures and sometimes relatively high temperatures) that made it unlikely that proliferation was a disturbing factor. Using such cells and such conditions we were able to demonstrate (1924) for the first time that a reversible equilibrium exists between succinate and fumarate, that fumarate acts as a "hydrogen acceptor" and may even replace oxygen as an oxidant. We also demonstrated the equilibrium, in *E. coli*, between fumarate and L-malate and that our failure to demonstrate that L-malate was a hydrogen donor was due to its formation of fumarate which acted as an oxidiser. This work had quite a variety of results so far as I was concerned. In the first place I immediately provoked the ire of a group of bacteriologists who objected to the term "resting" cells on the grounds that "resting" cells were "dying" cells and that what results I obtained with them were due to post-mortem phenomena. I may say that before I ventured to use this term, I had the encouragement of Hopkins, Harden (the discoverer of NAD and Editor of the *Biochemical Journal* at that time), and Donaldson who all told me to go ahead and use the term although I already foresaw the objections of many microbiologists. However, this claim about post-mortem phenomena was easily rebutted. Marjorie Stephenson joined us and together we showed (1925) that it was possible to grow a variety of microorganisms anaerobically, using fumarate instead of oxygen as oxidant. Thus we were able to develop for the first time "synthetic" media for anaerobic growth and we outlined the principles, concerning oxido-reduction processes in bacteria, that nowadays are referred to as electron-transport systems.

In our articles [3, 4] the statement was made that anaerobic growth of an organism might be expected to occur on a pair of organic compounds A and B (neither of which supports anaerobic growth alone) when these fulfil the following conditions:

"(1) That both A and B are 'activated' by the organism so that simultaneous oxidation and reduction may occur; (2) that the energy necessary for growth is liberated in the interaction; (3) that as a result of such an interaction some substance is produced capable of entering into the synthetic processes of the cell.

It was then shown that these conditions are fulfilled in the case of *B. coli communis* by lactate and nitrate, and by lactate and fumarate, the activations of these substances by the organism having been demonstrated by the methylene blue technique. The lactate behaves as a hydrogen donator and the nitrate and fumarate as hydrogen acceptors. It was noted that the interaction of both pairs of substances occurred with output of energy. No anaerobic growth was obtained on lactate alone or on fumarate alone. In this communication [4] an attempt has been made to extend these observations, in the first place, so as to include the growth of *B. coli* on other pairs of substances (which had been shown to be activated) and in the second place to compare the results obtained with *B. coli* with those of other organisms possessing different activating powers. Glycerol, for example, is activated by *B. coli* [5] and both glycerol and nitrate and glycerol and fumarate, if they interact, do so with output of energy, e.g.

$$CH_2(OH).CH(OH).CH_2.OH + COOH.CH{:}CH.COOH$$
$$= CH_3.CH(OH).COOH + COOH.CH_2.CH_2.COOH + 33\ \text{Cal.}$$

and

$$CH_2(OH).CH(OH).CH_2.OH + 2COOH.CH{:}CH.COOH$$
$$= CH_3.CO.COOH + 2COOH.CH_2.CH_2.COOH + 50\ \text{Cal.}$$

Experiment shows that whilst *B. coli* does not grow anaerobically on glycerol alone or on fumarate it grows luxuriantly under anaerobic conditions on a mixture of glycerol and nitrate or of glycerol and fumarate."

We had realised that oxygen may be replaced by other hydrogen acceptors, as long as these reacted, in the presence of appropriate enzymes, to yield energy for anaerobic growth.

Later, it was discovered that formates may greatly stimulate anaerobic growth of *E. coli* on a mixture of lactate and fumarate by enabling synthesis of formic hydrogenlyase to take place [6], the proton gradient thus formed, in the presence of formate, which is a powerful hydrogen donator [5], resulting, according to the now current Mitchell hypothesis, in the bacterial synthesis of ATP and consequent acceleration of growth.

Miss Whetham and I (1925) showed in 1925 that *E. coli* possesses dehydrogenases for a large variety of molecules such as formic acid, α-hydroxy acids, α-glycerophosphate, polyols, amino acids, and a variety of sugars. "Resting" cells became a rich field of investigation and although, nowadays, it is considered more respectable to refer to them as washed bacterial suspensions, etc., the term still persists in various quarters, over fifty years after its introduction. Our systematic work with *E. coli* launched this organism into the field of biochemical research, although it was not the first biochemical work with this organism. Today it is one of the most popular organisms of investigation in the field of molecular biology.

Our demonstration that succinic dehydrogenase was a reversible enzyme came as a surprise, for it was thought at the time that the enzyme acted on succinate only by activation of its hydrogen atoms; the fact that it apparently could activate fumarate to take up hydrogen was something quite new. Our conclusion as to the reversible nature of the enzyme was followed by its confirmation with muscle tissue by Thunberg in Sweden, and by the demonstration, in many laboratories, of other reversible dehydrogenase systems. The idea of activation of hydrogen being a significant property of dehydrogenases disappeared and today no one even discusses it. It gave place to modern concepts of the activation of the molecule when combined with its enzyme. I had the thought, at the time, that once the fumarate molecule was activated by its enzyme it would be in a condition not only to accept atoms of hydrogen to form succinate, but also to accept the elements of water to form malic acid and perhaps the elements of ammonia to form aspartic acid. It had been known for a long time that aspartic acid is reduced to succinic acid during bacterial growth and Stephenson and I had found that aspartate can act as a hydrogen acceptor with *E. coli,* enabling this organism to grow anaerobically in a glycerol-aspartate medium. Woolf and I then found [7], to my delight, that an equilibrium existed between fumarate, aspartate and ammonia, in presence of *E. coli*, the responsible enzyme being later termed aspartase. I thought that this idea of one activating source being responsible for the various conversions of fumarate must be correct. However, I soon learned that this was not the case, that succinic dehydrogenase, fumarase and aspartase were quite different enzymes and that they could be selectively poisoned. I realized that activation of the fumarate molecule was not enough; other factors must also be operating to confer specificity.

This period, the early and mid-twenties, was an unforgettable experience for me, full of rich promise and achievement. I was fortunate enough to be elected a Fellow of Trinity College, a coveted academic distinction that I have valued all my life. In those years, that I spent at Trinity, I met some of the greatest intellects in the world and although I was awed by a number of them I found such friendliness and understanding, even by those for whom the term "biochemistry" had little or no meaning, that I felt that there was no greater intellectual haven in which I could hope to live and work. I made many new friends whose thoughts greatly expanded my perspective and met many distinguished scientific visitors who were guests of various Fellows. Among these I remember was R. Willstätter, for whose work my admiration knew no bounds, who was a guest of Hopkins. I asked them both to my rooms after dinner on one occasion, and looked forward to hearing the words of wisdom that would come from the lips of those two great men. But all they would talk about was old age pensions and what they should do on retirement. Little did Willstätter realize what was in prospect for him later in Nazi Germany.

About this time Margaret Whetham and I founded the journal *Brighter Biochemistry* – the illustrated annual outpouring of the Biochemical Laboratory in Cambridge. It lasted eight or nine years but I only helped to edit it in the first two years. Much of the journal contained noteworthy articles, worth reprinting today; happy verses by F.G. Hopkins, excellent articles and rhymes by J.B.S. Haldane, many delightful cautionary tales (in verse) by Margaret Whetham, and cartoons and witticisms and stories of high merit by M. Dixon, J. Needham, B. Woolf and many others. All this reflected the expression of a laboratory in good heart and buoyant spirit, full of brightness and lively comradeship, due to the warmth and inspiration of its leader.

I was invited to quite a number of conferences but one that I remember particularly was a "Congrès des Fermentations" held in Bordeaux in 1928. I recall this because it was there that I delivered a paper in French – almost the only time that I have ever done so. I fear that my delivery was given with a pronounced Yorkshire accent, so that few people understood what I was saying. However, I was able to clear up satisfactorily all difficulties by discussions later on, in English, with interested participants. I decided there and then never to give a scientific paper in a language in which I was not quite fluent. Among other invited speakers were such noted investigators as Meyerhof, Neuberg, Wurmser, Aubel and we had a very enjoyable time. Perhaps the most remarkable aspects of the Congress were the delightful luncheons and dinners provided us. I must admit, however, that I have never been, since then, to any scientific conference where the culinary art was so magnificently displayed.

I met many remarkable men outside my sphere of work who nevertheless affected my future activities, for example Dr. C. Weizmann, who later became the first President of Israel, for whom I did a small service in Cambridge, a very great statesman and leader whose words impressed me more than those of any other in matters concerning events in the Middle East.

At the request of the Cambridge Natural History Society, with which I was closely connected, I met, in Vienna, Dr. Paul Kammerer the subject of a hot dispute concerning the inheritance of acquired characters and obtained photos of the famed nuptial pads of *Alytes*. I persuaded him to come to Cambridge to exhibit these specimens and lecture on his evidence and views. All this is written up in a fascinating manner by Arthur Koestler in his *Case of the Midwife Toad.*

Space alone prevents my writing further on my varied activities, both within and outside my laboratory life, during those years in Cambridge.

I carried on much work with *E. coli* suspensions. I found to my amazement that there were present in these cells at least 56 specific dehydrogenases, if the current opinion was correct that for every substance, that could be classed as a hydrogen donator in the Wieland-Thunberg sense, there was related to it a corresponding specific dehydrogenase. At this time [8] I had evidence that these dehydrogenations must be taking place, at any rate to some extent, at the cell surface. It must be recalled that, at that time, enzymes were not known definitely to be proteins. Sumner had only just shown urease to be a protein, but many objected to the idea that all enzymes were proteins. A popular view, upheld for example by Willstätter, who carried out his well known absorption and elution techniques in an endeavour to obtain the necessary evidence, was that an enzyme consisted of some active group or molecule attached to a colloid carrier. As late as 1925 Oppenheimer considered enzymes to be compounds of unknown chemical structure belonging to no known group of substances. At the time I was working, prior to 1926, I was wondering how it was possible for at least 56 specific dehydrogenases to have their activities located at cell surfaces [8]. Moreover, in considering specificity, how was it possible for the cell to possess specific enzymes to cope with substances to which it was not normally accustomed? Was there present in an organism a specific enzyme, for example, that could activate chlorate as a hydrogen acceptor, when the product chlorite was known to be a highly toxic substance and so far as I knew there was no enzyme to break it down. Yet many organisms, including *E. coli*, I found able to activate chlorate in this manner.

This line of thinking led me to suggest [8] that there might exist at the bacterial surface, or at cell interfaces, particular regions or centres that had the property of absorbing molecules (substrates), producing electric changes in them and so causing changes in their chemical activities. This idea was extended when, with the cooperation of W.R. Wooldridge, we concluded [9], from an examination of the effects of a considerable variety of toxic substances on bacterial dehydrogenase systems, that there existed on surface structures "active centres" that were the seats of enzymic activity.

"Enzymes, on this view," (i.e., active centres) "are themselves part of and cannot be dissociated from the architectural units of the cell."

This concept was not very different from that held today, if it be accepted that proteins are architectural units of the cell.

Concept of active centres as a possible explanation of enzyme action

The possibility of the existence of active centres at biological surfaces as being responsible for the phenomenon of enzyme activity occurred to me in 1925. Together with the discovery of malonate as a specific competitor (inhibitor) of succinic dehydrogenase, it is expressed in a number of articles by Wooldridge and myself [6, 10] and also by myself [8, 11]. The following quotation may suffice to illustrate our concept at that time [9]:

"The active centre in the molecule $\underset{\displaystyle NH_2}{-\overset{}{A}}-CO-S-B-$ we would imagine to be made up at least of the groups $-NH_2-CO-$ and $-S-$ and each of these groups will play its part in rendering a substrate accessible to the centre. We have given this illustration of a possible formation of an active centre in order to make clear the difference between the centre and the usual conception of an enzyme. The actual composition of an active centre will in all probability be much more complex than in the illustration given. We may regard the entire aggregate as the enzyme, or the particular centre as the enzyme. Each view is equally legitimate. But the residue of the aggregate, distinct from the region occupied by the centre, may be the seat of other active centres, so that the aggregate as a whole may have a much wider range of specificity than were the residue inert in this respect. Such an aggregate would be a relatively large colloidal particle and it would certainly be difficult to regard it as a specific enzyme. On the other hand, the residue may be of comparatively small dimensions and contain no other active centres. The specificity of action would be determined by the single centre and the range of specificity may be so small as to make the particle a highly specific enzyme."

Out of the series of substances adsorbed by a particular centre (or enzyme) only a limited number is, in this concept, actually activated, i.e., only a number dependent on their molecular constitution receives the critical energy required to transform them into "active" molecules. The reason therefore why an enzyme is considered so specific in its action is, in the first place, because only a limited number of substances – containing a certain type of structure – is accessible to or adsorbed by the enzyme, and in the second place, because out of this limited number of substances specifically adsorbed only a few are capable of being turned into the "active" molecules capable of the reactions under investigation. Thus, each enzyme has a limited and definite range of specificity of action. Or, as N.K. Adam puts it,

"Emil Fischer's comparison of the action of an enzyme on its substrate to the highly specific relation between a lock and key, now seems capable of being analysed further; the key must not only be capable of entering the keyhole (adsorption) but it must be capable, once inserted, of operating the mechanism inside (activation)."

The orientation of a molecule at the centre must play a very important part in the activation process. Differences between D- and L-form would be anticipated just as there are differences in the activation of *cis* and *trans* isomers. Bearing on this point, Murray and King have recently shown with the lipases differences in the inhibition of ester splitting action by D- and L-carbinols, the polar carbinol group being apparently largely responsible for the adsorption at the lipase. Another important point is that a molecule may have a constitution which will make it accessible to more than one centre, with subsequent differences in the chemical events occurring after activation. A number of enzymes would in this case be specifically related to one substance.

An enzyme (lactic dehydrogenase) extracted from the cell has much the same properties, when tested in the manner described above, as it has in situ; there are a few modifications. Clearly the constitution of the active centre itself is of fundamental importance [10]. In another article [12] it is stated:

"Just as carbon monoxide inhibits oxidations of the cell by competition with oxygen for an oxidase, so certain substances inhibit dehydrogenations by competition with hydrogen donators for their enzymes – the competition being quite reversible."

Examples of this are the effects of malonate on succinate oxidation and of oxalate or hydroxymalonate on lactate oxidation. The phenomena are best studied by using suspensions of bacteria which have been shaken with toluene and then well washed. Such an organism can still activate succinate, lactate and formate to an extent very little different from that of the normal cell, but other dehydrogenase activities have been suppressed. This "simplified" organism is particularly useful for quantitative work on dehydrogenases. The lactate dehydrogenase is studied by observing the effect, upon the rate of reduction of methylene blue by lactate of the mixture of various substances which are themselves inert in presence of the toluene-treated organism; adsorption or combination of any of these substances, at the lactic acid enzyme, will result in a competition with the lactate and hence, by reducing the effective concentration of lactate, will result in a decrease of the rate of reduction. The succinic and formic acid enzymes are investigated in the same way.

Original experimental data [10] are shown below.

The effects of malonic acid and substituted malonic acids on bacterial dehydrogenases

Table I shows reduction times by *E. coli* due to lactic, succinic and formic acids in presence of toluene-treated organism and the effects on these times of the admixture, with the normal substrate, of either malonic acid, hydroxymalonic acid, or ethylmalonic acid. Results with oxalic acid are also inserted for comparison. It will be seen how specifically effective malonic acid is on the succinic acid enzyme, and hydroxymalonic and oxalic acids are on the lactic acid enzyme. The formic acid enzyme is not affected. Ethylmalonic acid has little or no effect on the three enzymes [10].

TABLE I

Effects of the addition of certain substances upon the times (t) of reduction of methylene blue by succinic, lactic and formic acids in the presence of toluene-treated *B. coli*. Each vacuum tube contained 2 cc. phosphate buffer pH 7.4, 1 cc 1/5000 methylene blue solution. 1 cc toluene-treated *B. coli,* 1 cc of the donator at the head of each vertical column, 1 cc. of the substances, in a concentration of M/2, given horizontally and 1 cc. water. All the acids were brought to pH 7.4 with sodium hydroxide. The reductions were carried out in vacuo at 45°C. ∞ indicates that reduction was not complete in 3 hours.

Added substance, M/14	$t_{succinate}$, M/140	$t_{lactate}$, M/350	$t_{formate}$ M/70
NaCl	23'	10.7'	12'
malonic acid	∞	14.5'	11.7'
hydroxymalonic acid	21.3'	∞	10'
ethylmalonic acid	20.5'	18'	10.5'
oxalic acid	32.5'	∞	11.3

In summary, the experiment showed:

"(1) The enzyme (or active centre) of *B. coli,* which activates lactic acid as a hydrogen donator has the property of specifically adsorbing compounds characterised by the possession of a particular structure which seems to be –CO–COH*– or –CHOH–COH* where H* is mobile, the compound having acidic properties. The specificity of adsorption is very marked.
(2) The enzyme (or active centre) which activates succinic acid has also the property of adsorbing compounds characterised by the possession of a particular structure. This seems to be C–CH–COO or C–CH_2–COOH.
(3) The formic acid enzyme is independent of the lower fatty acids (other than formic acid itself) and of all the substances we have so far investigated.
(4) The activity of glucose as a hydrogen donator is not perceptibly inhibited by the presence of oxalic acid, or of hydroxymalonic acid. The reduction of methylene blue by glucose in presence of bacteria is independent, therefore, of the intermediate production of lactic acid [10,11]."

The investigations show that each dehydrogenase or centre adsorbs or combines with a particular type of structure, only one class of substance being taken up at each centre in the toluene-treated organism. The combination is reversible and the substances appear to act as "poisons" simply by competing with the substrate at the centres. A relatively large number of substances can be adsorbed or combined in this specific way but out of this large number only a few can be activated to function as donators of hydrogen. In this manner the principle of competition inhibition of an enzyme by a substrate analogue originated.

Although I considered at first that active centres were the seats of enzymic activity located on bacterial surfaces, the basic idea of the existence of such centres as responsible for enzymic activity was adopted almost without opposition by later enzymologists even at a time when enzymes were shown to be proteins that could be separated, crystallised, their molecular weights determined, and their structures unravelled. The concept was useful and capable of development. The practical importance of our observation on reversible enzyme inhibition by substrate analogues did not become clear until 1940 with Wood's work on sulfanilamide (1940) as a competitive analogue of *p*-aminobenzoate, found to be essential for bacterial growth, and with our work with Mann (1940) on competition between amphetamine and cerebral amines for monamine oxidase. These results stimulated much work in laboratories all over the world on the practical therapeutic aspects of enzyme inhibition by substrate analogues. It is the basis nowadays of much work on chemotherapy, not only in medicine, but also in some aspects of agriculture.

These ideas when put forward in 1928 and 1929 seemed to make very little impact, though I felt that they were important. I became very discouraged. Hopkins was only mildly interested. Marjorie Stephenson seemed to have no interest in them at all – she was, if anything, hostile. She was working with a group of able young men on intracellular bacterial enzymes, busy with her book on bacterial metabolism and turning her attention to other problems of microbiology. The only person, I knew, who seemed really interested in our work and thought it of considerable promise was J.B.S. Haldane whose very great intellectual ability and extraordinary retentive memory made discussions with him most enjoyable and fruitful.

I had few contacts with other members of the laboratory; notable ones were with David Thomson whose work was something of a mystery to me but who greatly influenced my life later on, and with the dynamic Albert Szent-Györgyi, who had recently come to the laboratory at the invitation of Hopkins and whose continued friendship I have greatly valued. I think I was among the very first to see the crystals of his "hexuronic acid" which later turned out to be vitamin C.

Hopkins at this time was becoming more and more involved in his important public duties and more and more remote so far as I was concerned. He continued to run his laboratory, that had become host to many young scientists of different nationalities, with the support of such old and loyal colleagues as Dixon, Needham, Stephenson and others who all helped him in every possible manner. I began to feel that my time there was coming to a close. My College Fellowship which was of limited duration was coming to an end. I suppose I could have stayed on had I been truly anxious to do so, but I had neither the personality nor the desire to live indefinitely the life of a College don. Moreover, I wanted not only a steady post but a home of my own. The possibility of my finding my life's partner, who would have a background in any way similar to mine, in Cambridge, was very remote and I had but few friends outside. I had no intention or desire to "assimilate" and so I started earnestly looking for a job outside Cambridge I found that this was no easy matter. I was considered unsuitable for one or two jobs, for which I applied, as I was thought to be over-qualified and therefore unlikely to stay long, for others because I was not medically qualified and for one or two others for non-biochemical reasons. I was beginning to despair of finding anything suitable in Britain. I had no intention of going anywhere outside Britain, though there were some good opportunities, one even in Canada, when my attention was drawn one day to an advertisement in *Nature* viz. "Biochemist wanted for Cardiff City Mental Hospital. Salary 800 pounds a year, dinners and teas included." This salary was quite generous and somehow the inclusion of dinners and teas, as part of the emoluments seemed to make the post even more attractive. I had not the slightest idea of what a biochemist in a Mental Hospital was supposed to do, but now my war-time experience of a hospital laboratory stood me in good stead and I thought that I might be able to carry out research as well as routine duties. Hopkins was a little surprised that I should think of such a post but thought I could do much worse. My mind was not finally made up until I had a call from Sir Walter Fletcher, Secretary of the Medical Research Council, who said that if I succeeded in obtaining the job, the Council would probably back me financially (for assistance and equipment) if I carried out investigations on the brain. So I applied, stating that I was anxious to carry out research work on the nervous system. Owing doubtless to the efforts of Dr. Goodall, the enlightened superintendent of the Hospital, who was dear to the hearts of the Cardiff City Council as he often spoke to them in Welsh, and who had shown outstanding initiative in being the first to substitute for the term "lunatic asylum" the term "mental hospital" and to introduce into such a hospital female nursing for the first time, I was duly short-listed.

The interview for the appointment, where I was on a list of three or four, left me somewhat apprehensive, for the first question that I was asked was about my ethnic background and the second was whether I was married, or intended to be married. Apparently my answers were satisfactory, and I was duly appointed. So I packed my few belongings, books and myself into my Morris-Oxford and left Cambridge for Cardiff with very mixed feelings, with sorrow that I should leave Cambridge and many friends behind me, with high hopes for the future, and with a sense of adventure into a realm of research new and quite unknown to me.

Cardiff

I found the Cardiff City Mental Hospital, in which I lived for one or two years until I married, rather different from what I expected. It was clean and quiet and well-run and I soon became accustomed to locking and unlocking doors as I found my way through various corridors to the laboratory. This was quite well equipped in a rather severe and trim manner, with not a speck of dust showing, the benches bright and shining and everything arranged tidily. I found that I had inherited there a wonderful colleague, Mr. A.H. Wheatley, who was nearly 20 years older than I was, whose passion for accuracy and cleanliness knew no limits. He soon fell into my way of doing things and I do not think that I ever had a more loyal and conscientious colleague. I was apprehensive, at first, when I saw that he had a Parkinsonian tremor, but I noted that when he used a pipette his hands were as firm as a rock and I came to trust him implicitly in all his work with me. He had no grounding or training in biochemistry but he was a first class analytical chemist. He was not particularly interested in the nature of our results, only in obtaining them, and I do not think that he ever read the papers that were published under both our names. He certainly never bothered much to look at my manuscripts, or the proofs, or the articles we published, but I knew he was delighted to see his name on the papers. He was a bachelor, a pillar of his church and when he retired, he married (I have been informed) a well-to-do and delightful lady and so far as I know they were happy ever afterwards.

Through the help of Donaldson, in London, I managed to acquire the services of an able young man, who, under my supervision, carried out the routine clinical work of the laboratory, urine analyses, blood sugar estimations, etc. and I turned my attention to research on the brain. One of my first studies was to show that the malonate inhibition of succinate oxidation, apparent with bacteria also applied to mammalian tissue viz. muscle and brain. Moreover, I showed at the same time, using malonate and oxalate as inhibitors of succinic and malic dehydrogenase respectively, that succinate is oxidised quantitatively to fumarate and L-malate and thence to oxaloacetate – a sequence of events now well known as part of the citric acid cycle. I had the fear previously that succinate and fumarate might undergo other metabolic changes. The following quotation from my article [13] indicates:

"This evidence (as to the mode of oxidation of succinate) may be secured from a quantitative study of the oxidation of succinate in presence of those tissues or organisms with which the oxidation of fumarate or malate is small or nil. In presence of such cells, one atom of oxygen should be taken up per molecule of succinate forming an equilibrium mixture of fumarate and L-malate in the ratio of 1:3. Thus, the uptake of 1 g-atom of oxygen should give rise to 3/4 g-mol of L-malate. If succinate suffered an oxidation other than that through fumarate, the uptake of g-atom of oxygen would lead to less than 3/4 g-mol L-malate. If, for instance, it proceeded first to DL-malate, and this to a mixture of D-malate, L-malate and fumarate the final rotation observed would not only be smaller than if fumarate were the first step but of the opposite sign."

For experiment, succinate at an initial concentration of M/15 was oxidised in presence of (1) muscle tissue, with which the fumarate oxidation was very small compared with that of succinate; (2) a suspension of *B. coli* which had been treated with toluene so as to eliminate the oxidation of fumarate.

The oxidation was carried out in the Barcroft apparatus and after a certain period the oxygen uptake was calculated, and the solution in the Barcroft cup made up to 5 ml with water. 1 ml glacial acetic acid was added and the whole mixed with 10 ml 14.2% ammonium molybdate solution. After filtering or centrifuging, the clear solution was examined polarimetrically. Using the mercury green line a rotation of 1° (2-dm tube) was equivalent to 9.65 mg L-malic acid. The experimental results were as follows.

(1) With muscle tissue:

Oxygen uptake at 37°C and 760 mm	357 mm^3
Rotation observed	0.30°

Rotation calculated on the basis of succinate being oxidised entirely through fumarate	0.30°

(2) With *B. coli* treated with toluene:

Oxygen uptake at 37°C and 760 mm	1222 mm^3
Rotation observed	1.06°
Rotation calculated	1.02°

These results clearly indicate that if succinate normally undergoes some other oxidation, besides that of proceeding through fumarate, the amount of this oxidation must be exceedingly small.

The entire evidence is now in support of the view that the normal course of biological oxidation of succinate lies largely, if not entirely, through fumarate and L-malate.

This was done in 1931 and the results contradicted a view, upheld at that time by Warburg, that a "respiratory enzyme" was wholly responsible for these oxidations. Such an enzyme, for example, that responsible for the oxidation of *p*-phenylenediamine, was quite unaffected by malonate. My observations with malonate took on considerable importance later (1935) in the hands of Szent-Györgyi in his formulation of the succinic acid cycle playing a catalytic role in the mammalian respiratory system and still later (1937) in the hands of Krebs when he formulated the now classic citric acid cycle, a mechanism whose dominant feature was the condensation of oxaloacetate with acetate derived from pyruvate. About this time I learned too, from studies of the actions of dyestuffs that not only a substrate, but a substrate analogue could protect the activating enzyme from the inhibitory effects of toxic substances.

Turning my attention specifically to brain, I found, among a variety of results [14] that glucose is the major fuel of the brain in vitro, that other sugars notably fructose and mannose are oxidised and that L-glutamate supports brain respiration. The following quotation demonstrates the evidence with L-glutamate:

"Glutamic acid (0.05 M) is oxidised at a comparatively slow rate by either fresh brain tissue or tissue which has been depleted of oxidisable material by previous oxidation for 3 hours. The following figures are typical for guinea pig brain. (Glutamic acid was always neutralised with sodium hydroxide.)

	mm^3
O_2 uptake of fresh brain (0.5 g) in 1½ h	770
O_2 uptake of fresh brain with glutamic acid (0.05 M) added initially in 1½ h	942
O_2 uptake by brain in 1½ h, after 3 h prior autooxidation	200
O_2 uptake by brain in 1½ h, with the glutamic acid added subsequent to the 3 h autooxidation	388

Polarimetric examination of the products of oxidation of glutamic acid by brain tissue showed no indication of L-malic acid formation. Iodoacetic acid (M/4000) had but little effect on the oxidation by brain of the amino acid."

Later evidence confirmed this; when much later labelled glutamic acid became available it was easy to show that, in the presence of rat brain, labelled CO_2 is formed from labelled glutamate even in the presence of excess glucose.

But whilst these results were of interest to me they made little impact on my clinical colleagues in the hospital. I remember well the new Superintendent, the successor to Dr. Goodall, coming into the laboratory, watching my experiments, done with pieces of brain, with a rather glum expression, and stating that he could not see how on earth such work could have the slightest influence in mental disorder problems. However, he went on, he had no strong objections to the work, as he had calculated that the total cost of the research laboratory to the hospital was less than that of the laundry and that the effect of stating, in advertisements for nursing staff, that research was being carried out in the hospital helped to secure more intelligent and able applicants.

It was the custom of the medical staff to meet in my laboratory every Friday afternoon and to discuss research problems and results. This was an excellent custom and I felt more and more that I had to do something to justify the existence of our laboratory in a mental hospital. It must be recalled that at that time there were very few such biochemical laboratories in the world and work on the brain had been done rather fitfully since the days of Thudicum over 30 years before. Certainly there had been no systematic studies on brain chemistry related to problems of mental disorders. The nearest such work was that of Peters and his colleagues in Oxford on brain metabolism in vitamin B_1 deficiency (1932). It took some time for me to become familiar with psychiatric nomenclature and with psychiatric thought, none of which was conducive to laboratory studies on the brain. In fact, there was not only doubt of the value of such studies (carried out with bits, or preparations, of apparently dead brain) by the majority of doctors in mental hospitals but an antagonism existed that made it difficult to find funds or space for such work in the limited facilities of a hospital. I was fortunate in that I was backed financially (for assistance) by both the Medical Research Council, U.K., and by the Rockefeller Foundation.

Being tired of facing the challenge and sceptical attitude towards our work by some of my medical colleagues, I decided on a major diversion of my line of research. I had been impressed, on going through the wards with my medical colleagues, with the success, on many occasions, of the treatment of certain classes of psychotic-patients with what was then termed prolonged narcosis therapy. Patients were put to sleep, or near sleep, with barbiturates for many days and I saw what seemed to me to be miraculous remissions. However the Superintendent, a most intelligent and humane man, was not satisfied, for some patients had not fared well under this treatment and indeed one or two had succumbed. He saw no reason why patients who had a fairly good prognosis, and might be discharged from hospital within a year or two, should be exposed to a toxic treatment, even if it reduced hospitalization to two or three months. I argued that this was no way to handle the situation and that the real problem was to discover why, with certain patients, the barbiturates used should be so toxic. Such thoughts prompted me to observe whether barbiturates could affect brain metabolism and I speedily found [15] that the barbiturates had powerful inhibitory effects on cerebral glucose metabolism in vitro, that the effects were reversible and that they had no effect on succinate oxidation, so that they were not just general tissue poisons. I hazarded the opinion, to the medical staff, that perhaps the toxicity of barbiturates to their patients was due to some interference with carbohydrate metabolism. I asked whether they had examined their patients on prolonged barbiturate anaesthesia, for the excretion of acetone bodies? This they had not done and they proceeded to do so, finding such excretion in quite a number of their patients. So Ström-Olsen and I [16] carried out experiments showing that glucose-insulin administration to such patients prevented ketosis, and much of the toxicity, due to barbiturate treatment. This form of therapy was, thenceforward, carried on under much more favourable circumstances. Later, it was found that even the insulin was not necessary and that large doses of glucose would suffice to diminish or abolish toxicity. The attitude of the medical staff towards our work changed after that time and there was much desire to cooperate with us in appropriate clinical work.

Although the barbiturates used at that time (1932) were at somewhat high concentrations, those that were clinically active showed much greater inhibitory effects than those with little or no pharmacological action. This is shown in some of the results given in Table II [15].
Later results showed that when cerebral respiration in vitro in the presence of glucose is stimulated, either electrically or by the presence of relatively high concentrations of potassium ions it becomes exceedingly sensitive to the clinically active barbiturates. The stimulated respiration

TABLE II

Barbiturate (0.12%)	Percentage inhibition of autoxidation	Percentage inhibition of O_2 uptake in presence of glucose	Hypnotic action
Isopropylbarbituric acid 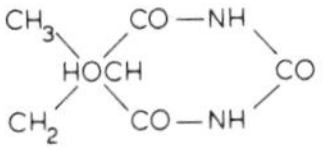 CH_3 CO—NH HOCH CO CH_2 CO—NH	10	6	0
Isopropylbarbituryl urethane 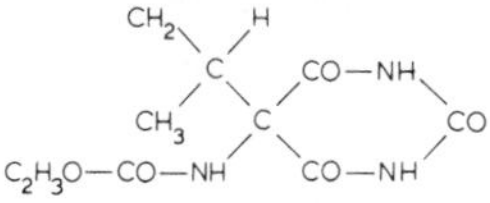 CH_2 H C CO—NH CH_3 C CO C_2H_3O—CO—NH CO—NH	9	4	0
Isopropylbrompropenylbarbituric acid 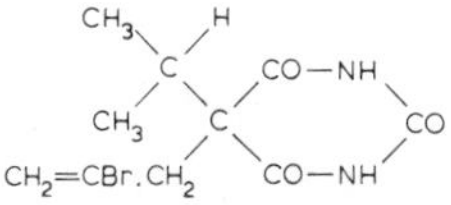 CH_3 H C CO—NH CH_3 C CO CH_2=CBr.CH_2 CO—NH	40	50	+ +
Isopropylbrompropylbarbituric acid 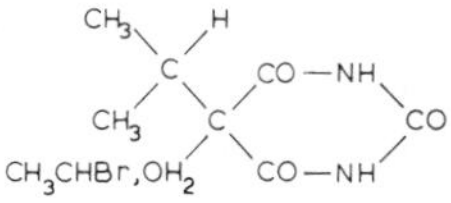 CH_3 H C CO—NH CH_3 C CO CH_3CHBr,OH_2 CO—NH	9	0	Very weak
Isopropylallylbarbituric acid (numal) 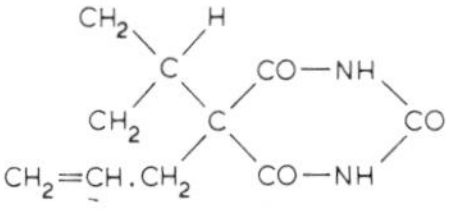 CH_2 H C CO—NH CH_2 C CO CH_2=CH.CH_2 CO—NH	34	40	+ +
Phenylallylbarbituric acid 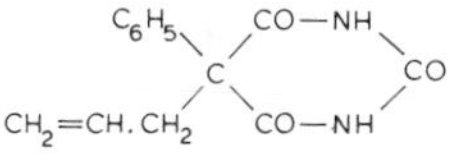 C_6H_5 CO—NH C CO CH_2=CH.CH_2 CO—NH	50	57	+ +
Phenylethylbarbituric acid (luminal) 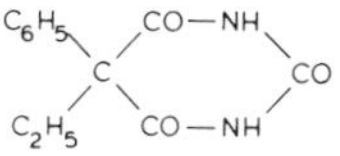 C_6H_5 CO—NH C CO C_2H_5 CO—NH	33	40	+ +
Diethylbarbituric acid (veronal) 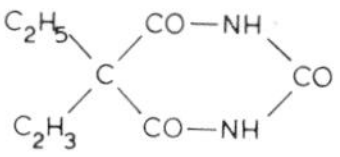 C_2H_5 CO—NH C CO C_2H_3 CO—NH	15	10	+

approximates that found in vivo. Amytal (0.25 mM) at a concentration approximately that present in the blood of rats under Amytal anaesthesia [17] brings about complete suppression of electrically or K^+-stimulated respiration of incubated rat brain cortex slices [18]. Pentothal (0.2 mM) suppresses electrically stimulated respiration and even at 0.1 mM shows a significant inhibition [18]; at somewhat higher concentrations it suppresses K^+-stimulated brain respiration. The barbiturates at the low concentrations that are effective in vitro [19, 20] have no significant inhibitory effects on the membrane-bound, ouabain-sensitive ATPase that is involved in the active transport of Na^+ and K^+ across the brain cell membranes [18].

Recent investigations on the effects of barbiturates and other neurotropic drugs on Na^+ influx into incubated rat brain cortex in vitro show that Amytal (0.25 mM) or pentothal (0.5 mM), either of which blocks rat brain cortex respiration stimulated electrically or by the addition, for example, of protoveratrine (5 μM), does not diminish the influx of Na^+ that occurs under these conditions [18, 21].

However, when there is a lack of Ca^{2+} in the incubation medium together with an augmented rate of respiration, e.g., by addition of protoveratrine or high K^+, barbiturates (Amytal 0.25 mM, pentothal 0.5 mM) bring about a diminished rate of entry of Na^+ into the incubated brain cells and an increased cell concentration of K^+ [21]. These results become understandable when they are considered in relation to the known effects of Amytal on the release of Ca^{2+} from cell mitochondria [22]. Mitochondria from a variety of tissues are able to accumulate Ca^{2+} either by a respiration-dependent process or by a process requiring the presence of ATP. This accumulation of Ca^{2+} by mitochondria is prevented by respiratory inhibitors, including Amytal, and by a variety of uncoupling agents. The suppression of mitochondrial metabolism, particularly when stimulated by ADP, by barbiturates at anaesthetic concentrations is presumably sufficient to release Ca^{2+} from its bound sites in the mitochondria, and the increase of cytoplasmic Ca^{2+} causes alterations in the membrane changes in permeability to Na^+ and K^+ brought about by Ca^{2+} deficiency in the incubation medium. Such a conclusion is compatible with the view [23] that anesthetic barbiturates may act by causing a rise in cytoplasmic Ca^{2+}, by suppression of mitochondrial metabolism, resulting in a changed membrane permeability to K^+. Thus changes in membrane permeabilities, or transport processes, may play an important role in development of anaesthesia.

The work on barbiturates in 1932 brought me into touch with Dr. Max

Guggenheim, the extremely able Director of Research for Messrs. Hoffmann-LaRoche, whose headquarters were in Basle. He was quite blind but he carried out his work with great success. He asked me to visit him and to pick up a series of barbiturates whose clinical effects were known and whose effects in vitro I could study. Some good friends of mine, whom I had first met in Cardiff, were visiting an important conference in Basle and they invited me to go with them. So, combining business with pleasure, I decided to visit Guggenheim and to see something of the conference at the same time, during 1931. At the conference I had the good fortune to meet someone who was to become my partner in life for the next 42 years. She and her mother had just visited Egypt and Palestine and were so enthusiastic about the events taking place in the latter country that they felt they must attend the conference in Basle. My wife and I were married in Sheffield a few months later and so began my happy home life that terminated with her death, by cancer, early in 1973 and only resumed on my remarriage in 1975.

Our work on the chemistry of the brain became increasingly oriented so as to bear as much as possible on the problems of mental disorder. This involved, of course, basic speculations on the interrelations between cerebral organic processes and cerebral behaviour. That such connections exist is largely taken for granted today but in those days, the 30s, this was far from the case.

The work with barbiturates started a train of thought that led to a very productive period of research. If it was possible that these drugs, by their potent effects on some aspects of cerebral metabolism, could affect behaviour and lead to severe mental disturbances, then there might also exist substances that were normally produced in the body but which, when present in more than ordinary amounts, would have similar effects on cerebral metabolism and thereby affect mental processes. This working hypothesis led me (1932) into lines of research which I have already briefly described [24] and which I do not think it necessary to repeat here at any length.

I was fortunate in having with me during the next few years very able colleagues, not only Wheatley, but Maurice Jowett, Philip Mann, Cecilia Pugh, Lionel Penrose, Moritz Michaelis and others who all played important roles in our work.

Jowett and I were able to show (1933) that the substrate of glyoxalase was not methylglyoxal itself but a complex of methylglyoxal and glutathione, as indicated in the following quotation from our article [25]:

"The work described indicates that the presence of glutathione is necessary for the activity of glyoxylase of red blood cells, confirming the work of Lohmann on tissue extracts. The results with lysed cells show that the red blood cell contains ample glutathione to account for its glyoxalase activity and that the disappearance of glyoxylase activity on completely lysing corpuscles is due to dilution of the cell contents. The thermolabile glyoxalase itself is probably fairly stable and unaffected by lysis of the corpuscle. The experiments recorded also show that methylglyoxal and glutathione combine reversibly. The following scheme therefore represents, we suggest, the course of transformation of methylglyoxal into lactic acid, catalysed by glyoxalase:

$$CH_3COCHO + HS.G \rightarrow CH_3COCHOH.SG \xrightarrow{H_2O} CH_3CHOHCOOH + GSH.$$"

This work was the forerunner of the investigations of a later colleague of mine, E. Racker, who carried out in the USA his well known work on triosephophosphate-thiol interrelationships and still later by Hopkins.

We were also able to show [26] that normal cerebral respiration must proceed by a mechanism independent of lactate, thought, at that time, to be a necessary intermediate in the oxidative breakdown of glucose. This is now known to proceed through phosphopyruvate and pyruvate.

Pugh and I discovered monoamine oxidase in brain early in 1937. Wheatley and I had, in fact, in 1933 [27] shown that a variety of amines (including β-phenylethylamine, thyramine) were highly inhibitory to brain respiration in the presence of glucose or of glutamate, but the explanation of this phenomenon was not made clear until our discovery of monoamine oxidase in brain early in 1937. The effective inhibitory agent was the corresponding aldehyde. The results of Pugh and myself [28] showing the marked effect of isoamylamine are given in the following table, indicating the formation of ammonia (Q_{NH_3}),

Amine added	Q_{NH_3}	
	Amine absent	Amine present
Methyl (M/60)	0.5	0.6
Ethyl (M/60)	0.6	0.9
Propyl (M/150)	0.7	0.9
Butyl (M/150)	0.5	1.3
Butyl (M/60)	0.8	1.4
Amyl (M/150)	0.3	0.9
*iso*Amyl (M/150)	1.0	2.2
Heptyl (M/150)	0.4	1.1

"Isoamylamine is broken down by brain and liver to yield a substance giving a hydrazone with 2:4-dinitrophenylhydrazine. The possible formation of isoamyl alcohol is indicated by the development of the characteristic odour. The oxidation of the amine appears to occur only under aerobic conditions. An amine-oxidizing system exists in brain and other tissues which is distinct from the α-amino-acid oxidizing system." [28]

Later in the same year, we published a second paper to show that the enzyme could bring about the oxidation of various aromatic amines. In the very same issue of the *Biochemical Journal* in which our later article appeared there appeared also, to our astonishment, a paper by Blaschko, Schlossman and Richter giving similar results. Our summarized results are quoted below [29].

"(1) Indolethylamine and tyramine, which greatly inhibit the respiration of brain cortex slices in a glucose medium, are oxidised by an amine oxidase in brain with liberation of ammonia and an aldehyde. Mescaline, benzedrine (phenylisopropylamine) and β-phenylethylamine which also inhibit brain respiration are attacked feebly or not at all. There is no evidence for deamination of histamine by brain.

(2) Rat liver slices and guinea pig kidney slices oxidize and deaminate indolethylamine, tyramine and phenylethylamine and their effects on mescaline, histamine and benzedrine are small or negligible. The order of activity of the tissues on the amines is rat liver>guinea pig kidney>rat brain. Rat kidney has relatively little activity.

(3) The presence of cyanide does not affect the activity of the amine oxidase. In presence of cyanide, 1 atom of oxygen is taken up for each mol of indolethylamine or tyramine consumed and 1 mol of ammonia is liberated.

(4) Oxidation of indolethanolamine by tissues is always accompanied by the formation of a melanin-like pigment. In presence of cyanide, although oxidation of the amine and deamination take place, pigment formation is greatly reduced. It appears therefore that indolethylamine is attacked by a cyanide-resistant mechanism (amine oxidase) and a cyanide-sensitive system resulting in pigment formation.

(5) Arsenite (M/1000) does not inhibit the activity of amine oxidase, nor does it inhibit appreciably pigment formation from indolethylamine.

(6) Well-washed liver extracts oxidize butylamine, isoamylamine, tyramine and indolethylamine. Dialysis of the extract has no apparent diminishing effect on its activity.

(7) From the products of oxidation of isoamylamine by liver extract isovaleraldehyde has been isolated as the dinitrophenylhydrazone (79% yield) and identified.

(8) Analyses of the aldehyde, ammonia and volatile acid formed during the oxidation of isoamylamine and butylamine by liver extracts show that the following reactions take place during the oxidation of the amines:

$$R.CH_2NH_2 + O = R.CHO + NH_3$$
$$2R.CHO + H_2O = R.COOH + R.CH_2OH$$

(9) Hydrogen peroxide is formed during amine oxidation by liver extracts as shown by coupled oxidation of ethyl alcohol in presence of the amine. Alcohol does not, however, double the oxygen uptake in presence of indolethylamine.

(10) When indolethylamine is oxidized by a liver extract, there is an uptake of between 3

and 4 atoms of oxygen for each mol of ammonia produced. It is concluded that the extra oxygen uptake, above that required for the aldehyde formation, is concerned with the pigment formation.

(11) Experiments carried out with mixtures of amines show that a common amine oxidase system is concerned with the oxidation of the higher aliphatic amines, tyramine, β-phenylethylamine and indolethylamine. It is proposed to term the enzyme amine oxidase, this being identical with tyramine oxidase."

This work led to our finding [30, 31] that amphetamine, well known for its pharmacological activity, is a structural inhibitor of monoamine oxidase and to our suggestion at that time, that amine metabolism in the brain may be linked with brain behaviour. Our results [31] are quoted below:

"(1) When brain respiration in a glucose medium takes place in the presence of tyramine, β-indolethylamine or isoamylamine a fall in respiration occurs. This fall in respiration is neutralized by the presence in the system of benzedrine (β-phenylisopropylamine sulphate). This phenomenon applies to the respiration of intact brain cortex slices or of minced brain tissue.

(2) The same phenomenon obtains in the case of tyramine when glucose is replaced by sodium succinate.

(3) The fall in brain respiration due to the presence of tyramine, etc. is not due wholly to the amine itself but to a product of oxidation of the amine, i.e., the corresponding aldehyde, or a product formed by further metabolism of the aldehyde.

(4) Succinate protects succinic dehydrogenase from the toxic influence of the oxidation product of tyramine (probably *p*-hydroxyphenylacetaldehyde).

(5) Benzedrine owes its stimulating influence to its ability to compete reversibly with amines for the amine oxidase of brain and of other organs, thereby reducing the rate of formation of inhibitory aldehyde.

(6) L-Ephedrine, 3:4-methylenedioxyphenylisopropylamine and 3-methoxy-4-hydroxyphenylisopropylamine also compete reversibly with tyramine for amine oxidase and they all show effects similar to benzedrine in their retarding actions on the fall of respiration of brain tissue in glucose or succinate media due to the presence of tyramine. Ephedrine and 3-methoxy-4-hydroxyphenylisopropylamine are much less active than benzedrine, and 3:4-methylenedioxyphenylisopropylamine has about the same activity as benzedrine. The equilibrium constants of these amines with amine oxidase have been determined.

(7) Aldehydes such as isovaleric aldehyde, and *p*-hydroxybenzaldehyde are inhibitory to the respiratory processes of brain and their inhibitory effects are not neutralized by the addition of benzedrine.

(8) Benzedrine has no retarding action on the fall of respiration of brain tissue in a glucose medium due to the presence of luminal, chloretone or bulbocapnine.

(9) The possible clinical significance of these results is discussed. It is pointed out that the relative stimulating actions of ephedrine, benzedrine and benzedrine derivatives on brain respiration in the presence of tyramine bear a parallelism to the relative stimulating actions of these substances on the central nervous system as observed in vivo."

Furthermore [32]

"The effects of benzedrine in vitro, in partially neutralising the inhibition of glucose oxidation by brain due to the presence of tyramine and other amines, take place at concentrations which are not markedly greater than those which have pharmacological effects in vivo. Moreover, since competition between benzedrine and other amines takes place at amine oxidase according to the laws of mass action, it follows that the influence of benzedrine may be exerted in the body at much lower concentrations than have been used in in vitro investigations. It seems not unreasonable to suggest that the action of benzedrine in vivo is linked with its ability to compete with amines which give rise by oxidation to toxic substances; the lower the concentration of such amines, the lower the quantity of benzedrine which will be required to compete successfully with them. Ephedrine is known clinically to be less powerful than benzedrine as a stimulant of the central nervous system and it has been shown that such a difference between their effects on brain respiration occurs also in vitro."

Similar results

"are suggestive of a possible connection between amine metabolism in the brain and the development of the clinical conditions known to be relieved by benzedrine (amphetamine) administration." [32].

Since that time it has been found that the amphetamines also suppress amine transport and doubtless this affects cerebral amine levels that play a basic role in animal behaviour [33].

There is no doubt, at present, of the importance of cerebral amines in the brain (though in 1940 very few had been found) in the control of cerebral behaviour. Even recent achievements in the field of encephalins and endorphins, the neutral peptides, which have potency as analgetics, have supported my long-held belief that endogenous substances may affect cerebral chemical or physicochemical processes and so affect cerebral behaviour. There has occurred much change in the attitude of psychiatrists to the biochemical study of the brain, so much so that the early scepticism towards in vitro studies, and their possible importance in the understanding of animal and human behaviour, has greatly diminished.

In the 30s, another line of research in my laboratory in Cardiff became prominent. This arose by a curious circumstance. During the early thirties I received letters from a number of scientific men in Germany who sought posts in the U.K. that would enable them to obtain exit-visas and the means to escape the dreadful conditions beginning to take place in Nazi Germany. I could do little for them in my small laboratory, but I must have helped in all perhaps a dozen families to escape. One of the first of these was a Michael Tennenbaum who wrote stating he had experience of enzyme assays under the well-known enzymologist Rona. I decided to find a place for him and obtained his living expenses from a charitable organisation and from private sources. When he arrived I was dismayed to find he could do little for us in our current work and in despair I asked him to outline everything that he could do. Almost as an afterthought he mentioned that he could estimate acetylcholine by the leech bioassay. I decided, there and then, to commence a line of work using this technique. In this way, our studies of acetylcholine synthesis in the brain commenced, studies that I have pursued for almost the rest of my life. We showed [34] for the first time that acetylcholine, already established by Loewi and Dale and his colleagues as a transmitter in the peripheral nervous system, was synthesised in the brain at the expense of glucose breakdown. This process proved, in fact, to be the first example of a biological acetylation in vitro. With Philip Mann, we discovered the existence of a "bound" acetylcholine. We suggested a scheme of the mechanism of acetylcholine formation in brain [35] and described the effects of potassium and other ions on acetylcholine biosynthesis and breakdown [36]. This was very exciting work and it was in full swing when it all had to stop because of the advent of the war.

Our provisional scheme for the formation of acetylcholine in brain sluggested in 1938 was as follows [35]:

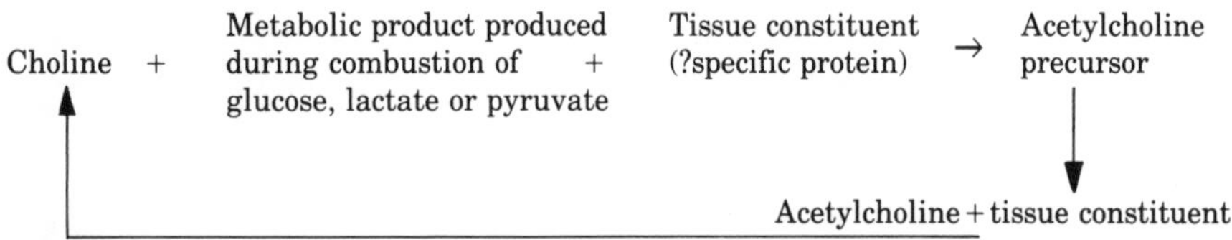

This scheme incorporated the results that had been found experimentally. It was observed that the "tissue constituent" behaved as a catalyst and resembled an enzyme, since it brought about the synthesis of acetylcholine from choline and was itself regenerated. The tissue constituent is referred to nowadays as cholineacetyltransferase. The marked effects of high [K^+] on acetylcholine synthesis are shown in the following quoted results [36]:

"The experiments with rat brain slices show that the addition of K^+ to the medium in which the tissue is respiring may bring about a very large increase in the rate of acetylcholine formation. Typical results are shown below. The effect of the addition of 0.027 M K^+ may result in the formation of acetylcholine to the extent of over 40μg/g.

TABLE III

Rat brain slices in glucose (0.01 M)-bicarbonate-Locke-eserine medium

KCl was added to the medium, with a corresponding decrease in NaCl to preserve isotonicity. 95% O_2 + 5% CO_2 1h, 37°C. Capital letters A–E refer to individual experiments.

	Total acetylcholine formed μg/g				
Exp.	A	B	C	D	E
Medium with no extra KCl	12.7	10.4	12.0	20.0	21.3
+ 0.0135 M KCl	16.5	–	–	–	–
+ 0.027 M KCl	37.5	41.0	34.0	41.0	45.0
+ 0.054 M KCl	–	–	15.0	–	–
+ 0.11 M KCl	–	4.0	–	–	–
+ 0.13 M KCl	4.3	–	–	–	–

The presence of a much greater [K^+] brings about, however, a fall in the rate of formation of acetylcholine. Concentrations of the order of 0.1 M K^+ inhibit the synthesis of the ester.

It is evident that the accelerating action of 0.027 M KCl on the rate of formation of acetylcholine is due to K^+ itself and not to a fall in the [Na^+]. A fall in the NaCl concentration to the extent of 0.027 M has little or no effect on the rate of formation of acetylcholine.

Analysis of the "total" acetylcholine formed in presence of excess K^+ shows that the increase of acetylcholine formation lies entirely with the free acetylcholine, the amount of the "combined" ester being usually decreased."

Work on acetylcholine biosynthesis and location, and on its function in the nervous system, is now a major field of investigation in many parts of the world. I still find enjoyment in working in this field but it is little compared with that in the pre-war days. Nevertheless, the subject still continues to occupy my thoughts and I try my best to keep up with the now massive literature on it and perhaps to introduce something original and useful.

At about this time (1937) I was carrying out a long-distance collaboration with Lionel Penrose, whom I had introduced into the Cardiff City Mental Hospital and with whom I had formed a friendship resulting in our collaboration. We were both much interested in the newly discovered Fölling's disease, known as phenylpyruvic imbecility or oligophrenia. Penrose moved to Colchester, about 200 miles away, whence he sent me samples of urines of patients having this disorder and I devised quantitative methods of estimating phenylpyruvic acid, after first isolating it and confirming that it was indeed phenylpyruvic acid. As the work went on I realized that the nomenclature for this disease was a poor one, at any rate for a biochemist, and I wanted a term descriptive of the biochemical abnormality, and not necessarily including a reference to the mental state. So I invented the term "phenylketonuria", in analogy with the well-known term "alcaptonuria". This term is now used universally and it has often been shortened to PKU. Penrose and I published an article in 1937 that was the first to describe the metabolic effects of phenylalanine administration in man [37]. We found that there was a diminished rate of oxidation of the benzene ring in phenylpyruvic acid. The disease is now known to be due to the absence of phenylalanine hydroxylase. I would have liked to pursue this work, but I found long-distance collaboration very time-consuming and unsatisfactory, and in any case the war stopped our further work.

Our small laboratory received many applications from investigators in Britain and overseas but we had not the space to accommodate them. However, we had numerous visitors from the USA and Continental Europe who were interested in our work on the brain and its possible relationship to problems of mental disorder. I do not doubt that we had some considerable influence on the development of neurochemical laboratories in various countries, both in hospitals and elsewhere, that began gradually at first and then rapidly to spring up in many parts of the world.

During the whole of this period in Cardiff, I was a member of various committees of the Medical Research Council and of the Department of Scientific and Industrial Research. I was a member of the Council of the Royal Institute of Chemistry and later, after the war began, of many boards and committees. This necessitated a lot of travelling but the experience, although time-consuming, was very useful and rewarding.

With the advent of war my thoughts turned increasingly towards some manner in which I could be of more direct service in the war. In 1941 when I was alone, for, after the Battle of Dunkirk, I had sent off my wife and children on the last evacuation ship to the USA, I received a cable from Fulton of Yale University asking me to join a group there to work on a problem that had war-time significance and I was ready to go. I had already received approval to do so from my hospital and from the Medical Research Council, when I received a midnight call from London, after a lengthy air raid, asking me to meet, next day, Dr. Topley, Secretary of the Agricultural Research Council. He asked me to drop my plans to go to the USA, to resign my job in Cardiff and to direct the first Unit of the Agricultural Research Council. I was to work on soil biochemistry with the objective of improving crop production. At that time Britain stood alone and many of us thought that we were in for a long siege. Naturally I felt that I had no choice and I accepted Topley's invitation. And so commenced another of those steps into the unknown that have marked my life. I knew nothing about soil and, at that time, I had not the faintest idea what I would do with it. Nevertheless I still felt the urge for scientific adventure and feeling that I might be lucky in my new job I entered a new line of work with all the confidence that comes from ignorance and high hopes. So I rented my house in Cardiff, and had great difficulty later on, when my wife and children returned, in getting it back, and went to the Rothamsted Experimental Station, where I was to be located, presumably as long as I held my war job.

Rothamsted Experimental Station

I was not particularly happy in the Rothamsted Station. The atmosphere depressed me. The director, Sir John Russel, had promised to give me a laboratory there but when I arrived he had forgotten all about it. When we toured the completely filled laboratories there was nothing available except an old potting shed which was, however, supplied with electric light and water. There were no windows but there was a large door, opposite a glasshouse. I determined to use this space and, in a relatively short time and with the aid of Government permits, had it equipped with laboratory benches and a tiny balance room. There was very little space but three or four of us managed to work there and we had easy access to the glasshouse. I recall being in a complete quandary as to how I should start the work. Placing me in that station was probably a good thing, not only because of certain friendships that I formed, but because it gave me an insight into the nature of soils that I never knew or suspected before. One of the few who took some interest in my problems was Thornton (later Sir Gerard) who was Head of the Department of Soil Microbiology. Together we thought of a scheme for improving crop production by increasing the process of nitrogen fixation in the roots of legumes and decided to see if a plant hormone, indoleacetic acid, might facilitate the formation of the rhizobia nodules wherein the process of nitrogen fixation occurred. To our astonishment we, with the cooperation of Philip Nutman, found that this hormone was extraordinarily toxic to root growth, and later, that this toxicity did not occur in normal soils but that it did occur in sterile soils or in sands. As our thoughts then began to turn towards the practical use, both in war and in peace, of substances which in low concentrations would kill plants, we sought other substances that would be as effective as indoleacetic acid but which would withstand bacterial attack in natural soils for relatively short periods of time. I knew, after a short war-time visit to the USA in 1942, that phenoxyacetic acid and some of its derivatives were being used there as plant hormones. I visualized them as analogues of indoleacetic acid. I felt that chlorinated phenoxyacetic acids would be fairly resistant to bacterial attack. A friend of mine, who worked in St. Albans, nearby, synthesized and gave me samples of monochlor-, dichlor-, and trichlor-phenoxyacetic acids, and Thornton and I, together with Nutman, found, by the fall of 1942, that 2:4-dichlorphenoxyacetic acid (2:4D) was highly toxic in natural soils, in very small concentrations, to many dicotyledonous plants. However, it was much less toxic to various other plants including cereals. In this manner we discovered the selective herbicidal effects of 2:4D [38]. My initial report on the subject, and its possible implications, was sent to Topley on November 17, 1942, less than a year after we started work on the subject. We were not allowed to publish our results for security reasons and we were asked to collaborate after November 1942 with the scientific staff of the Imperial Chemical Industries who had been working, unknown to us and to our astonishment, on rather similar lines. All this is now past history, and described in various places, but I will never forget the excitement I felt, at the termination of large field experiments in 1943, carried out with the collaboration of the I.C.I., when we saw the effects of application of the synthetic herbicide on a large scale. Abolition of weeds by the herbicide increased the yields of grain crops, unscathed by the herbicide, as there was less competition for the soil nitrogen. I could claim that my original goal, to make the two blades of grass grow where one grew before, was to some extent accomplished. The work on 2:4D, not only by ourselves but by the I.C.I. and somewhat later by an American group, has been described briefly, and so far as I know fairly accurately, in C. Kirby's *The Hormone Weed Killers* published by the British Crop Protection Council, 1980. Thus commenced an agricultural revolution whose far-reaching effects are only now being realised.

Our work, the results of which we were allowed to publish in 1945, is summarized in the following quotation from our article [38].

"The action of 2:4-dichlorophenoxyacetic acid on the plant varies according to concentration from complete inhibition of germination to a stunting of the root system and dwarfing of the leaves. This effect on the roots is accompanied by a marked thickening of the cortex and also, in certain cases, by the development of numerous short laterals. The macroscopic symptoms are similar to those produced by β-indolyl- and α-naphthylacetic acids.

Problems of great interest are raised by the fact that these compounds, which are known to stimulate extension growth and to have profound developmental effects, are also toxic to the whole plant in doses of 0.1–1.0 part per million. 2:4-Dichlorophenoxyacetic acid, however, differs from such compounds as β-indolyl and α-naphthylacetic acids in that it persists long enough in unsterilized soil to produce marked toxic effects. These effects show at a concentration of 1 part per million of soil solution, equivalent to 1 part in 4 millions of soil at 25 per cent moisture, representing on a field scale about 1/2 lb. per acre of soil to 6 in. depth. It would thus seem that this compound should be of use in controlling plant-growth. It is not readily lelached from soil but yet possesses the advantage of ultimately losing its toxicity; thus, it would be unlikely to poison the land. The fact that different types of plants vary greatly in susceptibility to its toxic action may here be of considerable practical importance".

After I was informed that the Imperial Chemical Industries Ltd., had independently worked on possible weed killers for several years, there arose some confusion concerning priority of the discovery of the herbicidal action of 2:4D, mainly due to the independent secret work of the two British groups – a confusion worsened by the death of Topley in 1944. In order to clarify the situation I give below a copy of a letter written by Lord Mc Gowan, Chairman of the Imperial Chemical Industries Ltd.

NOBEL HOUSE,
2. BUCKINGHAM GATE, LONDON, S.W.1.

FROM LORD Mc GOWAN. TELEPHONE: VICTORIA 4444.

14th May, 1952.

Dear Dr. Quastel,

I thank you for your letter of January 15th, on the subject of the reference, in my Dalton Lecture, to the discovery of 2,4-D.

I was abroad when this arrived and have only just been able to complete my enquiries here.

I find that you are correct in your statement that the properties of this material, as a selective herbicide, were discovered independently by Templeman and Sexton, and by Nutman, Thornton and yourself; and that publication was not permitted during the war. Also it is correct to say that, at the request of the Agricultural Research Council, the results of the investigations by the two groups of workers were pooled.

I suppose that, in strict logic, it would be legitimate for you to claim discovery for your group and for us to claim it for ours.

On the other hand I must agree with you that, when either of us makes reference to this discovery, it is desirable that the fact of separate but simultaneous discovery by the two groups should be mentioned.

I am sorry, therefore, that I did not make reference to this in my Lecture but I can assure you that it was an oversight. I do not think that there is anything that I can do now in connection with the Lecture itself but I will do my best to see that, in any future reference to this work, both groups of workers are mentioned.

Yours sincerely,
(signed Mc Gowan)

J.H. Quastel, Esq., D.Sc., F.R.S.,
Professor of Biochemistry,
McGill University, and
Director, Research Institute,
Montreal General Hospital, Montreal.

While this work was going on in great secrecy, my colleagues Philip Mann and Howard Lees and I were also at work on what is perhaps best called soil metabolism. We decided to perfuse soil, as though it was a living organ, using a circulatory system, devised by the ingenuity of Lees and modified later by my colleague Leslie Audus. With this apparatus Mann and I studied the conditions underlying manganese metabolism in soils [39], and Lees and I embarked on a series of studies of soil nitrification that were published and culminated eventually in an article by Scholefield and myself entitled "Biochemistry of nitrification in soil" [40]. Further papers on the subject were published by us [e.g.41]. Another thought I had at this time, in 1945, that bore fruit later on, was that it might be possible to improve the structure of a soil, so that it would become more suitable for crop growth, by the addition to it of synthetic substances, later to be known as synthetic soil conditioners. Webley and I demonstrated [42] the effects of the addition of alginates, and of other organic substances, on soil crumb stability and on soil aeration and water holding power, by a manometric technique. These results led to a series of experiments by the scientific staff of the Monsanto Chemical Co., in which I was associated that led to the use of synthetic substances much more stable in soil than alginic acid or polyuronides, that were water-soluble polymeric electrolytes of high molecular weight, one of the most active being made by the hydrolysis of polyacrylonitrile. This substance became known as Krilium [43]. The study of synthetic soil conditioners is now, I have been informed, as I am now completely out of this field, an important aspect of soil science.

After the war ended, I felt a desire to leave the subject of soil science to those more personally interested in soils than I was. Although I knew the subject of soil biochemistry to be a fascinating one, capable of great extension by those interested in the inter-relations of soil microorganisms and in problems of soil ecology, and in the influence of microbiological changes in soil on plant nutrition, I had a yearning to return to my normal fields of endeavour. I had been out of touch with animal biochemistry for over six years, at a critical time in my scientific life, but I felt that I would be able to pick up again, at least where I had left off. I did not grudge the loss of those years, so many people had lost far more than that, but I knew that if I was to return to the normal problems of biochemistry I must do it at once. It was a misfortune to many of us that Topley died of a heart attack before the end of the war. I think that he alone might have persuaded me to stay on. I had great admiration for his shrewdness and insight and he was a source of great encouragement to me in those gloomy periods when I detested soils and the stubborn conservative outlook of some agriculturists, which I felt impeded the development of soil science.

I had, in 1946, various options. I was invited to go back to my old job in Cardiff but I refused. I was also invited to go to the Maudsley Hospital but this I refused, too, as I did not want at that time to be confined to problems of the nervous system. I was invited by Weizmann to help found the new buildings of the Weizmann Research Institute in Rehovot, Israel and it was touch and go as to whether I might eventually settle there, but an offer came from an unexpected quarter, Montreal. My old friend, David Thomson, wondered whether I would like to go to McGill University as Professor of Biochemistry and the Montreal General Hospital asked me if I would like to help direct their proposed new research institute, both institutions to collaborate in paying my salary. The offer attracted me as it would combine teaching with research. Finally Cyril James, Principal of McGill University, met me in the Athenaeum in London and formally invited me to full professorship in McGill. I was hesitant at first because I had heard of the manner in which persons of my ethnic background had been treated in Montreal but James assured me that I would be welcomed by all, as was, in fact, the case. So I accepted the invitation (1947). My pay was not much greater than I had in England and I had trouble later on with my pension owing to the split in my salary between a hospital and a university. I went because I considered McGill to be an important university in the British Empire, and because the conditions for teaching and research, in fields of my own choosing, were far better than I was likely to obtain at that time in Britain. Moreover, I considered that the future prospects for my children would be very good in Canada. Also my wife, who was born in New York, wanted the change, though she would certainly have stayed in England if I had wished to do so. So in view of these, and other, considerations, I refused a very tempting offer of a professorship in London University, made to me much too late, at a time when I had already given my word to McGill and the Montreal General Hospital. We packed up our home in Cardiff and, leaving our many dear friends in London, Cardiff and Sheffield behind, we embarked (1947) on a new adventure that led to my present life in Canada.

Montreal

We received a great welcome in Montreal. Within a few weeks we seemed to have made a host of new friends who were all anxious to help us settle down. We bought a house within 48 hours of our arrival and there we lived happily for the next 19 years. We fell in love at once with Montreal which we thought to be one of the most interesting and beautiful cities in North America. I was treated in the most friendly manner by all my colleagues in McGill and in the Montreal General Hospital.

My laboratories were located in a large house, that formerly belonged to the owner of a big department store, and every room in the house was converted into a laboratory or an office. Every nook and cranny was used. The cost of the house and of its conversion into laboratories was borne by the hospital which was also responsible for heat, light and power. My pay and that of my secretary was borne by both the Hospital and the University. I embarked on my teaching duties – I was responsible for teaching the advanced course in biochemistry to medical and other students – almost at once and I enjoyed them. Applications from students who wished to take a Ph.D. in biochemistry at McGill under my supervision began to flow in immediately. Dr. Peter Scholefield, who came with me from Cardiff, was the first of my students to take a Ph.D. and I found him to be of great assistance, a little later on, in running the laboratory. He subsequently became Assistant Director of the Institute and later Professor of Biochemistry in McGill and still later Assistant Director of the National Cancer Institute of Canada.

My chief problem was to find the money for the research. Unlike the conditions in the U.K. where I had a fixed budget and I could conduct research as I pleased within this budget, I found, at first to my consternation, that I had to find every penny to pay for all equipment, all chemicals, books and journals, animals and animal care, all technical and laboratory assistance, grants for the subsistence and welfare of my students and of the post-doctoral investigators, except for those who were fortunate enough to have their own scholarships, who began to appear in increasing numbers. I was not allowed to apply for funds from private sources as this would interfere with the appeals for funds made annually, by the hospital, and by McGill. So I had to obtain funds from various granting agencies in Canada and the USA. I had to devise research projects that would appeal to these agencies and which would be suitable and interesting topics for my students, each of whom had to work on a different line of investigation. However, I tried to arrange that these lines would have something in common so that the men took an interest in each other's work. For example, I received a small but useful grant from the Sugar Research Foundation with which I was able to start work on the mode of sugar transport through the isolated surviving intestine. This became a most fruitful field of investigation and led to all our work on transport processes at the cell membrane. Riklis and I were able, for example, to show for the first time that sodium ions were needed for the energy-assisted glucose transport across the isolated surviving intestine [44]. This was the forerunner of widespread investigations on the need for Na^+ in energy-assisted transport systems. Another grant enabled us to work on D-glucosamine, which we discovered to be a substrate of hexokinase [45]; another on the mode of action of anaesthetics and so on [e.g. 46, 47]. I had to exercise considerable ingenuity in finding different but useful projects for the students, and for which I could find the funds. The trouble with most of these grants, at one time, I had about 14 or 15 of them, was that I had to present annual and even semi-annual reports and this took up a great deal of time. Moreover, I had to travel extensively, lecturing in many places in the USA and Canada and elsewhere. I used to enjoy these journeys but I found them rather tiring. I have calculated that the total funds I received by way of grants, during the 19 years I was in Montreal was about three million dollars, and this was considered by people in my line of work, in Canada, in those years, quite a large sum of money. Over 300 publications, many of them of high quality, came out of the Institute and I think that the money was well spent. About 70 students took their Ph.D. under my supervision and most of them, I am happy to say, are now professors of biochemistry in many universities in the world or directors of research or occupy senior scientific posts. I enjoyed these years of hard but rewarding work. Unlike the situation in the U.K., I was never asked to sit on Government committees and only rarely on others, so that my time was almost wholly spent in the laboratory or in teaching.

The fact that there were always graduate students in the laboratory – at one time there were more than 15 at work in the total Institute population of about 35 – resulted in my having to think along a number of different lines instead of concentrating on one or two of them. This was inevitable owing to the multi-grant system which I believe to be a weakness in present-day university research in this country. Health-directed scientific research is essentially a public service and investment in it, and in the country's best brains, should be a first priority of any competent government. It is a profession of great actual and potential value, and properly qualified workers in it should have sure salaries with adequate research funding and periodic reviews. The present system of grant application is time-consuming, and quite often unjust. I know that it is sometimes in the hands of committees, whose members know far less of the subject at issue than the applicant. They have to rely on the opinions of referees who may be biassed or almost ignorant of the subject at issue or both. I personally have been treated well but I know of others who have fared badly and perhaps unfairly. I know even the case of a man who was not allowed to apply for a grant because he had not a faculty position and he was not allowed to be on the faculty because he had no grant! This absurd state of affairs can only block scientific progress.

I was able eventually to delegate many of my students to the care of the senior members of the laboratory and the load on me personally was diminished. We had about twelve post-doctoral investigators present at any one time in the Institute.

Biochemistry was, by 1947, a rapidly developing science. The advent of labelled isotopes and of chromatographic techniques and the increase in the amounts of funds available for research were causing a marked acceleration in almost all phases of the science. We, at the Montreal Institute, played some part in these rapid developments and I have referred to some of these in an earlier publication [48].

Darlington and I devised a simple apparatus for the study of the transport of glucose in isolated surviving intestine, showing [49] among a variety of results, that this substance is transported by an energy-assisted process and this led eventually to the discovery I have already mentioned: that the active transport of glucose across the intestinal membrane requires the presence of sodium ions. J.E. Guthrie and I also showed [50] that experimental shock causes inhibition of the active transport of sugars and amino acids in the intestine.

Our transport work was extended to tumour cells, to brain and other tissues. Together with my students and colleagues (O. Gonda, A. Tennenhaus, S.K. Sharma, R.M. Johnstone, A. Vardanis), I demonstrated in a variety of publications, that many substances including creatine, choline, ethanolamine, amino acids such as glycine and serine, thiamine and ascorbic acid and acetate were taken up by isolated brain against a concentration gradient by sodium-dependent, but very specific, processes as they could be selectively poisoned either by substrate analogues or other substances and moreover that multiple carriers exist for amino acid transport, at the cell membrane. Many of these results were summarised in an invited review [51] delivered before the Royal Society (London) 1965. I had over 2500 requests for reprints of this review. At about this time (1958–60) I began to see the implications of possible linkages between neurochemistry and neurophysiology.

Kini and I investigated carbohydrate-amino acid inter-relations in brain showing, for the first time [52], that labelled glutamine was formed from labelled glucose in brain, and that the conversion of glucose into cerebral amino acids is a process very sensitive to the effects of neurotropic drugs and the ionic conditions surrounding the brain cells.

With Bragança, I showed [53] that mitochondrial oxidative enzymes (e.g., succinic dehydrogenase) are attacked by phospholipase A and introduced the idea that phospholipids may form part of the structure of various oxidising systems. Some of our summarized results are quoted below:

"(1) Investigations have been made on the effects of snake venoms heated at 100°C for 15 min on a variety of enzyme systems. Such venoms still retain toxicity to animals.

(2) The enzyme systems concerned with the oxidation of glucose, pyruvate, L-glutamate, succinate, alpha-ketoglutarate and fructose by brain homogenates are inhibited by small quantities of the heated venoms.

(3) Pyruvic dehydrogenase of brain, succinic dehydrogenase of brain and heat, cytochrome oxidase of brain, and choline oxidase of liver are all vigorously attacked.

(4) The following systems are not attacked: anaerobic glycolysis by rat brain, glucose fermentation by yeast extract, pyruvic acid dismutation by pigeon liver, glucose oxidase of liver, and glucose oxidase of fungi (notatin), soluble pyruvic dehydrogenase of pigeon-breast muscle lactic and malic dehydrogenases, and urease. Cytochrome *c* is not broken down.

(5) The enzyme systems that are not attacked by heated venom are distinguished from those that are attacked by being active in aqueous cell-free tissue extracts and are apparently independent of cellular structures for their activities.

(6) Heated snake venom contains a heat-stable enzyme (probably a lecithinase) that attacks brain phospholipids or lipoproteins. Its inhibiting effect on a susceptible enzyme system is enhanced by calcium ions which activate lecithinase.

(7) It is suggested that heated venoms exert their inhibitive effects by enzymic degradation of mitochondrial structures upon whose integrity or particular spatial configurations, the enzyme systems concerned depend for their optimal activities."

This led to fairly recent work by others on the reconstitution of mitochondrial oxidative enzymes. Moreover, we showed [54] that venom phospholipase A greatly accelerates acetylcholine synthesis in brain in vitro perhaps by affecting the bound form of this ester.

My late colleague, I.J. Bickis, and I were, I think, the first to demonstrate the profound effects of Actinomycin C on nucleic acid metabolism in mammalian (Ehrlich ascites cells) cells and discussed these at a meeting of the Canadian Biochemical Society in March 1962. But I made the mistake of not publishing immediately. So we lost our priority, though I thought that our publication as an abstract, was enough to secure it. I should have realised that the abstracts of the Annual Meeting of the Canadian Biochemical Society have a very restricted circulation in the world. I give a copy of our abstract [55] below:

"Effects of actinomycin D on nucleotide and nucleic acid metabolism of Ehrlich ascites cells.

Incorporation of labelled adenine and guanine into the nucleotides and nucleic acids of Ehrlich ascites cells during incubation, aerobically and anaerobically, in a glucose medium was measured in the absence and in the presence of actinomycin D. This antibiotic (0.3 μg./ml.) has little or no effect on the rate of respiration or of glycolysis of Ehrlich ascites cells. It has no inhibitory effect on the incorporation of adenine-8-C^{14} or guanine-8-C^{14} into the nucleotides nor does it inhibit the incorporation of labelled glycine into the proteins. It inhibits markedly, however, the incorporation of adenine-8-C^{14} or guanine-8-C^{14} into RNA. This amounts to 70% at a concentration of 0.3 μg./ml. and 90% at a concentration of 2 μg./ml. Incorporation of labelled adenine into DNA is stimulated by the presence of actinomycin D, the stimulation not being as large as the inhibition of incorporation into RNA. Incorporation of labelled adenine or guanine into *total* nucleic acids is inhibited by actinomycin D. Stimulation of incorporation of adenine into DNA by actinomycin D reaches an optimum at a concentration of approximately 1 μg./ml., after which the stimulation diminishes. Isolated DNA combines with actinomycin D and its addition to a suspension of Ehrlich ascites cells diminishes the inhibitory effect of the antibiotic on adenine incorporation into DNA of the cells. RNA has no such effect and treatment of isolated DNA with DNA-ase diminishes its protective effect on Ehrlich ascites cells against actinomycin inhibition."

An important result, that resulted in much further work elsewhere, was our demonstration [56–58], in 1961, that hydrogen peroxide is formed during phagocytosis and that NADH and NADPH oxidases are released in the leukocyte, accounting for some of the unique biochemical changes accompanying phagocytosis. We suggested that hydrogen peroxide is partly responsible for the bactericidal effects accompanying phagocytosis and this suggestion has been amply confirmed. We, in fact, indicated the importance of peroxidase reactions in phagocytosis and in bactericidal activities. With Roberts [59], we studied the conditions underlying particle uptake by leukocytes and this has proved to be a useful field of investigation. I was very pleased with these results as I had always wondered how phagocytes, our first line of defence against bacterial parasites, killed their prey. Some of our results [56] are quoted below:

"Evidence that hydrogen peroxide is actually formed during phagocytosis comes from experiments the results of which are shown in Table IV. In the first place it is evident that the addition of catalase more than doubles the rate of oxidation of formate in the presence of resting or phagocytizing leucocytes, whereas the addition of heated catalase has no effect. Such an accelerating effect must be attributed to the action of catalase on hydrogen peroxide leaking from the leucocyte, as we have found no evidence that the catalase content of the leucocyte is increased on exposure to catalase. Secondly, the addition of sodium nitrite (2 mM) which will act as competitor to formate brings about an inhibition of oxidation of formate by resting or phagocytizing leucocytes. Moreover, the system oxidizing formate is found in the leucocyte homogenate, which is capable of extensive oxidation of formate in the presence of a system producing hydrogen peroxide such as is afforded by a mixture of glucose oxidase (notatin) and glucose (Table IV).

TABLE IV

FACTORS AFFECTING AEROBIC OXIDATIONS OF FORMATE BY LEUCOCYTES AND LEUCOCYTE HOMOGENATES

Formate-^{14}C (2 m*M*) present throughout. Glucose = 5 m*M*.
Notatin = 0.01 per cent

	$10^4 \times \mu$moles formate oxidized/hr./mgm. leucocytes
Resting leucocytes	
Leucocytes only (*p*H = 6.0)	36
Notatin + glucose	0
Leucocytes + notatin + glucose	715
Leucocytes + notatin + glucose + sodium azide (1 m*M*)	1
Leucocytes only (*p*H = 7.4)	18
+ soldium azide (1 m*M*)	0.8
+ potassium cyanide (1 m*M*)	6
+ catalase (1/750)	40
+ heated catalase (1/750)	18
+ sodium nitrite (2 m*M*)	3
Phagocytizing leucocytes	
Leucocytes only (*p*H = 7.4)	86
+ catalase (1/750)	198
+ heated catalase (1/750)	87
+ sodium azide (1 m*M*)	2
+ sodium nitrite (2 m*M*)	12
+ potassium cyanide (1m *M*)	9
Leucocyte homogenate	
Homogenate only (*p*H = 6.0)	65
+ glucose + notatin	529
+ glucose + notatin + catalase (1/15,000)	708
Glucose + notatin + catalase (1/15,000)	105
Homogenate + glucose + notatin + sodium azide (1 m*M*)	5
Homogenate + sodium azide (1 m*M*)	3
Homogenate + sodium nitrite (2 m*M*)	6

The evidence indicates that the systems oxidizing glucose (by the hexose monophosphate shunt pathway) and formate are present in the non-particulate cytoplasmic fraction of the leucocytes, a fact in agreement with observations concerning the localization of glucose-6-phosphate and 6-phosphogluconic acid dehydrogenases in many other tissues."

Together with my colleagues (O. Gonda, S. Lahiri, J.J. Ghosh, C.T. Beer, M.M. Kini) we carried out considerable work on the effects of drugs on brain metabolic processes; for example, the inhibitions brought about by ethanol and higher alcohols on stimulated brain respiration; the notable effects of fluoracetate on ammonia and amino acid metabolism of the brain, an investigation which has stimulated much work on brain compartmentation; the highly potent effects of ouabain in suppressing active transport of amino acids and other substances into the brain, indicating the dependence of brain transport reactions on brain ATPase activity; and the powerful effects of protoveratrine on cerebral carbohydrate–amino acid interrelations.

Some results showing the effects of fluoroacetate on brain metabolism are given in Tables VI an VII [60].

TABLE V

Effects of sodium fluoroacetate (1 mM) on the formation of ^{14}C-labelled amino acids from [$^{14}C_6$] glucose in rat-brain-cortex slices

Rat-brain-cortex slices from adult hooded rats of about 150 g. were placed in ice-cold Krebs–Ringer phosphate medium [of the following composition: NaCl, 128 mM; $CaCl_2$, 3.6 mM; KCl, 5 mM; $MgSO_4$, 1.3 mM; Na_2HPO_4, 10 mM (brought to pH 7.4 with N-HCl)] containing glucose (5 mM; 2 x 10^5 counts/min., i.e. 4 x 10^4 counts/min./μmole). Extra KCl was added, when required, to the main vessel at the start of the experiment. Fluoroacetate was tipped into the main vessel after gassing with O_2 and temperature equilibration, or added as indicated (*) to the main vessel with the other components before gassing. The average wet wt. of slices was 60–80 mg. The final volume of medium in each Warburg manometric flask was 1 ml. The temperature of incubation was 37°. Incubations were carried out in O_2 for 1 hr. The results are means ± S.D. of six independent determinations.

Additions (final concn.)	Q_{O_2}	Incorporation of C derived from glucose into amino acids (μmg.atoms of C/hr./100 mg. wet wt. of tissue)				
		Glutamic acid	Glutamine	γ-Amino-butyric acid	Aspartic acid	Alanine
K^+ ions (5 m-equiv./l.)	9.9 ± 0.5	828 ± 15	227 ± 12	238 ± 11	378 ± 18	109 ± 5
K^+ ions (5 m-equiv./l.) + fluoroacetate (1 mM)	9.6 ± 1.1	1090 ± 60	142 ± 16	291 ± 12	350 ± 20	196 ± 15
K^+ ions (5 m-equiv./l.) + fluoroacetate (1 mM)*	10.4 ± 0.2	1120 ± 75	20 ± 3	303 ± 11	365 ± 19	201 ± 14
K^+ ions (105 m-equiv./l.)	15.1 ± 1.1	918 ± 18	420 ± 14	380 ± 13	260 ± 17	176 ± 14
K^+ ions (105 m-equiv./l.) + fluoroacetate (1 mM)	16.6 ± 0.5	1473 ± 150	46 ± 5	376 ± 11	241 ± 12	102 ± 6
K^+ ions (105 m-equiv./l.) + fluoroacetate (1 mM)*	16.0 ± 1.0	1640 ± 100	0	381 ± 18	251 ± 13	106 ± 6

"Sodium fluoroacetate, at concentrations of 1 mM or less, which have no effect on the respiration of rat-brain-cortex slices incubated in Krebs–Ringer phosphate medium containing glucose, suppresses the formation of ^{14}C-labelled glutamine from [$^{14}C_6$]glucose, the effect being greater in a medium containing 105 m-equiv. of K^+-ions/l. than in one containing 5 m-equiv. of K^+-ions/l. There is a concomitant increase in the amount of ^{14}C-labelled glutamate formed.
Sodium fluoroacetate (1 mM) suppresses the accelerating action of NH_4^+-ions on the rates of oxygen consumption and of the formation of ^{14}C-labelled glutamine (from [$^{14}C_6$]glucose) by the brain cortex slices."

TABLE VI

Effects of sodium fluoroacetate at various concentrations on the formation of ^{14}C-labelled amino acids from [$^{14}C_6$]glucose in rat-brain-cortex slices

The experimental conditions were as described in Table I, the glucose concentration being 5 mM (2×10^5 counts/min., i.e. 4×10^4 counts/min./μmole). Fluoroacetate was tipped in after gassing with O_2 and temperature equilibration. The results are means ± S.D. of at least four independent determinations.

Additions (final concn.)	Q_{O_2}	Incorporation of C derived from glucose into amino acids (μmg.atoms of C/hr./100 mg. wet wt. of tissue)				
		Glutamic acid	Glutamine	γ-Amino butyric acid	Aspartic acid	Alanine
K^+ ions (105 m-equiv./l.)	16.0±1.0	918± 21	439±16	340±18	250±14	185±17
K^+ ions (105 m-equiv./l.)+ fluoroacetate (0.01 mM)	17.1±1.2	792± 19	430±20	328±16	236±11	147± 3
K^+ ions (105 m-equiv./l.)+ fluoroacetate (0.03 mM)	16.5±0.5	1084± 30	235±14	262±13	220±16	98±13
K^+ ions (105 m-equiv./l.)+ fluoroacetate (0.1 mM)	16.7±0.7	1590±150	66± 7	298±17	263±12	68± 6
K^+ ions (105 m-equiv.l.)+ fluoroacetate (1.0 mM)	16.0±1.0	1538±100	36± 8	381±21	249±18	84± 9

Some results showing the effects of ouabain [61] are quoted below:

"Ouabain at the low concentrations that have little or no effect, within experimental error, on rates of respiration of rat-brain-cortex slices brings about the following phenomena: (*a*) inhibition of formation of labelled glutamine from either labelled glucose or labelled glutamate, this inhibition being diminished by the addition of NH_4^+ ions or of glutamate; (*b*) increased efflux of amino acids (glutamine, glutamate, alanine, γ-aminobutyrate and aspartate) from the brain-cortex slices into the glucose–Krebs–Ringer phosphate medium in which the tissue is respiring; (*c*) diminished influx of glutamate into brain slices from the medium, the percentage diminution increasing with increase of glutamate concentration; (*d*) diminished influx of creatine from the medium, though labelled-creatine phosphate formation in the brain is unaffected by ouabain; (*e*) increased inhibition of influx of glutamate on addition of an increased concentration of K^+ and NH_4^+ ions.

It is postulated that ouabain combines with carrier sites for K^+ (or NH_4^+) ions, for glutamate and for creatine, such sites requiring ATP for their activities. This hypothesis makes it possible to link adenosine-triphosphatase activities with transport phenomena."

TABLE VII

Effects of ouabain on radioactive-amino acid formation from radioactive glucose by rat-brain-cortex slices in the presence of 105 m-equiv. of K^+ ion/l.

Rat-brain-cortex slices were incubated in Krebs–Ringer phosphate medium (see Table V) for 1 hr. in oxygen at 37° in the presence of [$^{14}C_6$]glucose (5 mM), and ouabain was added where indicated. After 1 hr. the slices were removed from the incubation medium and the labelling of amino acids was determined in the tissue and medium as described in the Materials and Methods section. Results are means ± S.D. of four independent determinations.

	Amount of amino acid (incorporated μmg.atoms of carbon derived from glucose/hr.)							
Concn. of ouabain (μM)...	0		0.1		1		10	
Amino acid	Tissue (100 mg.)	Medium (1 ml.)	Tissue (100 mg.)	Medium (1 ml.)	Tissue (100 mg.)	Medium (1ml.)	Tissue (100 mg.)	Medium (1 ml.)
Glutamate	965 ± 62	35 ± 15	960 ± 57	38 ± 12	1030 ± 72	163 ± 12	1101 ± 62	234 ± 23
Glutamine	341 ± 17	40 ± 23	350 ± 15	40 ± 18	60 ± 5	105 ± 17	0	0
γ-Amino-butyrate	325 ± 25	15 ± 9	320 ± 21	15 ± 6	280 ± 15	90 ± 8	260 ± 12	117 ± 9
Aspartate	264 ± 15	15 ± 11	270 ± 13	22 ± 9	282 ± 12	40 ± 6	163 ± 12	80 ± 9
Alanine	162 ± 12	38 ± 15	170 ± 15	50 ± 11	105 ± 8	86 ± 9	92 ± 7	110 ± 12

We had a flourishing research institute that attracted considerable attention in various parts of the world and I had very many applications from scientists overseas, as well as from the USA and Canada, to spend some time with us. In fact, at one time we had representatives of over 12 nationalities working in the laboratories. But lack of space and funds prevented me accepting any but a small fraction of the applications.

Perhaps I should mention some other work in which I was involved. Early in the 50s, K.A.C. Elliott of the Montreal Neurological Institute and I decided that the time was ripe to launch a book on neurochemistry, for the new science was developing rapidly. With the long-distance collaboration of I.H. Page, for whom I had written many years before (1937) a chapter on cerebral oxidation in his book *Chemistry of the Brain* we, together with the invaluable cooperation of a number of co-authors, published *Neurochemistry* (1955). This book, I believe, was of considerable help in the growth of this subject and it ran into a second edition in 1961. Nowadays the science of neurochemistry has reached such massive proportions that a *Handbook of Neurochemistry* composed of many volumes and written by many experienced investigators has been published and is about to go into a second edition with even more volumes and authors.

In the early '60s my old student and colleague, Rolf Hochster, whose death a few years later came as a shock to all of us, and I decided that another subject, of much importance in various aspects of biochemistry, was at a stage when a suitable book concerning it might be written. The first volume of *Metabolic Inhibitors* appeared in 1963. This has run into four volumes and I daresay it will be extended by those younger than I.

The work of the Montreal Institute had to stop because I reached the age of 65. It was considered that the Institute was no longer "viable" because I had reached that milestone in our lives. The real reason, though I was not told this officially, was that it was thought unlikely that a successor would be able to secure the necessary funds for the Institute's maintenance and growth. My budget at the time was about $250 000 per annum. I was never consulted nor was my advice seriously sought. The laboratory was ultimately split into two parts, the larger part being converted to a Cancer Institute, and supported financially by the National Cancer Institute under Professor Scholefield, and the smaller part devoted to problems of cell metabolism, under my direction and supported to a considerable extent by the Medical Research Council of Canada. We were housed in beautiful laboratories in the McGill-McIntyre Medical Science Building. I had been promised years before that our Institute would be fully incorporated there. But I had also been promised, many years before that, that our Institute would be incorporated into the new Montreal General Hospital and then I was told that sufficient money was not available. My old Institute, for which I had done much to create and maintain a high scientific standing, sank into oblivion. I was due to be fully retired in a couple of years and I saw no prospect of carrying on research work in Montreal beyond that time. I was, however, in no mood for retirement and I began to look around to see what were the possibilities for the future.

Several options came along. In 1966 I was generously offered, through the kind offices of Professor W.C. Gibson, a post as Professor of

Neurochemistry, in the University of British Columbia, to help in the research work and teaching in the neuroscience there. I gladly accepted this excellent offer. My children all being married and independent of us, my wife and I once again packed up our home and we moved to the beautiful city of Vancouver, whose scenery, temperate climate, and freedom from ragweed allergens became a source of great comfort and delight to us. We were sorry to leave our many friends in Montreal. But we both looked forward to making new friends and living a new life in one of the most attractive areas of the world, British Columbia.

Vancouver

My conditions for work in Vancouver were altogether different from those that I had experienced in Montreal. Quite a group of us left Montreal for Vancouver. Many of my old students accompanied me and there was my ever loyal friend, Dr. S.C. Sung. The party of eight or so greatly strengthened the composition of the Kinsmen Laboratory which became the Division of Neurological Sciences, a part of the Department of Psychiatry. I came as a sort of elderly bride with a not inconsiderable dowry. The grants in my name and my equipment together amounted to well over $100 000. I became a post-retirement professor, to be reappointed each year so that I never knew which year would be my last. This greatly interfered with my planning for research, hire of post-doctoral students, etc. I took care not to interfere with the general running of the laboratory. My own students all got their Ph.D.s but I had a small and able group of senior investigators who had decided to join me. I had few responsibilities, and my entire energies were devoted to carrying out research work on the brain. I feel deeply grateful to the University of British Columbia, and to the Department of Psychiatry for giving me the opportunity to carry on research on the basic lines that I like and to the Medical Research Council of Canada without whose assistance I could do nothing.

I have now been here over fourteen years and in this period of time we have tried to throw more light on the mode of action in the brain of neurotropic drugs, on the interrelations between glia and neurons in brain metabolism and function, to understand more clearly compartmentation phenomena in the brain, a subject so very popular with neurochemists today, to study the chemical changes in the brain that occur during development from infancy to maturity and, above all, to try to integrate neurochemical and neurophysiological phenomena, for I believe that such integration is essential for progress in this field in the future.

I owe much to discussions with my son David, Professor of Pharmacology in U.B.C. whose extensive knowledge of physiology and pharmacology has been of enormous help to me. It was he who first told me of the powerful effect of tetrodotoxin in abolishing action potentials under the conditions used by physiologists. I then decided to see whether it would affect the neurochemical events in brain cortex slices which were favourite objects of my studies. I have been able to use this drug with some success in our neurochemical studies from 1967 onwards and the results only served to indicate to me the great importance of bringing neurochemical and neurophysiological phenomena together. They are both of equal importance but only recently is this being realized by all parties. We demonstrated [62–65] the marked effects of tetrodotoxin on stimulated brain respiration and transport processes and that this drug affects both water uptake and sodium influx into the brain under a variety of conditions. We concluded that action potentials are generated in brain slices under a variety of incubation conditions, including the absence of glucose or at the onset of anoxia. This conclusion (because of the extraordinary patience and expertise of K. Okamoto who learned much of the technique of measuring action potentials in David's laboratory) we have now fully confirmed [66]. We have used guinea pig cerebellum slices to demonstrate the development of spontaneous action potentials in vitro. We [21] have shown that local anaesthetics, in common with tetrodotoxin, block the sodium influx into brain cells when action potentials are generated, but that barbiturates must act, at least in part, indirectly by suppression of mitochondrial metabolism with resultant changes in cell calcium content and membrane permeabilities. Itoh and I have been able to demonstrate the quantitatively important role of acetoacetate in the infant brain as compared with that in the mature brain [67]. I have had a most fruitful collaboration with a former student and colleague, Dr. A.M. Benjamin. We have been able to demonstrate [68] that during the generation of action potentials in brain slices, many amino acids are released and by the appropriate use of the neurotropic drugs, tetrodotoxin, protoveratrine and ouabain, to show definitely that the neurons are the sites of the major pools of glutamate, aspartate, γ-aminobutyrate, glycine and serine but that the glia are the site of the major pool of glutamine. The amino acids are now the subjects of great endeavour throughout the world of neuroscience. Some of them have proved to be transmitters and thus they have obtained a significance not even dreamed of in earlier days.

Some sort of instinct must have kept me in close touch with cerebral amino acids for very many years – since 1930 in fact – not simply as parents of the biogenic amines or their relations the neural peptides, but because I sensed in them the means by which some of the great problems of neuroscience may ultimately be approached.

Benjamin and I [68, 69] have given evidence to indicate the existence in the brain of a cycle of events in which neurons and glia are coupled so that part of the glutamate released from the neurons during excitation is withdrawn from the extraneural space and is returned to the neurons eventually in the form of glutamine. We have suggested a positive function of ammonium ions in the brain and by both metabolic and transport studies have thrown more light on the role of amino acids in the brain. This subject has grown to such massive proportions that no one man can handle it. In fact, the subject has drawn into it, as into an intense vortex, behaviourists, anatomists, physiologists, neurologists, people who call themselves molecular this or molecular that, that I feel more or less completely lost among them. They will grow in size and numbers as they search for opportunities that will give them fame and fortune – and grants. Publications galore is the watchword today, publish or perish is the motto, and this will be so until there is a radical change in the system of providing funds for research.

I am at work now (1981) with at most one colleague and still hope to make a significant contribution. I find, as always, the work to be fascinating and I will carry on as long as I am allowed to do so.*

I have referred in this article to about one-fifth of my scientific publications omitting, through consideration of space, much that has seemed to me to be of importance (see also [48]).

The road of science to me, over a period of nearly 60 years, has been full of adventure. There have been lots of ups and downs, fair weather and foul, entrancing by-ways which I have often entered for I never could resist trying to see what lay beyond. If this constitutes enjoyment of life, I have enjoyed it. I know that my chemical background and instincts, rather than the biological, have dominated my thoughts and activities, but I have never allowed these to underrate the significance of results and

* In revising this script (1983), I think that I should mention that I am now on the staff of TRIUMF, the Inter-University Meson Facility, situated on the campus of the University of British Columbia, Vancouver, and concerned with the neurochemical aspects of Positron Emission Tomography.

conclusions in neighbouring sciences. There should be no class distinction in science – all phases of it are of equal importance. There has always been a pull on me, ever since I was a child, called scientific curiosity and I have never been able to resist this pull though maybe it is a little weaker than it was in earlier days. I am getting old now; but old scientists, like old soldiers, just simply fade away.

REFERENCES

1 J.H. Quastel, Biochem. J., 18 (1924) 365.
2 J.H. Quastel, Biochem. J., 19 (1925) 641.
3 J.H. Quastel, M. Stephenson and M.D. Whetham, Biochem. J., 19 (1925) 304.
4 J.H. Quastel and M. Stephenson, Biochem. J., 19 (1925) 660.
5 J.H. Quastel and M.D. Whetham, Biochem. J., 19 (1925) 520.
6 D.J. Kushner and J.H. Quastel, Proc. Soc. Exp. Biol. Med., 82 (1953) 388.
7 J.H. Quastel and B. Woolf, Biochem. J., 20 (1926) 545.
8 J.H. Quastel, Biochem. J., 20 (1926) 166.
9 J.H. Quastel and W.R. Wooldridge, Biochem. J., 21 (1927) 1224.
10 J.H. Quastel and W.R. Wooldridge, Biochem. J., 22 (1928) 689.
11 J.H. Quastel, Trans. Faraday Soc., 26 (1930) 853.
12 J.H. Quastel, Ergeb. des Enzym-Forschung, 1 (1932) 209.
13 J.H. Quastel and A.H.M. Wheatley, Biochem. J., 25 (1931) 117.
14 J.H. Quastel and A.H.M. Wheatley, Biochem. J., 26 (1932) 725.
15 J.H. Quastel and A.H.M. Wheatley, Proc. Royal Soc. B., 112 (1932) 60.
16 J.H. Quastel and R. Ström-Olsen, Lancet, 1 (1933) 464.
17 H.W. Reading and J. Wallwork, Biochem. Pharmacol., 18 (1969) 2211.
18 S.L. Chan and J.H. Quastel, Biochem. Pharmacol., 19 (1970) 1071.
19 J.J. Ghosh and J.H. Quastel, Nature, 174 (1954) 28.
20 H. McIlwain, Biochem. J., 53 (1953) 403.
21 R. Shankaran and J.H. Quastel, Biochem. Pharmacol., 21 (1972) 1763.
22 J.B. Chappell and A.R. Crofts, Biochem. J., 95 (1965) 378.
23 J.M. Godfraind, H. Kawamusa, K. Krnjevic and R. Pumain, J. Physiol., 215 (1971) 199.
24 J.H. Quastel, Canad. J. Biochem., 52 (1974) 71.
25 M. Jowett and J.H. Quastel, Biochem. J., 27 (1933) 486.
26 M. Jowett and J.H. Quastel, Biochem. J., 31 (1937) 275.
27 J.H. Quastel and A.H.M. Wheatley, Biochem. J., 27 (1933) 1609.
28 C.E.M. Pugh and J.H. Quastel, Biochem. J., 31 (1937) 286.
29 C.E.M. Pugh and J.H. Quastel, Biochem. J., 31 (1937) 2306.
30 P.J.G. Mann and J.H. Quastel, Nature, 144 (1939) 943.
31 P.J.G. Mann and J.H. Quastel, Biochem. J., 34 (1940) 414.
32 J.H. Quastel, Trans. Faraday Soc., 39 (1943) 357.
33 J.H. Quastel, in L.L. Iversen, S.D. Iversen and S.H. Snyder (Eds.), Handbook of Psychopharmacology. Vol. 5, Plenum, New York, 1975, p.1.
34 J.H. Quastel, M. Tennenbaum and A.H.M. Wheatley, Biochem. J., 30 (1936) 1668.
35 P.J.G. Mann, M. Tennenbaum and J.H. Quastel, Biochem. J., 32 (1938) 243.
36 P.J.G. Mann, M. Tennenbaum and J.H. Quastel, Biochem. J., 33 (1939) 822.
37 L. Penrose and J.H. Quastel, Biochem. J., 31 (1937) 266.
38 P.S. Nutman, H.G. Thornton and J.H. Quastel, Nature, 155 (1945) 498.
39 P.J.G. Mann and J.H. Quastel, Nature, 158 (1946) 154.
40 J.H. Quastel and P.G. Scholefield, Bacteriol. Rev., 15 (1951) 1.
41 J.H. Quastel, Lecture on Soil Metabolism, Royal Institute of Chemistry, London 1946.
42 J.H. Quastel and D.M. Webley, J. Agricult. Son., 37 (1947) 257.
43 J.H. Quastel, Nature, 171 (1953) 7.
44 E. Riklis and J.H. Quastel, Canad. J. Biochem. Physiol., 36 (1958) 347.

45 R.P. Harpur and J.H. Quastel, Nature, 164 (1949) 693.
46 W.J. Johnson and J.H. Quastel, Nature, 171 (1953) 602.
47 J.J. Ghosh and J.H. Quastel, Nature, 174 (1954) 28.
48 J.H. Quastel, Canad. J. Biochem., 52 (1974) 71.
49 W.A. Darlington and J.H. Quastel, Arch. Biochem. Biophys., 43 (1953) 194.
50 J.E. Guthrie and J.H. Quastel, Arch. Biochem. Biophys., 62 (1956) 485.
51 J.H. Quastel, Proc. Royal Soc. B., 163 (1965) 169.
52 M.M. Kini and J.H. Quastel, Nature, 184 (1959) 252.
53 B.M. Bragança and J.H. Quastel, Biochem. J., 53 (1953) 88.
54 B.M. Bragança and J.H. Quastel, Nature, 169 (1952) 695.
55 I.J. Bickis and J.H. Quastel, Proc. Canad. Fed. Biol. Soc., 5 (1962) 13.
56 G.Y.N. Iyer, M.F. Islam and J.H. Quastel, Nature, 192 (1961) 535.
57 G.Y.N. Iyer and J.H. Quastel, Canad. J. Biochem. Physiol., 41 (1963) 427.
58 J. Roberts and J.H. Quastel, Nature, 202 (1964) 85.
59 J. Roberts and J.H. Quastel, Biochem. J., 89 (1963) 150.
60 S. Lahiri and J.H. Quastel, Biochem. J., 89 (1963) 157.
61 O. Gonda and J.H. Quastel, Biochem. J., 84 (1962) 394.
62 S.L. Chan and J.H. Quastel, Science, 156 (1967) 1752.
63 S.L. Chan and J.H. Quastel, Biochem. Pharmacol., 19 (1970) 1071.
64 K. Okamoto and J.H. Quastel, Biochem. J., 120 (1970) 37.
65 R. Shankar and J.H. Quastel, Biochem. J., 126 (1972) 851.
66 K. Okamoto and J.H. Quastel, Proc. Royal Soc. B., 184 (1973) 83.
67 T. Itoh and J.H. Quastel, Biochem J., 116 (1970) 641.
68 A.M. Benjamin and J.H. Quastel, Biochem. J., 128 (1972) 631.
69 A.M. Benjamin and J.H. Quastel, J. Neurochem., 23 (1974) 457.

G. Semenza (Ed.) Selected Topics in the History of Biochemistry: Personal Recollections (Comprehensive Biochemistry Vol. 35)
© 1983 Elsevier Science Publishers

Chapter 7

A Biochemist's Approach to Autopharmacology

HERMANN BLASCHKO

University Department of Pharmacology, South Parks Road, Oxford OX1 3QT (U.K.)

1. Introduction

Biochemistry as an independent subject has a very recent history. My own scientific development, as well as that of my contemporaries, has run a course parallel to that of our science. I hope to illustrate this by an account of what I still remember. I shall try as far as is possible to base myself on my own experience and my own recollections. This means that I unavoidably shall expose myself to certain risks. However much I still remember, some uncertainty is bound to remain. Where it was unavoidable I have consulted the literature in order to reinforce what has been kept alive in my memory.

For my own field of study, that of the substances active in the nervous system, it can safely be said that it has become a legitimate field of study only in my own lifetime. It was a field that practically did not exist when I started research.

A special aspect of my personal experience is one of education in Germany and independent research in Britain. I owe much to the German tradition that I experienced before I chose my own field of research. I can therefore look at my early experiences from a new standpoint; I can see advantages and defects, in the light of experience of the system into which I was transplanted. Fortunately, this transplantation was gradual as I had already worked in England before I moved there for good in 1933 [1].

Plate 7. Hermann Blaschko.

2. Early influences

Of these influences, the most important one was my father's, Alfred Blaschko (1858–1922), dermatologist, who was a leader in the fields of medicine and social hygiene. His position in the field of social medicine has recently been fully discussed in a biographic essay [2]. By the time I appeared on the scene this field was his main interest, but he never lost his interest in natural science, and he continued to take a lively interest in modern scientific achievements and scientific thought. Early in his career he had worked in the laboratories of Rudolf Virchow (1821–1902) and Wilhelm Waldeyer (1836–1920). Shortly after I was born he had published, for the German Dermatological Society, an atlas of the linear diseases of the skin [3]. There he described the lines that are preferentially taken by linear naevi, lines that are still called the "lines of Blaschko" [4].

At this time also my father was drawn into activities that from then on determined his attitude to international co-operation. It began with his attendance at Brussels of the first two international leprosy conferences, and these were followed by an international conference on venereal diseases, also at Brussels. He became a convinced internationalist and attended many congresses, the last of these being the International Congress of Medicine, held in London in 1913, to which he contributed a review [5]. When the first World War broke out he was for a while intellectually isolated. Invited to add his signature to the notorious nationalistic declaration of German intellectuals in the autumn of 1914, he not only refused to do so, but as a reply he wrote an editorial in the journal he edited on the value of international congresses and expressing the hope that the bridges between different nations would not for long be broken; this editiorial has recently been reprinted in full (see [2]).

In 1916, before our family holidays in the Bavarian Alps, my father gave me a book to read that had been sent to his editiorial office for review. It was entitled *Grundriss der Organischen Chemie* and the author was Carl Oppenheimer (1874–1941), about whom I shall have to say more below. I learn from the publishers, Messrs. Georg Thieme, now at Stuttgart, that the ninth and last edition of this book came out in 1916 [6]. I presume this book was written mainly for medical students. At any rate, it had a profound effect on me. I was attending a school strictly based on classics, Latin and Greek. We had reasonably good teachers of mathematics and also learnt some physics, but of chemistry we had only, I believe, a semester's course in inorganic chemistry. I had never before heard anything of organic chemistry, and I was equally ignorant of the implications of chemistry for medicine. As I became increasingly interested in what I was reading I thought that here there might be a field in which I could eventually take an active interest.

As a matter of fact, my introduction into practical work of this kind followed soon afterwards: nine months later, in the spring of 1917, I had left school and was working in the Institute of Animal Physiology at the Agricultural Academy in Berlin. My work there, which I have recently described [1], was related to nutritional problems. The director of the Institute, Nathan Zuntz (1847–1920), was a pupil of Eduard Pflüger; he was an expert in the field of respiration and metabolism, and a pioneer in high altitude physiology. In the institute there was a collection of respirometers for large farm animals, and although they were not in use, they had to be kept in good order. It was there that I first heard the names of Reignault and Reiset and of Atwater and Benedict. I was lucky to have started in the laboratory of Dr. C. Brahn, who was an authority on nutrition, and especially on bread.

In that laboratory I stayed until the end of the war, the autumn of 1918, but in the autumn of 1917 already I was able to enroll as a student of medicine at Berlin University, since that was the time when I should normally have left school. Thus, during my first year at the university, I really had two jobs: in the mornings I worked in the institute, and in the afternoons and evenings I became a medical student.

3. Medical studies and the role of biochemistry

During my time as a medical student, from 1917–1922, teaching of biochemistry at German universities was in a stage of transition. In contrast to anatomy and physiology, biochemistry was not a compulsory subject and it did not appear in the "Physikum", the examination that terminated preclinical studies after five semesters, or two-and-a-half years. I remember that in Berlin I once went to the biochemistry lecture advertised by H. Steudel (1871–1967) who had a laboratory in the Institute of Physiology of the University. At that time he held the position of Professor extraordinarius; there was no full professor of biochemistry in Berlin. I found there about four students, and I was deterred by these small numbers from attending the course because I was afraid I might make myself too conspicuous if I decided not to attend regularly. At that time Berlin had a very big medical faculty; that is to say, only a negligible fraction of the undergraduates attended the biochemistry lectures. I think the position at Berlin was similar to that to be found at many other German universities, although the German university of Strasbourg, reorganised after 1871, had a distinguished department of biochemistry founded by Hoppe-Seyler. There were universities in which the professor of physiology was what we would call today a biochemist. One such place was Heidelberg, which I remember visiting in the summer semester of 1918. There the professor of physiology was A. Kossel (1853–1927), Nobel Laureate and discoverer of the amino acid lysine. However, the lecture that I attended was on physiology of vision!

A different arrangement was in operation in Frankfurt a.M. where there was a recently founded university. Here there were two chairs of physiology, one of "Animal Physiology", held by A. Bethe, the other of "Vegetative Physiology", held by G. Embden.

I moved in 1919 from Berlin to Freiburg i. Br. and it was at this university that I took my "Physikum". In Freiburg biochemistry, or rather, physiological chemistry, had a more favourable position than it had held in Berlin. There was a professorship of physiological chemistry, which was held by F. Knoop (1875–1946). He gave a course of lectures that was not compulsory but was well attended. In the only lecture list in my possession that has survived I see that Knoop lectured three times a week for one hour. Also, there was an arrangement between Knoop and the Professor of Physiology, J. von Kries (1853–1928): they had divided the obligatory practical course in physiology between the two laboratories. I seem to remember that the greater part of the course was in the Institute of Physiology. The smaller part of the course was held in the Institute of Physiological Chemistry, under Knoop's supervision; I believe he had one assistant demonstrator. The institute was small but well organised; Hans Krebs, one of my contemporaries [7], has recently reminded me that it was situated in the old Botanical Garden; the building had served as the Botanical Institute, but the latter had been moved recently to bigger and more modern quarters.

Knoop was a good teacher; through him we were initiated into what were at the time the modern problems of biochemistry. Knowledge of intermediate metabolism was restricted, but we heard him discuss ideas like oxidative deamination of amino acids and, of course, β-oxidation of fatty acids. The practical course was well run. The students were given mimeographed sheets in which the practical work was described and discussed. Knoop was always ready to talk to us about any question that we wanted to ask. I am sure that I am not the only student who became personally known to him, and he always remembered me when we met in later years, for instance, when he was one of my oral examiners when I took my M.D. viva in January 1923 (L. Aschoff, Pathologist, and W. Straub, Pharmacologist, were the other two examiners). In 1932, I met Knoop in the Mediterranean summer resort of Forte dei Marmi, where together with Alex von Muralt and Hans Krebs I spent a week after the 14th International Physiological Congress at Rome. My last meeting with Knoop was in the summer of 1938, at the 16th International Physiological Congress in Zurich. After a paper by E.A. Zeller (then Basel, now Chicago), I had made a discussion remark in English, when he got up and queried my evidence. All I could do in reply was to repeat what I had said, but this time in German. He then said: "Now I understand; why did you not say it in German the first time?" Well, there was little that I could add. Soon afterwards a letter from him reached me in Cambridge in which he said he hoped I had not been offended, but in view of our long acquaintance he had probably been rather informal; he sent me his best wishes. This was my last experience of F. Knoop.

Biochemical thinking had not penetrated very deeply into our curriculum. The physiologist in Berlin was Max Rubner (1854–1932), and from him we learnt about metabolic quotients, and the isodynamic action of foodstuffs, and the specific dynamic action of protein. I never heard from him anything about vitamins, and his one-sided attitude was later criticized by Hopkins [8]. In fact, it was in Freiburg that I first heard anything about the "accessory foodstuffs" (Hofmeister); this was in the lectures given by our professor of physiology there, J. von Kries. J. von Kries was a pupil of Karl Ludwig and Hermann von Helmholtz, and his chief field of research was in physiology of vision. One of his doctorandi, W. Trendelenburg (1877–1946), had shown that the efficiency spectrum of bleaching of visual purple coincided with the intensity spectrum in skotopic vision [132]. Eventually, I did the experimental work for my M.D. thesis in von Kries' laboratory.

In Freiburg our most outstanding teacher was the professor of pathology, Ludwig Aschoff (1866–1942). He was an eminent representative of the German school of pathology, which excelled in what one would today call "morbid anatomy". However, he was a man of wide interests, and the problem of function was never far from his thoughts. It was in his laboratory that the "Aschoff–Tawara" node of the mammalian heart had been discovered. His work on what he called the "reticulo-endothelial system", aroused his interest in biochemical problems, and particularly in lipid metabolism, an interest that led Aschoff to install a biochemical laboratory in his institute; this laboratory was directed by R.

Schoenheimer (1898–1941), whose importance in the development of modern biochemistry is based chiefly on his later work carried out in the U.S.A.

When I returned to Berlin during my clinical studies I attended in the winter semester of 1921/1922 a course of lectures entitled "colloid chemistry" by P. Rona (1871–1944).

The significance of P. Rona as a teacher of biochemists has been often acknowledged in recent years [9–11]. When I had my interview with Otto Meyerhof in December 1924, he told me that he was only accepting German students who had had their training in Rona's laboratory. Rona's outlook had been shaped by his long association with Leonor Michaelis (1875–1949), one of the pioneers of modern biochemistry. Rona's importance lay in his ability to convey to his students the principles of modern biochemistry, especially the properties of proteins and enzymes. To Rona's lectures I owed my first acquaintance with terms like pH, isoelectric point and many other concepts familiar to the biochemistry student of today.

At the time I attended Rona's lectures, I was already a clinical student. He was at that time in charge of the chemical laboratory of the Institute of Pathology at Berlin. Some of the medical clinics also had chemical laboratories, but I had no contact with these as a student. Of course, the German clinicians included a number of experts on metabolism. In my last semester as a clinical student, I once, on a visit to Breslau, attended a lecture by Oscar Minkowski (1858–1931), one of the discoverers of pancreatic diabetes.

I have already described the influence that the book of Carl Oppenheimer had upon me in 1916. On the whole I was not a very assiduous attender of lectures, and only went to them when I thought the lecturers were interesting. I preferred to get my information from books. I managed, for little money, to collect quite a number of textbooks, and I used to enjoy reading and becoming aware of the differences in emphasis in the treatment of the material. I had some special preferences for out-of-the-way kinds of books. For instance, I remember I learnt much from an introduction into organic chemistry for students of medicine by G. Bunge (1844–1920), Professor of Physiological Chemistry at Basel. Bunge had come from the Baltic provinces of Russia, where after a rather alcoholic period as an undergraduate of the University of Dorpat he had become a very active fighter for the Blue-Cross temperance movement. My own copy of Bunge's book dates from 1913; so it must have been a bit out of date, but this did not really matter [11a]. He gave an unusually large amount of space to the discussion of intermediate metabolism, but also to proteins, amino acids and their metabolites.

There was another work that was not on our reading list, but that I acquired after I had graduated. These were the two volumes of lectures on physiological and pathological chemistry by Otto Fürth [12], of Vienna. I notice the two volumes carry the dates 1925 and 1927, respectively; by that time I was already with Otto Meyerhof in Dahlem. It was a rather uneven book but I found it useful, chiefly on account of the very extensive list of references. Since Fürth was also very good on metabolism of nitrogenous compounds, I still made much use of his work in my Cambridge period.

I have left the discussion of pharmacology, my own field of interest, to the last. Naturally, in pharmacology chemical considerations were of some importance. In Berlin the chair was held by Heffter (1859–1925), best known at present as the founder of the *Handbook of Experimental Pharmacology*. At the time I was a student he was elderly and not very stimulating as a lecturer; so I do not have many vivid recollections. Of course, at an earlier time he had made important contributions. For instance, he had first made the suggestion that the toxicity of mercury might be due to an interaction with sulphydryl groups of cysteine. Also he was a pioneer in the study of mescaline. All these things I learnt only much later. In Freiburg I often attended the lectures by W. Straub (1874–1944); as a lecturer he was a little uneven, but when he was in good form he was worth listening to.

During the first World War, while working in the Institute of Zuntz, I first encountered Wolfgang Heubner (1877–1957). He was one of the few pharmacologists who took an interest in biochemical problems. His work on toxic substances that formed methaemoglobin found its recognition when eventually Otto Warburg took note of it after he had moved to Berlin. I encountered him during my time in Göttingen and later on again in Heidelberg. In England he became a familiar figure at early meetings of the British Pharmacological Society; at these he used to express his resistance to Nazi ideas and practices. Heubner is remembered as a steadfast and courageous opponent of the Nazi regime, and also as the founder of a German school of "biochemical pharmacology" which has produced a number of well-known researchers, e.g. H. Herken, H. Kewitz and H. Remmer.

Some time ago I wrote down a few recollections of a meeting of the German Pharmacological Society held at Innsbruck in 1924 [13]. Much of what I experienced there I have forgotten, but I do remember a demonstration given by Julius Geppert (1856–1937), Professor of Pharmacology at the University of Giessen. In order to demonstrate the mode of action of cyanide he took a suspension of goose erythrocytes, aerated it well and poured it into two stoppered glass cylinders; to one of these he had added a little potassium cyanide. After a few minutes the colour in the control cylinder was seen to change from that of oxyhaemoglobin until it eventually took the dull colour of reduced haemoglobin. In the other cylinder the cyanide had inhibited respiration, with the result that the bright colour of the oxygenated haemoglobin persisted. I was impressed by the neatness and simplicity of the demonstration. I presume that this demonstration has remained alive in my memory because when I joined Otto Meyerhof in the following year, I did some work on the mechanism of cyanide inhibition [14]. Much more recently, when I read David Keilin's [15] book on the history of cell respiration, I learnt that Geppert had been interested in the mechanism of cyanide inhibition at a much earlier stage [16]; this was when he was still with Eduard Pflüger (1829–1910) in Bonn. According to Keilin, it was directly to Pflüger that we owe the conclusive demonstration that the seat of oxygen consumption was not in the bloodstream but in the living cells of the tissue; the nucleated avian red cells, with their high respiratory activity, served as paradigms for the tissue cells in Geppert's experiment.

4. The status of biochemistry as an academic subject

I have already mentioned that at the time I entered research the position of biochemistry at the different German universities was extremely uneven. On the whole its status was low, in contrast to that of anatomy, physiology, pathology and hygiene. This state of affairs was given prominent treatment by Sir Frederick Gowland Hopkins (1861–1947) when he gave the Opening Lecture at the 12th International Physiological Congress held at Stockholm in the summer of 1926. I was fortunate to attend this congress as well as the 14th Congress held at Rome in 1932, because I earned part of my travel expenses by reporting on these two for the *Vossische Zeitung*, the old and respected Berlin liberal daily newspaper, which folded up early in the Hitler period. Hopkins devoted the first part of his lecture to the status of biochemistry [8]. Re-reading this lecture today one is struck by his foresight and the strength of his argument. His claim that biochemistry was a distinct academic subject, has been fully vindicated in the 55 years that have passed.

Into his script Hopkins had inserted some comments on the unsatisfactory state of affairs at German universities. He paid tribute to the historic achievements of men like Liebig and Hoppe-Seyler, but he warned his German colleagues, physiologists mainly, that the treatment of biochemistry at their universities was detrimental to the development of biological science in Germany. I seem to remember that I was told at the time that Hopkins had inserted these remarks upon the initiative of some German biochemists, notably Knoop, but I have no positive information on this. It seems to me plausible, since I know that Knoop, at the occasion of my oral examination for the doctorate of medicine in January 1923, had deplored the low status of biochemistry in my own native city of Berlin and had asked me if something could not be done about this!

I might add that the attitude taken by Hopkins was not shared by everyone. Otto Meyerhof was of the opinion that physiology should be considered as one indivisible whole; he feared an impoverishment from the splitting off of a part. It was for this reason that when he moved into the new Kaiser Wilhelm Institute for Medical Research at Heidelberg in 1929, he insisted upon calling his own institute "Institut für Physiologie".

When the idea of the First International Biochemical Congress was mooted, I at first shared Meyerhof's misgivings. However, one can confidently say that time has justified Hopkins' claim for biochemistry to be considered as a subject in its own right. In retrospect, it seems fitting that the first of the International Biochemical Congresses was held in Cambridge, Hopkins' old university, in 1949.

In my own subject, pharmacology, international recognition came even later. I think it was at the 16th International Physiological Congress, held in Zurich in 1938, that the pharmacologists for the first time had one afternoon's session set aside specifically to consider the needs of the pharmacologist. This session, I believe, was held at the suggestion of W. Heubner. I remember a contribution given by Sir Henry Dale in which he made a plea for the field of pharmacology to be drawn as wide as possible, in regard to method as well as to scientific scope.

The example set at Zurich in 1938 was followed at the three subsequent international congresses that I attended: in 1947, at Oxford, the Congress was followed by a day's session at University College London. Similarly, I attended a very successful day after the 18th International Physiological Congress held at Copenhagen in 1950; this pharmacological session was organised by Eric Jacobsen, whom I first met in Heidelberg in the nineteen thirties, when he worked with Meyerhof. At Montreal in 1953, after the 19th International Physiological Congress, the pharmacologists had a day at MacDonald College; after the Montreal Congress I was no longer a regular attender of these congresses and can thus not speak from personal knowledge. The first independent international meetings of pharmacologists began at Stockholm in 1961, followed by the Prague Congress in 1963; after that congress the present three-year rhythm was introduced, with the biochemists, the pharmacologists and the physiologists meeting in successive years.

5. Biochemistry in Dahlem and Heidelberg

I have recently described [1] how by good fortune as well as by sound advice from trusted friends, I began my apprenticeship with Otto Meyerhof (1884–1951) in January 1925. This was only a few months after Meyerhof's move from Kiel, where he had held a post as an assistant professor in R. Hoeber's Institute of Physiology. There he had only a very small number of co-workers, among them H.H. Weber, Rolf Meier and two Americans, Loebel and H.E. Himwich (1894–1975).

Himwich had just left when I arrived; I only met him once, many years later when he organised a meeting at Galesburg, I.L. Loebel was still there but just leaving. There was only one established position in the laboratory; that was held by K. Lohmann (1898–1978). When I arrived there was also the technician, W. Schulz, who had been a friend of Kubowitz, O. Warburg's technician at the Siemens works at Spandau near Berlin, and there was Schroeder, the "Laborant". Lohmann, Schulz and Schroeder stayed with Meyerhof until he left Germany in 1938, and so did W. Moehle who joined the team some time before the move to Heidelberg in 1929. The number of co-workers steadily increased during my stay with Meyerhof. Among the first were K. Meyer, later the discoverer of hyaluronic acid and for many years at the Columbia University College of Physicians and Surgeons in New York, and J. Suranyi of Budapest. On the whole the number of Germans was small, because there was not much money for research workers at the time. I only remember Paul Rothschild (1901–1965), Karl Meyer, David Nachmansohn and Fritz Lipmann. However, there were many visitors from abroad, Hungarian, French, American and Japanese; I think that Severo Ochoa also came to Dahlem already.

When Meyerhof first accepted me, I had no salary. It was quite common at the time for young men to work for several years without a salary. I remember that Paul Rothschild also had no salary or grant; he was going to specialise in internal medicine later, and it was understood that a training in a scientific laboratory was essential for one who was contemplating an academic career in medicine. As a matter of fact, towards the latter part of 1925 Meyerhof was able to offer me a very modest salary. If I remember correctly he had been given some funds from the big German banks, and my payment came from this source. It was only after the move to Heidelberg towards the end of 1929 that Meyerhof acquired two additional paid positions; one of these went to H. Laser (1899–1980), the other one to me.

I am not going to dwell on my experience in Meyerhof's laboratory in any detail; there have been several recent contributions especially by Nachmansohn [17, 18], but also by others [1, 10]. I have already described some of my earliest experiences. The main task that I had in Meyerhof's laboratory was to contribute to the laboratory's equipment in the area of thermometry and calorimetry. My own output of publications was very modest; this was due to the fact that my work was at least twice interrupted by prolonged periods of illness.

During the time that I stayed in Meyerhof's laboratory the energetics of muscle contraction was the main topic. The period covers the successful isolation of the lactic-acid forming system in cell-free suspension from skeletal muscle, a feat that was essential for the isolation of the various inorganic and organic co-factors of glycolysis, and also for the subsequent elucidation of the various steps of the so-called Embden–Meyerhof pathway. The first report on this work that I remember was at the Stockholm Congress of 1926.

I do not want to go into detail on the work that was carried out at Dahlem at the time, because I feel certain that it will be much more fully gone into by others. My own recollections cover essentially the first successes in the preparation of the cell-free system. In the early days, this was a task in which the whole laboratory took part. Everyone helped with the cutting up of the rapidly dissected rabbit muscles with scissors in ice-cooled containers. The same procedure was still in use when a year or two later Lohmann first prepared adenosine triphosphate, ATP, in the laboratory.

The story of the development of muscle biochemistry, and particularly of the contribution made by Meyerhof's laboratory, has recently been told by Nachmansohn [18] (see also [19]). The important historical break was Einar Lundsgaard's (1899–1968) discovery of the a-lactacid contraction of the muscle poisoned with iodoacetic acid. This important event almost exactly coincides with the move of Meyerhof's laboratory from Dahlem to Heidelberg (see [1]).

In Dahlem Meyerhof's laboratory still had a second research interest which stemmed from the early links between Otto Meyerhof and Otto Warburg. This was the problem of cell respiration.

The history of modern concepts of cell respiration has been the topic of the classical monograph by Keilin (1966). He describes the development of our knowledge of the subject, beginning with the observation by Spallanzani and the work of Moritz Traube (1826–1894) and E. Pflüger (1829–1910). It appears from Keilin's account that the recognition of oxygen consumption being a phenomenon due to the participation of all – or almost all – cells of the body, was a relatively modern one. The role of Pflüger was recognized by Keilin, who also refers to the work of Pflüger's pupil, Geppert, mentioned earlier.

The modern arguments started with the observations by T. Thunberg (1873–1952) on dehydrogenases and their interpretation by H. Wieland (1877–1957). According to the view of these authors cell respiration was the outcome of the action of this group of enzymes, in which hydrogen atoms were removed, a process in which a constituent of the enzyme was reduced. This constituent was autoxidisable, and the catalytic cycle was completed by the oxidation of the enzyme by atmospheric oxygen.

This theory of "hydrogen activation" was opposed by Otto Warburg, who saw as the primary event in cell respiration the oxidation of an organic molecule by ferrous iron. His view was expressed by the term "oxygen activation".

There is no doubt that Meyerhof was strongly influenced by the work and the ideas propagated by Warburg. For a while, experiments on metal-catalysed oxidation reactions and their inhibition by potassium cyanide were carried out in Meyerhof's laboratory [1].

In the year after my arrival in Dahlem Otto Warburg demonstrated the inhibition of cell respiration by carbon monoxide. It has, I believe, been told already that A.V. Hill on his visit to Otto Meyerhof, en route for the 12th International Physiological Congress in Stockholm, saw Warburg's results; Hill told Warburg of the thirty-year-old observation by J.S. Haldane on the reversal of the CO–haemoglobin interaction by light. This piece of information immediately led Warburg to his work on what he called the "Wirkungsspektrum" of the respiratory ferment. I remember the first public announcement of Warburg's work particularly well; I reported on this lecture which was held in the rooms of the Kaiser Wilhelm Gesellschaft in the old Berlin Imperial Palace. Immediately after the lecture I had to go to the office of the "Vossische Zeitung", to write my report for the next morning's edition. I gather from the Archives of the

Max-Planck-Gesellschaft that the date of Warburg's lecture was 22nd February, 1928, that is about nineteen months after A.V. Hill's visit.

At the Stockholm Congress of 1926, Sir Frederick Gowland Hopkins had made the theories of cell respiration his main scientific theme. The position that he took was more favourable to the ideas of Wieland and Thunberg, which was in accord with the general attitude at Cambridge at that time; however, he conceded that the cyanide inhibition of succinic dehydrogenase was presumably due to an interaction of the cyanide with iron in the enzyme. In his lecture Hopkins mainly reported on the work of his own laboratory and that of his colleagues in the Biochemistry Department at Cambridge; these included Malcolm Dixon and J.H Quastel. In re-reading his lecture today, I am struck by one important omission: no mention is made of a paper, that was also published from Cambridge in 1925 and that has since made scientific history. This is David Keilin's first paper on cytochrome [20]. In this first publication we find a description of the wide distribution of cytochrome and also of the spectral changes that it underwent when substrate was oxidized. McMunn's old contention of the haemin nature of cytochrome was fully substantiated. Hopkins' listeners at Stockholm did not hear of the work that had been carried out also at Cambridge, but in the Molteno Institute. As far as I remember, the reaction in Dahlem was different; here Keilin's papers were immediately read with great interest and the relation between cytochrome and the "Atmungsferment" was fully discussed.

Hopkins mentions that Meyerhof shared Warburg's ideas. This on the whole is a fair statement. However, Meyerhof was probably taking a somewhat more detached view. It was in 1925 that Carl Oppenheimer began discussing ideas in which the opposing positions taken by the two schools were somewhat reconciled. At that stage Oppenheimer came to Dahlem and submitted his ideas and his writings to Meyerhof. I am certain the latter did not fully agree with Oppenheimer, but I presume he considered the text reasonable, a position that would not have been shared by Warburg. Subsequently Oppenheimer's ideas were repeatedly re-stated, most fully in a book, published jointly with K.G. Stern in 1939 [20a]; by that time, mainly through the work of Keilin and of Warburg, a more balanced picture of the mechanism of cell respiration was emerging.

However, this time was still to come when Meyerhof's laboratory moved from Dahlem to Heidelberg in the autumn of 1929. Here we were at once exposed to the ideas expressed by the Munich school of H. Wieland. Richard Kuhn (1900–1967), the head of our institute of chemistry, had been brought up in the Munich tradition of Willstätter and Wieland. The members of the two institutes shared common-room facilities; so the contacts, particularly between the younger members, were quite close and all problems of joint interest were constantly discussed.

Soon after our arrival in Heidelberg, Kuhn, Hand and Florkin [21] made an observation that strongly influenced Kuhn's attitude: they found that purified preparations of horse-radish peroxidase showed the spectral lines of an iron-porphyrin compound. The presence of iron in peroxidase had earlier been discussed by Willstätter and his colleagues; they had dismissed the iron content as not essential to the chemical nature of the enzyme since in their stages of purification they found no parallelism between enzymic activity and iron content. Kuhn and his co-workers showed that even in Willstätter's purest preparations only about one-seventh of the total iron had been present as iron-porphyrin.

The observations from Heidelberg followed closely on analogous findings by Zeile and Hellström [22], from H. von Euler's laboratory at Stockholm, on the presence of haemin in the enzyme catalase.

Kuhn considered it essential that the adherents of the two opposing groups should meet. He therefore arranged a meeting that took place in the Heidelberg Institute on March 20–23, 1932.

The proceedings of this meeting were not reported in print; however, a list of the papers read was published in *Naturwissenschaften*. Moreover, Professor W. Hasselbach (Heidelberg) has kindly sent me a photocopy of a page from the Institute's visitors book, in which some of the visitors have put down their signatures. The visitors from abroad included R. Wurmser (Paris), Victor Henri (Liège), L. Genevois (Bordeaux), J.B.S. Haldane and D. Keilin (Cambridge). The list is not complete: Malcolm Dixon (Cambridge) was among the participants.

It is clear that the meeting was a very representative one; however, neither Thunberg nor Wieland were among the participants. The latter's laboratory was represented by A. Bertho.

I think it is fair to say that the Heidelberg meeting had the result that Kuhn had hoped for. The personal meetings of some of the chief protagonists and the full discussions helped to make the participants better acquainted with the positions taken by their scientific opponents. Polemics did not cease thereafter, and a final resolution of the points at issue had to wait until new discoveries made some of the old arguments obsolete. One of the personal gains that I had from the meeting was that I here met some of the personalities that I encountered later on again, in Cambridge, notably David Keilin and Malcolm Dixon.

At the end of my Heidelberg period came the 14th International

Plate 8. Snapshot, taken in March 1932, outside the Heidelberg Institute. From left to right: K.W. Hausser (1887–1933), Director of the Institute of Physics, D. Keilin, R. Kuhn, Director of the Institute of Chemistry. The picture was kindly provided by Mrs. Joan Keilin–Whiteley, Ph.D.

Physiological Congress, held at Rome in the late summer of 1932. I travelled to Rome by car, with one of my Heidelberg friends, Alex von Muralt, and with Edwin Cohn (1892–1953). Before coming to Heidelberg von Muralt had been at Harvard, in Edwin Cohn's laboratory; the outcome of his stay had been the famous joint paper with John Edsall, on the birefringence of flow of myosin solutions [23].

Biochemistry was well represented at the Rome Congress. I spent most of my time with Lundsgaard who read a paper on the delayed phosphagen breakdown in cooled muscle, an observation that cleared the way for the recognition that ATP was the ultimate energy donor for muscle contraction.

I have already recently described my post-Congress experiences in the company of Hans Krebs [7].

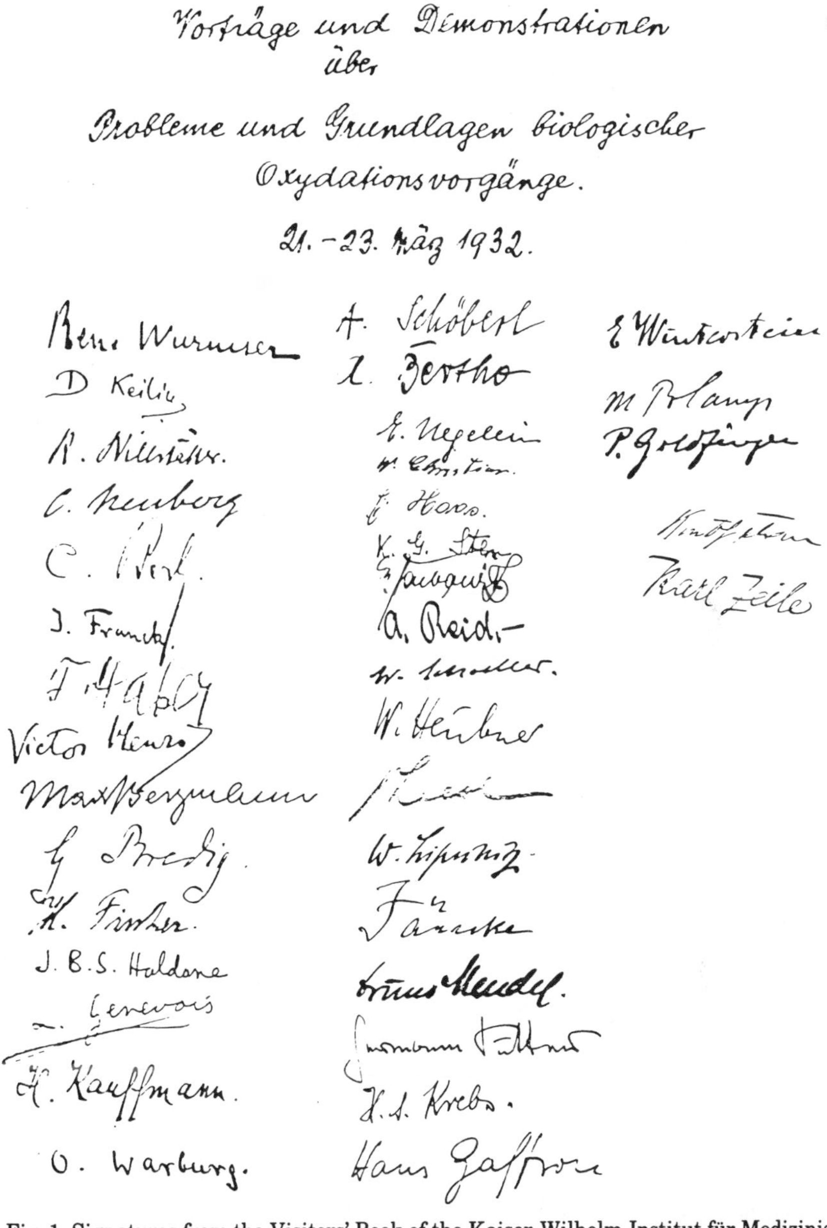

Vorträge und Demonstrationen über

Probleme und Grundlagen biologischer Oxydationsvorgänge.

21.–23. März 1932.

Fig. 1. Signatures from the Visitors' Book of the Kaiser-Wilhelm-Institut für Medizinische Forschung, Heidelberg. The heading, handwritten by the late Professor R. Kuhn, says: "Lectures and Demonstrations on Problems and the Foundations of Biological Oxidations. 21–23 March 1932". The photocopy was kindly given by Professor W. Hasselbach, Heidelberg.

This may be the place to add a comment on the scientific meetings I attended in Germany. While I was at Dahlem, all the members of Meyerhof's laboratory regularly attended the famous Haber Colloquia presided over by F. Haber (1868–1934) and dominated by his personality. These colloquia, held in the Kaiser-Wilhelm-Institute für Chemie, have often been described (see e.g. [18]). I have derived enormous profit from these colloquia. We also had colloquia in our Institute of Biology, and these, too, were often very exciting. I remember interesting talks, e.g. by Jollos, on relative sexuality, and I still have a reprint on non-mendelian inheritance that Carl Correns, the director of our Institute, sent me after I had left Dahlem [24]. I think that talk was a "try-out" of a lecture he was to give at the First International Congress of Genetics, held in Berlin.

Meyerhof's group also had very occasional informal seminars for the few people who worked in his Dahlem group; they took place in his small office. Here Karl Lohmann for the first time described to us his method of determining ATP and other phosphates by following the time course of acid hydrolysis. At another occasion at which I believe Severo Ochoa was present we pressed Otto Meyerhof to tell us something of his ideas on the oxidation of glucose and/or lactic acid. He said: "Well, it goes to CO_2 and water, doesn't it?"

At Dahlem there was no contact with the scientific activities that went on at Berlin University. Occasionally I attended meetings of the Berlin Physiological Society of which I was a member (as my father had been before me), but I was the only member of the Dahlem Institute who went there. Meyerhof actually encouraged me to do so; he regretted the lack of contact with the university laboratories.

At Heidelberg the Institute had regular colloquia, and here we had visitors not only from the University, but also from the industrial laboratories at Ludwigshafen (e.g. K.H. Meyer (1883–1952)), from Mannheim (W.S. Loewe, 1884–1963), and also from the technical university at Karlsruhe (H. Mark).

Meyerhof also regularly sent invitations to G. Embden's laboratory at Frankfurt and, although I do not think Embden ever came to these colloquia, some of his younger colleagues were present at these occasions. On the whole, these meetings were free of the formal atmosphere that one often experienced at meetings in Germany at that time.

Relations with the university were not close, but we did take part in the meetings of the seminars and societies at the university. One year George Barger (1878–1939) was a Visiting Professor in the University Department of Chemistry; we all attended his lectures regularly.

6. England in the Nineteen Twenties and Thirties

My move to England eventually led to a complete change in my own research interests. The move was a gradual one, since I had been working there before I definitely settled in England in 1933.

My earlier visits came about through the close contacts between the laboratories of Otto Meyerhof and A.V. Hill (1886–1977). I first visited University College London in the spring of 1926. I was most hospitably received by A.V. Hill and his colleagues. W.K. Slater, later Sir William Slater (1893–1970), had measured the heat of combustion of glycogen; this had immediate relevance for the work that I had done at Dahlem in the preceding year. Slater eventually had a distinguished career: he became the first director of the research laboratories at the Dartington estate in Devonshire, where I visited him with A.V. Hill in June 1929; later he became Secretary of the Agricultural Research Council. I also met for the first time Philip Eggleton (1903–1954) and his wife Grace (1901–1970), the co-discoverers of phosphagen; they became friends of whom I saw much in later years. An American visitor to Hill's laboratory was R.W. Gerard (1900–1974); he was engaged in measuring the heat production in stimulated nerve, a study that he followed up subsequently in Dahlem with that of the metabolic events during nerve stimulation. I was introduced to E.H. Starling (1866–1927); he invited me to visit his laboratory to watch him set up a heart-lung preparation.

On A.V. Hill's invitation, I attended, on my last day in England, my first meeting of the Physiological Society. That was on 5th June 1926, at Cambridge. In the chair was Joseph Barcroft (1872–1947), who seven years later was to become my host in Cambridge. I also have recollections of W.E. Dixon (1870–1931), an impressive figure with an eyeglass on a broad ribbon of black silk of his pioneering work in the field of humoral transmission I was entirely unaware at the time. My chief recollection of that occasion is meeting a number of young Cambridge biochemists, Eric and Barbara Holmes and Joseph and Dorothy Needham, all eager to get the latest news of the work being done in Warburg's and Meyerhof's laboratories. The friendly reception and the informality of the meeting are what stand out most clearly in my memory, an impression that has been reinforced on innumerable later occasions.

Later in the same year A.V. Hill visited our laboratory in Dahlem, en route to the 12th International Physiological Congress at Stockholm, and it was then that he invited me to come and work with him in London. It was only in 1929 that the opportunity came for me to work in Hill's laboratory. Prior to his move to Heidelberg, Meyerhof had offered me a post in his new institute at Heidelberg but he insisted on my going to A.V. Hill first in order to become familiar with myothermic methods. Things did not look too promising at first because the Rockefeller Foundation did not give me the hoped for fellowship; I presume that my medical record of pulmonary tuberculosis did not help; that was at the time a severe handicap in the search for jobs. However, Hill said he was quite willing to have me, even without the fellowship, so Meyerhof raised some money, chiefly from the "Notgemeinschaft der deutschen Wissenschaft" which enabled me to come to London in the spring of 1929. From the autumn of 1929, when Meyerhof's Heidelberg budget became operative, he supplemented my modest grant, from what was to become my salary after I started work in Heidelberg in 1930.

The year spent in A.V. Hill's laboratory at University College London became a very valuable introduction into my subsequent life in Britain. I have already described the debt I owe to Hill [1] who took great care to make me acquainted with British physiology and physiologists. One of the great educative experiences was to take part in the meetings of the Physiological Society. It was at the meetings of the Society that one was able to meet all the great figures of the day: Sharpey-Schafer, J.S. Haldane, Sherrington, Barcroft, Dale, Adrian, A.J. Clark, Verney and their younger colleagues. The level of the papers read was high and the discussions were interesting. A.V. Hill insisted that I should go to as many of the meetings of the Society as possible.

I cannot remember having been at meetings of the Biochemical Society at that time. Later on, when I used to attend Biochemical Society meetings, I did not think that the average level of the papers and the discussions was similarly high. One of the differences was that the physiologists had a rule that did not allow speakers to read their contributions from a script. Such a rule did not apply to the biochemists, and in consequence the presentation often became monotonous and occasionally degenerated into a recital of cookery recipes.

Hill also occasionally took me to meetings of the Royal Society, then still located at Burlington House. In 1929 Rutherford was president, and after a paper read by Hill, Rutherford initiated a brief discussion himself, by asking some questions, but the atmosphere was more formal than at the Physiological Society. Also, the attendance was poor. That was on June 13th, 1929. A week later I attended again, at a more formal occasion when Sir Charles Sherrington (1857–1952) delivered the first Ferrier Lecture [25]. I have since re-read this lecture; it eventually became the basis of the book *The Integrative Action of Central Nervous System.* It was in his lecture that he introduced the terms "central excitatory state" and "central inhibitory state". I suspect that my recollection of his important lecture stems from my later reading rather than from what I have retained from listening to the lecture.

A lasting gain from my year in Britain was the first experience of work in the Marine Biological Laboratory at Plymouth. A.V. Hill was in the habit of taking his whole laboratory down to Plymouth for the summer months, although in 1929 he was absent for much of the time in connection with the 13th International Physiological Congress held in Boston, MA. I returned from Heidelberg to Plymouth for a short stay in 1930 only, in order to complete my experiments. In 1930 I travelled with my friend Einar Lundsgaard who used the stay to carry out experiments on the effect of iodoacetate on crustacean muscle. Some time earlier Meyerhof and Lohmann had shown that in crustacean muscle the phosphagen present was arginine phosphate. Lundsgaard showed that the typical effect that he had described for vertebrate muscle could also be demonstrated, provided suitable concentrations of iodoacetate and adequate periods of exposure were chosen [25a].

The great experience at Plymouth was to see for the first time the enormous variety in form and function that could be found in life and life processes in marine animals. In later years I have often returned to Plymouth and have also introduced my younger colleagues to the same experience. At various times I have visited other marine biological laboratories, at Arcachon in France, at Naples in Italy and at Espegrend near Bergen, in Norway.

7. Humoral transmission and the catecholamines

My own research interests crystallised a few months after my move to the Physiological Laboratory at Cambridge. I went there upon the invitation of Joseph Barcroft, and it was through Barcroft that I was led to my new field of interest [1]. Accidental factors became decisive for my choice; it was the experimental experience gained in Meyerhof's laboratory that played an important part. However, I think it was not entirely fortuitous that I decided to do some work on the biological inactivation of adrenaline. I should like to give a brief sketch of the historical background to my own research.

When I was a student we were taught the following about the messages passing between different cells of the body. There were two different means of communication: one was by way of hormones, most clearly defined by Bayliss and Starling; according to this concept chemical substances made in one organ were released and transmitted messages to other parts, chiefly by way of the blood stream. In other words, the chemical messenger was delivered in an indiscriminate fashion, reaching all parts of the body. The specificity that resided in the message was in the presence, in the effector cell, of the necessary equipment for receiving the message. The second mechanism, long known, was communication by means of nerves. Here, we were told, there was no specificity in the kind of message; specificity was ensured by the anatomical arrangement: the nerve fibre terminated at an effector cell, and it was this cell alone that was activated.

Revolutionary changes in these views were on the way by the time I began my own work. These changes brought about, within one generation, new concepts that dominate our ideas on communication in living organisms today. In consequence, the boundary line between endocrine and nervous function has become blurred.

It is convenient to date the beginning of these changes to the year 1921 when the theory of humoral transmission of nerve impulses was first formulated. I have elsewhere described [13] my early encounters with Otto Loewi (1873–1961), who first gave experimental support for the transmitter concept. By the time I returned to England in 1933, the meetings of the Physiological Society served as the forum where the new findings in support of the idea of humoral transmission were presented, mainly from the laboratory of H.H. Dale (1875–1968). I also had many discussions with the younger pharmacologists of my generation: O. Krayer, W. Feldberg, M. Vogt, J.H. Gaddum (1900–1965), and many other workers in the field. At the Physiological Society meetings D. Nachmansohn, then in Paris, presented many of his early observations on acetylcholinesterase. I also frequently went to Paris to discuss with him the progress of his work.

The historical background of the transmitter concept has been the subject of several recent reviews [26, 27].

My own work from 1934 onwards concerned biochemical aspects related to adrenaline and the catecholamines in general, and this has from then onward remained my main theme. Eventually, it was concerned with problems of biological inactivation, of biosynthesis, and of transmitter storage.

I scarcely knew at the time how fortunate I had been in the choice of my field of work, which has enormously expanded since I started in 1934. At that time adrenaline alone of the catecholamines was known to occur in the body. The most significant fact that was beginning to emerge was its dual role: it was the secretion product of the chromaffin tissue of the adrenal medulla, a typical hormone; however, it was also suspected of acting as the "accelerans substance" of Otto Loewi, as a transmitter.

In this dual role adrenaline was unique, and it was to remain unique for a very long time. Then there came serotonin, and in the past decade an entirely new class of substances has been discovered that seem to have a similar dual role. These new substances are the neuropeptides, and it is interesting that in the analysis of their function one can discern many parallels with the early work on the catecholamines.

However, the only thing that was known for certain in 1934 was that adrenaline was the secretion product of the adrenal medulla. Its tissue was called chromaffin tissue, because, when incubated with potassium dichromate, it gives the chromaffin reaction, a brown colour. The chromaffin reaction had been described in 1865 by Jacob Henle (1809–1885), but already in 1856, Vulpian (1826–1887) had seen the appearance of a pink colour when he added to the adrenal medullary tissue iodine, ferric chloride, or alkali [28]. Many years later, I accidentally obtained iodo-adrenochrome, the pigment formed in the reaction between adrenaline and potassium iodate, in crystalline form and the structure of this compound was elucidated by my friend, Derek Richter [29].

Vulpian's classical paper still makes exciting reading today [28]. I have quoted from it already once [30]. Based on the observation that the typical pink colour with iodine can be seen not only in the adrenal medullary tissue but also in the lumen of the adrenal vein he suggested that the specific material responsible for the colour reaction was released as a secretion product into the blood stream.

The term "chromaffin tissue" was introduced in the eighteen nineties by A. Kohn (1867–1959); this terminology is more widely accepted than that used by H. Poll (1877–1939); he proposed the term "phaeochrome", which has mainly survived in the word "phaeochromocytoma", denoting a tumour of the chromaffin tissue.

More than a generation elapsed until Vulpian's prophetic suggestions were followed up. In the wake of the classical observations by Oliver and Schäfer [31] there followed a decade of intense activity in which the chemical structure of adrenaline was established and its synthesis was achieved. This enabled the chemical manufacturers to make many substances chemically related to adrenaline, and their pharmacological properties were fully investigated.

The physiological and pharmacological studies were paralleled by histogenetic investigations. Balfour (1851–1882) had drawn attention to the close relationship of the adrenal medullary tissue and the sympathetic nervous system. This work was continued by many histologists, notably by W.H. Gaskell (1847–1914) in Cambridge. These studies became significant when Lewandowsky (1876–1918) first pointed out that the pharmacological effects of adrenaline resembled those seen when the sympathetic nerves were stimulated. Lewandowsky's [32] suggestion was taken up and extended by Langley (1852–1925) and his co-workers in Cambridge, and particularly by T.R. Elliott (1877–1961). Elliott was the first to discuss the possibility of humoral transmission in the sympathetic nervous system. This idea was clearly expressed in a communication read to the Physiological Society at its meeting on May 21st, 1904, where he said

"... the point at which the stimulus of the chemical excitant is received, and transformed into what may cause the change of tension of the muscle fibre, is perhaps a mechanism developed out of the muscle cell in response to its union with the synapsing sympathetic fibre, the function of which is to receive and transform the nervous impulse. Adrenalin might then be the chemical stimulant liberated on each occasion when the impulse arrives at the periphery." [33]

In 1903, Poll and Sommer [34] discovered chromaffin cells in the ganglia of a number of annelid worms. These observations were extended by Biedl and J.F. Gaskell [35, 36]. The latter showed that an extract of ganglia from the leech was pharmacologically active. In his paper we read:

"The physiological test decided upon was the inhibition of the virgin uterus of the cat. I was fortunate in obtaining the assistance of Dr. H.H. Dale in carrying out the test."

(For a more recent study of active amines in annelids the reader is referred to a paper by Rude [37].)

Commenting on his son's first paper, W.H. Gaskell [38] emphasises that not only the nerve cells but also the nerve fibres emerging from them contain the material that give the chromaffin reaction. He says:

"In the case of the medullary cells of the adrenal gland, the adrenaline is discharged from the cell into the surrounding fluid by the action of the splanchnic nerve; it is just possible that in the case of the leech the adrenaline passes from the cell to the periphery by way of the motor nerve itself. If such a suggestion proves to be true it opens out a new and most important chapter in our conception of the nature of nervous action."

This passage leaves little doubt that Gaskell had the transmitter concept in mind when he wrote these lines. He must have been aware of Elliott's earlier suggestion. Elliott did not refer to it again when he published his full paper in 1905 (see [53]). One can only surmise what the reasons were for this omission. Probably the idea had met with some criticism when he read his paper to the Physiological Society in 1904. This is similar to Dale's interpretation in the biographical memoir he wrote on J.R. Elliott [39]. He refers to Langley's unwillingness to accept speculations. Elliott made only one further explicit reference to the earlier note; that was in his Sidney Ringer Lecture of 1914. There he writes:

"The theory which I at first held to explain these facts was that the nervous stimulus consists in a liberation of adrenaline itself from the ending of the nerve on the muscle. We have seen how the ganglion cell and the adrenal cell are both derived from what is almost a common cell with power to transmit a nervous impulse or to excrete adrenalin. It is very conceivable that as the nervous cell developed its peculiar outgrowths for the purpose of transmitting and localizing the nervous impulse it might lose its power of producing adrenalin and come to depend altogether on what could be picked up from the circulating blood and stored in its nerve endings." [40]

Apart from part of the last sentence – we know today that the adrenergic neuron has the ability to produce catecholamines – this statement substantially conveys the picture that we consider as valid today. The original idea of Elliott was criticised also in the well-known paper of Barger and Dale [41], which introduced the term "sympathomimetic". Over forty years later, when the paper was reprinted, Dale [42] adds this comment:

"Doubtless I ought to have seen that noradrenaline might be the main transmitter – that Elliott's theory might be right in principle and faulty only in this detail."

I was very interested to see that in Elliott's lecture of 1914 the idea of amine storage is already discussed. This is of relevance not only in view of the much later work by J.H. Burn (1892–1981) and last discussed by him fairly recently [43], but also in regard to another piece of work carried out in Cambridge in the first decade of this century and that is of historical interest for the history of the transmitter concept: W.E. Dixon (1870–1931), Professor of Pharmacology at Cambridge, was engaged in a study of the action of the vagus nerve on the heart. Dixon and Hamill [44] discussed the possibility that the action of the vagus on the effector organ might be mediated through the release of some muscarine from the nerve endings upon stimulation of the nerve. Like Elliott's earlier suggestion, this idea did not find favour with Barger and Dale [41]. In particular, they criticised an additional proposal: Dixon and Hamill thought that if their idea was true, there might be some substance that exerted a pharmacological action not by virtue of any intrinsic muscarine-like properties, but indirectly, by displacing some of the natural "transmitter" from its storage site. Barger and Dale dismissed this idea; they found it difficult to draw the dividing line between what we would today call "directly acting" and "indirectly acting" substances. It took almost half a century until support for such an idea came from Burn and Rand [45]; they provided evidence for directly acting and indirectly acting sympathomimetic amines.

There is one additional group of early papers in which a biochemical approach was taken and in which humoral transmission was foreseen. This work I was not aware of until Annica Dahlström and David Smith both referred to it, about ten years ago [46]. It is the work of F.H. Scott [47, 48]. His approach was a cytochemical one. He made a study of the material usually called the Nissl body, present in nerve cells. He found it not only in nerve cells, but also in secretory cells such as the exocrine cells of the pancreas and the secretory cells of the fundus of the stomach. He convinced himself that the chemical properties of his material were those of a nucleoprotein and he postulated a close analogy between the function of the glandular cells and nerve cells:

> "And it seems simpler to suppose that the nerve cells secrete a substance the passage of which from the nerve endings is necessary to stimulation." [48]

It seems remarkable that Scott's prophetic words were overlooked for so long. (See [48a].)

Scott's work, I suppose, remained unknown to the group that a generation later introduced the term "neurosecretion". We owe the term to Ernst Scharrer (1905–1965) and Berta Scharrer and to W. Bargmann (1906–1978) [49–51]. These authors described the material that accumulated in certain nerve cells, in the form of granules, and these granules were secreted. The material was characterized by certain histochemical properties; it was positive with Gomori's stain. These authors believed that there were nerve cells which had two functions: one, ill-defined, they called nervous, and another was secretion. In vertebrates, they saw the formation and release of the hormones of the neural lobe of the pituitary gland as such a neurosecretory process, but they found it a common occurrence in invertebrates. Scharrer and Scharrer [49] were aware of the theory of humoral transmission but they expressly did not equate the phenomena that they had seen as nervous activity but considered neurosecretion as a thing apart. This is remarkable since, like Scott, they had clear evidence of an involvement of the Nissl body in the elaboration of the neurosecretory material [50].

It may be added that more recently the term "neurosecretion" has been used slightly differently, and more in line with the modern views of the function of the nervous system, especially by De Robertis [52] in his monograph.

8. Amine oxidases

Speculations on the biological inactivation of adrenaline began at an early date, but the literature is confusing. Elliott [53] has an often quoted passage:

"Adrenalin disappears in the tissues which it excites."

The evidence in favour of his statement was slender, but it contains a truth that we would agree with today: tissues that receive adrenergic innervation would display amine uptake in the nerves, with subsequent destruction of the amine by monoamine oxidase (MAO); thus, in this respect the statement is valid.

The early history of MAO has been told before [54]; the story begins with names famous in the history of biochemistry and pharmacology, like Oswald Schmiedeberg (1838–1921) and Oskar Minkowski (1858–1931). However, the first enzyme study we owe to Miss Hare [55], later Mrs. Bernheim [56], a Ph. D. student in Hopkins' laboratory at Cambridge. She described a tyramine oxidase in the mammalian liver; it is this enzyme that is now called MAO. She tested adrenaline as a possible substrate, but with a negative result. One suspects that autoxidation of adrenaline proved a complicating factor in her manometric experiments. Only after it became possible to exclude autoxidation, by the use of potassium cyanide, a cyanide-resistant oxidation of adrenaline (and also of noradrenaline and dopamine) was revealed [57—59]. Fortunately, the enzyme, MAO, turned out to be cyanide-insensitive.

There exist a number of reviews on MAO [54, 60–62]. Since these reviews were written, two new developments have been added. First, there is the elucidation of the prosthetic group of the oxidase as a flavin adenine dinucleotide, covalently linked to the enzyme protein through a thio-ether bond to a cysteinyl residue [63]. This work accounts for early observations by Hawkins [64], that MAO activity is reduced in riboflavin deficiency and is restored after adding riboflavin to the diet. Second, there has been the discovery of several molecular forms of the enzyme [65], each with its own pattern of substrate and inhibitor specificities (see Ciba Foundation Symposium No. 39 on Monoamine Oxidase and its Inhibition [66]).

It might be added here that the possibility of different isoenzymes was foreseen at an early stage by Alles and Heegaard [67]. In their paper they write:

> "The considerable species variance that is notable from the data of previous workers and those here presented make it questionable that amine oxidase preparations from various sources may be viewed as but a single enzyme."

A long time ago it was found [68] that a suitable way of obtaining MAO was to prepare a so-called "Körnchensuspension" as described by O. Warburg in 1913 [69]. After the 2nd World War, when modern methods became available, it was shown that MAO was a mitochondrial enzyme [70, 71]. This was a finding that soon acquired significance for the pharmacologist: if it were true that the catecholamines were broken down by an intracellular enzyme this obviously implied that the amines were able to pass into the cell's interior. Uptake and storage of catecholamines had been discussed after Elliott's statement of 1914 [40], already quoted (see, e.g. [43]). It followed that the potentiation of the response to adrenaline by a drug like cocaine was not necessarily due to the substitution of an enzyme, such as MAO, but might be due to an interference with amine uptake, thus maintaining an effective concentration of amine in the extracellular space. Since this possibility was first discussed [72], amine uptake inhibitors have been widely studied (see also [73]).

However, our knowledge of the catabolism of the catecholamines remained incomplete until it was found that two enzymes are chiefly involved: MAO is the enzyme responsible for amine breakdown after intracellular (e.g. intraneuronal) uptake; in addition there is also the enzyme catechol-*O*-methyltransferase, discovered by Axelrod in 1957 [73a], which seems to act on extraneuronal amine.

The complexity of catecholamine catabolism, intracellular and extracellular, intraneuronal and extraneuronal, explains why it took so long for the true role of MAO to be established. As late as 1949, in a review entitled *The metabolism of adrenaline,* Bacq writes:

"Its deamination by amine oxidase is unlikely." [74]

I think that the chief reason for the delay in recognising the true physiological role of MAO was that pharmacologists were tempted to look for analogues between the biological inactivation of acetylcholine by acetylcholinesterase and that of the catecholamines. There is no doubt that one principal site of the biological inactivation of acetylcholine is at the level of the cell membrane.

In our early work [58] we had already shown that MAO was distinct from the enzyme histaminase discovered by C.H. Best (1899–1978), best known as one of the discoverers of insulin. However, it is to E.A. Zeller that we owe the classification of the amine oxidases still in use at present. He was the first to distinguish between monoamine oxidase and diamine oxidase, the enzyme that acts not only on histamine but also on the short-chain aliphatic diamines, such as putrescine or cadaverine. Zeller first showed that all these substances were oxidised by an enzyme that was inhibited by carbonyl reagents; today it is assumed that this enzyme contains pyridoxal although this is not yet securely established. Another important feature, and one that distinguishes it from MAO, is that diamine oxidase is a copper-containing protein.

Diamine oxidase is only one of a number of copper-containing enzymes, all of which are chemically closely related and which subserve a variety of functions, not in every case established as yet. One of the most interesting members of this family of enzymes was discovered in recent years, following the elucidation of the chemical structure of desmosine, a compound isolated from hydrolysates of elastic tissue [75]. We now know that in the formation of both collagen and elastin an enzyme is at work that catalyses the oxidative deamination of a lysyl residue in the protein, and that this reaction is a necessary preliminary to the formation of interchain cross-links that are present in both collagen and elastic fibres. The enzyme responsible for the deamination has received the name "lysyl oxidase" (for references, see [61]). Lysyl oxidase belongs to the group of enzymes that includes diamine oxidase; it contains copper and possibly pyridoxal.

One aspect of the early work on MAO deserves comment. The functional role of the enzyme has been mainly discussed in regard to its role in the disposal of neuro-hormones, like the catecholamines or 5-hydroxytryptamine. However, at the beginning of the biochemical studies, the source of enzyme was usually the liver or some other parenchymatous tissue. There is one notable exception: Pugh and Quastel [76] made a study of the enzyme present in mammalian brain that acted upon the aliphatic monoamines. The identity of his enzyme with MAO was soon established. This was the first link between MAO and the nervous system.

Early work established that ephedrine and other derivatives of phenylisopropylamine were inhibitors of MAO [77], and it is fitting that J.H. Quastel, the discoverer of the phenomenon of competitive inhibition, should have been instrumental in proving the competitive nature of the inhibition of brain MAO by phenylisopropylamine, at the time better known as benzedrine [78].

9. Biogenesis of adrenaline

Metabolic pathways were not discussed very much when I was a medical student, but we were fortunate: we became familiar with what was known when we listened to F. Knoop over sixty years ago. I do not remember having heard of A. Garrod until I came to England. My own understanding owes much to an extensive article by O. Neubauer (1874–1957) on protein metabolism. In it he touches also on the biogenesis of adrenaline and gives some possible pathways, but these do not include the pathway that is generally accepted now as the main route of formation [79].

In 1919 Rosenmund and Dornsaft [80] synthesised 3,4-dihydroxyphenylserine, and in their paper they suggested that this compound might be a precursor of adrenaline. The very complicated pathway that they proposed immediately attracted the spirited criticism of Knoop [81] who pointed out the lack of experimental evidence in favour of their proposal. After a reply by Rosenmund and Dornsaft [82], Knoop [83] reaffirmed that he saw no reason why tyrosine might not be the ultimate precursor; he quoted examples of decarboxylation of amino acids, for instance in the formation of taurine from cystine, and that of putrescine and cadaverine from ornithine and lysine. At the time these papers were written I was at Freiburg, listening to Knoop, but I only read them many years later, when my own work involved the decarboxylation of tyrosine and cysteic acid [84, 85].

In the nineteen thirties observations on an enzymatic formation in micro-organisms of amines like histamine and tyramine began to accumulate. However, the evidence of the enzymes that formed these amines in animals remained unsatisfactory. The breakthrough came with the work of P. Holtz (1902–1970). In 1938, Holtz, Heise and Lüdtke [86] discovered an enzyme in the guinea pig kidney that decarboxylated 3,4-dihydroxyphenylalanine (DOPA). The authors proposed that enzymic decarboxylation might be the normal pathway of catabolism of the naturally occurring amino acids of the L-series. The idea was that MAO would then be the catalyst of the oxidative deamination of the amines formed in this metabolic pathway.

This suggestion was soon disproved. The observations by Holtz et al. [86] on the enzymic decarboxylation of DOPA were confirmed, but it was not possible at the time to see any decarboxylation of the other L-amino acids tested. The method used was manometry; later work using more sensitive methods has revealed the decarboxylation of other amino acids, e.g. of L-histidine or ornithine.

It was the high degree of substrate specificity of the L-DOPA decarboxylase that was so challenging. It led to the suggestion that the enzyme might be the catalyst of one of the steps that led from L-tyrosine to adrenaline [87].

In order to establish the exact sequence of metabolic steps in the formation of adrenaline it was necessary to test amino acids related to L-DOPA. I went to discuss my problem with Charles Harington (1897–1972) who had always been interested in my work. He proved co-operative and promised to make me some of the amino acids required. A few weeks later he sent me the first two compounds; these were D-DOPA and *N*-methyl-DOPA. D-DOPA was not a substrate; this was a result already foreseen by Holtz et al. [86], who had found in their manometric experiments that only one half of the theoretical amount of carbon dioxide was formed when they used DL-DOPA as substrate.

Harington's other compound proved of greater interest: *N*-methyl-DOPA was not decarboxylated. The fact that the N-methylated compound was not a substrate, seemed of particular importance, in view of certain speculations that had been put forward as to the nature of sympathin. As far as I remember, Otto Loewi never committed himself as to the exact nature of the transmitter, but the identity of his Accelerans substance with adrenaline was generally implied. This idea was questioned by W.B. Cannon (1871–1945), a pioneer in the study of humoral transmission in mammals. Eventually he summarised his evidence in a book written jointly with A. Rosenblueth [87a]. The ideas put forward in this monograph are today only of historical interest, but the summary given of the differences seen between the actions of adrenaline and the response to sympathetic stimulation is still valid, a witness to Cannon's remarkable qualities as an observer of physiological phenomena.

It was one of Cannon's earlier co-workers, Z.M. Bacq [88], who first discussed the possibility that noradrenaline might be a mediator in the autonomic nervous system. In a review article he proposed two possible explanations for the differences between adrenaline and Cannon's "Sympathin E". He writes:

"On peut s'imaginer qu'il existe deux intermédiaires adrénergiques. Là où le sympathique est excitateur, la substance M serait la noradrénaline, c'est-à-dire de l'adrénaline non méthylée."

Bacq actually favoured a second interpretation; he believed that an oxidation product of adrenaline might be the mediator. However, the suggestion quoted above was taken up by a number of pharmacologists who published observations in which the responses to sympathetic stimulation were compared with those to noradrenaline [89—91]. All these authors provided evidence in favour of the identity of sympathin with noradrenaline.

In the light of these observations the inability of the decarboxylase to act on *N*-methyl-DOPA seemed particularly interesting. If it were true that the enzyme catalysed one of the steps in the biosynthesis of adrenaline, it seemed obvious that the appearance of a primary amine was an obligatory step in this pathway. The enzymic decarboxylation of L-DOPA led to dopamine; we now know that this step actually occurs; it is followed by the formation of noradrenaline, catalysed by the enzyme dopamine-β-hydroxylase. This is the now generally accepted main pathway of catecholamine formation in the animal body [92]. The pathway has not only been confirmed by the isolation and biochemical characterization of all the enzymes involved, but also by the fact that all three catecholamines, dopamine, noradrenaline and adrenaline, have been found in chromaffin cells as well as in nervous tissue. All three amines serve as mediators.

In 1950, H. Langemann [92a], demonstrated at Oxford the presence of large amounts of DOPA decarboxylase in the bovine adrenal medulla. This finding supported the metabolic pathway first proposed before the second World War. The bovine adrenal medulla has since been the material from which all the enzymes active in the pathway have been isolated. The enzymes also occur in neuronal tissue.

In 1939, evidence in favour of the proposed pathway of adrenaline biosynthesis was still incomplete. What had been made likely by our first observations was that decarboxylation preceded *N*-methylation; however, it still remained uncertain at what stage in the pathway the introduction of the hydroxyl group in β-position of the side-chain occurred. When he gave me the first two substances, Harington also promised to make me the 3,4-dihydroxyphenylserine (DOPS) first made by Rosenmund and Dornsaft in 1919 [80], as well as its *N*-methyl derivative. He told me he was expecting a Rockefeller Fellow from Switzerland in the autumn, and making these two substances would be his first task. Alas, the expected Rockefeller Fellow, A.E. Zeller from Basel, never arrived, because by the autumn of 1939 the war had started and Harington's laboratory had been closed. Still, Harington had not forgotten, and some time later he told me that he would discuss the making of these substances with a friend, F.G. Mann, of the Chemistry Department in Cambridge. Towards the end of the war Dalgliesh and Mann [93] completed the synthesis of the two compounds. By that time I had moved from Cambridge to Oxford. The *N*-methyl-DOPS, a new compound, was not decarboxylated, as expected. As a matter of fact, in our experiments, we were also unable to demonstrate a decarboxylation of DOPS by the mammalian DOPA decarboxylase; however, we found that it was a substrate of the L-tyrosine decarboxylase of *Streptococcus faecalis* R [94]. Some months later, Dr. K.H. Beyer, of the Research Laboratories of Messrs. Sharp and Dohme, wrote to tell us that he had observed the formation of a pressor substance when DOPS was incubated with preparations of DOPA decarboxylase. Further investigation showed that DOPS was indeed decarboxylated by the mammalian enzyme, but at a very much slower rate than L-DOPA [95].

The much more rapid decarboxylation of DOPS by the bacterial decarboxylase made it possible to resolve the configuration of our sample of DOPS. Dalgliesh and Mann [93] had already noted that the sample they had prepared was a simple racemate. Measurement of the CO_2 formation during incubation of the sample showed that fifty per cent of it was decarboxylated, and bioassay using the blood pressure of the spinal cat proved that the amine formed was only the naturally occurring laevorotatory form of noradrenaline. Subsequent work has shown that the compound made by Mann and Dalgliesh had the threo-configuration [96, 97].

The work on the substrate specificity of the mammalian decarboxylase was subsequently extended, when it was shown that the presence of the catechol group was not a necessary prerequisite for decarboxylation. The presence of one hydroxyl group in position meta (or ortho) was sufficient.

The implications of these findings were discussed in an article dedicated to Otto Meyerhof on the occasion of his 65th birthday [98]. Also, the inability of the *N*-methylated amino acids to be decarboxylated became clear when it was established that DOPA decarboxylase was a pyridoxal enzyme [99]: the interaction between the carbonyl group of the enzyme and the amino group of the substrate required the presence of an unsubstituted amino group.

I should like to say a word on Pamela Holton (1923–1977), who was involved in much of the work on the substrate specificity of DOPA decarboxylase. Our joint work owes much to her enthusiasm and spirit of enterprise. It was Pamela Holton who in 1948 suggested to the late W.O. James (1900–1978) the problem of separating adrenaline and noradrenaline by paper chromatography. The method devised by James [100] was widely applied in the years that followed. That noradrenaline was in fact the main transmitter in mammalian adrenergic neurons was eventually established, chiefly through the work of U.S. von Euler [101]. The final elucidation of the pathway of adrenaline biosynthesis had to await the introduction of radioactive precursors. Gurin and Delluva [102] were the pioneers in this field; they showed incorporation of radioactive label into adrenaline from radioactive phenylalanine. Within a few years the ideas first fully formulated in 1939 had been confirmed. There then followed a period of intense work in which all the enzymes involved in the catecholamine pathway were recognised, isolated and at least partly purified. Following Langemann, the bovine adrenal medulla served as starting material in these studies. To describe all the results obtained lies outside the scope of this essay. However, it should be mentioned that there has been one important outcome of all this work: amino acids have since repeatedly been used as medicinal agents. Although intrinsically without pharmacological action, they are converted in the body, and by the body's own enzymes, into substances with pharmacological actions that are of therapeutic interest.

10. Intracellular localization of the catecholamines

The very high concentration in which adrenaline is present in the chromaffin tissue had occasionally invited comment. However, a report [103] that a substance more active than adrenaline was present in the adrenal medulla, was soon disproved [101a, 104]. These last authors confirmed that the concentration of adrenaline amounted to as much as

10–15 mg of adrenaline per gramme of fresh medullary tissue. There was a report from Feldberg's laboratory in Australia by Trethewie [105] according to which an extract of adrenal medullary tissue contained traces of adrenaline in sedimentable form; these amounts of adrenaline could be released in hypotonic media or by lysolecithin [106].

Parallel to these observations there were reports on granules in chromaffin cells [28, 107], but whether these can still be considered valid is uncertain.

Two new techniques eventually led to the discovery of the chromaffin granules. One of these was a biochemical technique, the use of centrifugation of homogenates of adrenal medulla in isotonic sucrose. This technique was used at the same time in two laboratories, at Oxford [108, 109, 110] and by N.Å. Hillarp (1916–1965) and his colleagues in Lund, Sweden, [111, 112]. Both groups showed that at least two-thirds to three-quarters of the amine present was sedimentable in these homogenates. The second method that was applied at the same time was electron microscopy [113, 114]; the anatomical findings reinforced what had been discovered by biochemical techniques.

The exact nature of the chromaffin granules remained uncertain until the introduction of a new technique: the centrifugation of homogenates over sucrose density gradients. Using this method it was possible to prove that the amine carrying particles were distinct from mitochondria: in a sucrose density gradient the chromaffin granules equilibrated in a zone of greater density than the mitochondrial marker enzymes [115]. We were also able to show that the extremely high concentration of ATP, discovered by Hillarp, was present in the amine-carrying particles. Using the same method again, it was possible a few years later to demonstrate that the chromaffin granules were distinct form the lysosomes [116, 117].

The introduction of a new method allowed many new observations to be made in quick succession: not only adrenaline but also noradrenaline and dopamine were found to be present in a sedimentable form [118]. Eade also was the first to notice that there was a difference in the sedimentation of adrenaline and noradrenaline in the density gradient: the noradrenaline-carrying granules tended to settle near the bottom of the gradient, whereas the adrenaline-carrying particles were recovered at lower sucrose concentrations (see also [119]).

Early in the studies of chromaffin granules it was found that they were rich in soluble protein [120, 121]. A more systematic characterization of the soluble protein was begun in the nineteen sixties [122—124]. The soluble protein, or at least its more or less homogeneous main constituent, has received the name chromogranin. The function of this protein is obscure.

During the work on his Oxford Ph. D. Thesis, P. Banks noted that perfusates of the adrenal gland looked more opaque when a secretagogue had been added to the perfusion fluid and pressor amine appeared in the perfusate. Subsequently he followed up this observation in Sheffield and discovered an increased release of soluble protein. In the meantime the first antibodies against chromogranin had been prepared, and in joint work with Dr. K.B. Helle it was established that stimulation of the gland by a secretagogue, e.g. carbachol, led to a release of chromogranin [125]. This discovery led to a great number of publications, too numerous to be quoted here; the reader may be referred to a Royal Society Discussion Meeting, held in 1970 and edited by Blaschko and Smith [46]. The simultaneous release of the catecholamines and the soluble proteins held in the chromaffin granules was a weighty argument in favour of the idea that release of hormone occurred by exocytosis. It is now known that also other soluble granule contents, e.g. the soluble form of dopamine-β-hydroxylase and the more recently discovered peptides related to the enkephalins, are released when the chromaffin cell sets free amine under the influence of a stimulus. Similarly, at the endings of adrenergic neurons soluble protein is released when the nerves are stimulated and release catecholamine; this strengthens the idea that a similar process of exocytosis is at work at the nerve endings.

11. On methods

The work described in the preceding section has been greatly helped by the development of new methods in the period under review. Almost my first experience at Dahlem was the introduction into manometry. I had already become familiar with the so-called "Barcroft" differential manometer when I joined Adolf Loewy in Switzerland in 1924; at that time the manometer was used in order to determine the oxygen capacity of arterial blood. At Dahlem, where we used the simple "Warburg" manometers in Meyerhof's laboratory, they were used to follow the time course of chemical reactions, usually enzymic reactions. My experience with these methods was strengthened when early in 1933, as a patient and then as a convalescent in the Medical University Clinic at Freiburg i.Br., I spent many hours in Hans Krebs's laboratory. When a few weeks later I left Germany I bought a Warburg manometer bath and manometers from the firms of Warburg's glassblowers and mechanics. This equipment accompanied me from 1933 for about twenty years. By the time I arrived in Cambridge in 1934, Krebs was already established there and had introduced Warburg manometers. Until his arrival the instrument in current use had been the differential "Barcroft" type. Some of the scientists in Cambridge, e.g. Keilin, remained faithful to his instrument, but I think on the whole most people followed the lead given by Krebs and changed to the "Warburg" type of manometer. I do not think one can necessarily claim any particular merit for this type of instrument, but it so happened that the Warburg manometer was better designed and simpler. I remember discussing the relative merits once with Keilin, and he praised the advantage of the sloping position of the Barcroft type, whereas I pointed out that in the calculation of the vessel constants the tangent of the angle of inclination had to enter as a complication. When I told Barcroft of this argument he was amused. He said that all his manometers had been upright, with one exception. That one tilted manometer was there because he used for it a manometer bath that had originally served another purpose and that had a block added onto the outside; in order to avoid that obstacle the manometer had to be tilted. As it happened, Barcroft was not in Cambridge when the draftsman came to make the drawing of the manometer for his book [126], and the technician showed him that particular instrument. Thus it became the odd one that developed into the prototype.

In the work that I did in the two decades following my move to England manometry proved extremely useful. Not only the amine oxidases but also the amino acid decarboxylases were conveniently studied by manometry and when I came to Oxford and started a course in biochemical pharmacology we also used manometry for the study of the cholinesterases and their inhibitors.

As already mentioned, I used Otto Warburg's recipe for "Körnchensuspension" in my early work on MAO. After the war, when we became interested in the study of the intracellular localisation of the enzyme, improved methods were available. These had been mainly developed by A. Claude and his colleagues at the Rockefeller Institute. Miss Hawkins used centrifugation of homogenates of rat liver as described by Hogeboom, Schneider and Palade [127]. The same method proved useful when in the autumn of 1952 Dr. A.D. Welch and I started to examine the intracellular distribution of the catecholamines. Early on in these experiments we noted that the sediment formed was not homogeneous but showed at least two distinct layers. We thereupon used what was essentially a discontinuous sucrose density gradient, but soon went over to proper sucrose density gradient centrifugation, a method that was being developed at that time.

12. Conclusions: the amine-forming cells

The early observations on MAO concerned the biological inactivation of the chemical messenger, an event that occurs after the latter has been released and exerted its effects. However, very early in the work of our laboratory interest shifted to happenings in the cell in which the messenger is made, stored and set free: biosynthesis and storage. These are functions of the amine-forming tissue, both in endocrine cells and in nerve cells. In the old days these two systems were considered as functionally distinct, but since I first began to take an interest in this field much has happened to blur the line of demarcation between the two systems. Of course, the beginnings of this development go back a long way. At first it was the embryologists and anatomists who spoke of the existence of a relationship, but beginning with the work of T.R. Elliott, fully quoted above, there came the discovery of the dual role of the catecholamines as both hormones and mediators. This relationship between neuronal and endocrine tissue is one that the catecholamines are now known to share with quite a number of other systems (see e.g. [128]).

The biochemical discoveries of the past three decades have contributed much new material that has underpinned the idea of a relationship between the two types of tissue: both chromaffin cells and catecholaminergic neurons contain the host of biosynthetic enzymes that are generally absent from other cell types.

These discoveries represent a challenge to the biochemist. What do we mean when we say: "two different cell types are related"? This question was raised some time ago [129], and it was pointed out that, when talking of a relationship between different cells, we mean the fact that the same biochemical abilities are expressed. The intriguing problem is why the cells of the two tissue types that are ontogenetically related are also biochemically related. The reason for this is still obscure. There is another relationship that has frequently been pointed out by J.H. Burn, most recently in 1977 [130]: that is the relationship between cholinergic and adrenergic neurons. Such a relationship has recently also been demonstrated in tissue culture experiments: sympathetic ganglion cells, grown in pure culture, synthesise the enzymes of the catecholamine biosynthetic pathway. However, the same cells, grown in the presence of ganglionic satellite cells, synthesise acetylcholinesterase and choline acetyltransferase, enzymes characteristic of cholinergic neurons [131]. Thus, the factors responsible for gene expression must be composite: they are determined not only by the ontogenetic history of the cell but also by influences exerted upon it by other cells present in its environment. So far the nature of these influences, ontogenic and environmental, can only be described in words. To describe their exact chemical nature, and their mode of operation, will be the task of future historians of biochemistry.

Acknowledgements

My memory has been helped by friends and colleagues, too numerous to be named. Also, the author is greatly indebted to the Librarians of our Department of Pharmacology, the Radcliffe Science Library, Oxford, the Royal Society, the Royal Society of Medicine, of University College London, the Max-Planck-Institut für Medizinische Forschung, Heidelberg, and to the Director of the Archives of the Max-Planck Society, Dr. Rolf Neuhaus, and his staff.

Thanks are due to Miss Sharon Derbe for secretarial work. Financial help with expenses incurred from the Glaxo Research Group is gratefully acknowledged.

REFERENCES

1 H. Blaschko, Annu. Rev. Pharmacol. Toxicol., 20 (1980) 1–14.
2 F. Tennstedt, Sozialreform, 25 (1979) 513–523; 600–613; 646–667.
3 A. Blaschko, Die Nervenverteilung in der Haut in ihrer Beziehung zu den Erkrankungen der Haut. Beilage zu den Verhandlungen der Deutschen Dermatologischen Gesellschaft, VII Kongress, Breslau (quoted after Jackson, 1976), 1901.
4 R. Jackson, Br. J. Dermatol., 95 (1926) 349–360.
5 A. Blaschko, Syphilis: its Dangers to the Community and the Question of State Control. 17th Int. Cong. of Medicine, London, 1913. Section XIII, pp. 37–55, Oxford University Press, London, 1913.
6 C. Oppenheimer, Grundriss der Organischen Chemie, 9th ed., Thieme, Leipzig, 1916.
7 H. Blaschko, FEBS Lett., 117 (1980) Suppl., pp. K11–K15.
8 F.G. Hopkins, Skand. Arch. Physiol 49 (1926) 33–39.
9 R. Ammon, Arzneimittelforsch., 10 (1960) 321–327.
10 H.A. Krebs and F. Lipmann, in Lipmann Symposium: Biosynthesis and Regulation in Molecular Biology, De Gruyter, Berlin, 1974, pp. 7–27.
11 H.W. Kosterlitz, Annu. Rev. Pharmacol. Toxicol 19 (1979) 1–12.
11a G. von Bunge, Lehrbuch der Organischen Chemie, 2nd ed., Barth, Leipzig, 1913.
12 O. Fürth, Lehrbuch der Physiologischen und Pathologischen Chemie, 2 Vols., Vogel, Leipzig, 1925–1927.
13 H. Blaschko, TINS, 1 (1978) IX–X.
14 H. Blaschko, Biochem. Z., 175 (1925) 68–78.
15 D. Keilin, The History of Cell Respiration and Cytochrome, Cambridge University Press, Cambridge, 1966.
16 J. Geppert, Z. Klin. Med., 15 (1889) 208–242; 307–369; (quoted after [15]).
17 D. Nachmansohn, Annu. Rev. Biochem., 41 (1972) 1–28.
18 D. Nachmansohn, German-Jewish Pioneers in Science 1900–1933, Springer, Berlin, 1980.
19 D.M. Needham, The Biochemistry of Muscular Contraction in its Historical Development, Cambridge University Press, Cambridge, 1971.
20 D. Keilin, Proc. Roy. Soc. (Lond.) B., 98 (1925) 312–339.
20a C. Oppenheimer and K.G. Stern, Biological Oxidation, Junk, The Hague, 1939.
21 R. Kuhn, D.B. Hand and M. Florkin, Z. Physiol. Chem., 201 (1931) 255–266.
22 K. Zeile and H. Hellström, Z. Physiol. Chem., 192 (1930) 171–192.
23 A.L. von Muralt and J.T. Edsall, J. Biol. Chem., 89 (1930) 351–386.
24 C. Correns, Ind. Abstammungs- und Vererbungslehre, 1928, pp. 131–168.
25 C. Sherrington, Proc. Roy. Soc. (Lond.) B., 105 (1929) 332–362.
25a E. Lundsgaard, Biochem. Z., 230 (1931) 10–18.
26 Z.M. Bacq, Chemical Transmission of Nerve Impulses. A Historical Sketch, Pergamon Press, Oxford, 1975.
27 W. Feldberg, in The Pursuit of Nature, Cambridge University Press, Cambridge, 1977, pp. 65–83.
28 Vulpian, C.R. Hebdom. Acad. Sci., Paris, 43 (1856) 663–665.
29 D. Richter and H. Blaschko, J. Chem. Soc., I (1937) 601.
30 H. Blaschko, Experientia, 13 (1957) 9–12.
31 G. Oliver and E.A. Schäfer, J. Physiol. (Lond.), 18 (1895) 231–276.

32 M. Lewandowsky, Zbl. Physiol., 12 (1898) 599–600.
33 T.R. Elliott, J. Physiol. (Lond.), 31 (1904) XX–XXI.
34 H. Poll and A. Sommer, Verhandl. Physiol. Ges., Berlin, 10 (1903) 549.
35 J.F. Gaskell, Phil. Trans. Roy. Soc. (Lond.) B., 205 (1914) 153–212.
36 J.F. Gaskell, J. Gen. Physiol., 2 (1919) 73–85.
37 S. Rude, J. Comp. Neurol., 136 (1969) 349–372.
38 W.H. Gaskell, The Involuntary Nervous System, Longmans, Green, London, 1916.
39 H.H. Dale, Biographical Memoirs of Fellows of the Royal Society, 7 (1961) 53–74.
40 T.R. Elliott, Br. Med. J., I (1914) 1393–1397.
41 G. Barger and H.H. Dale, J. Physiol. (Lond.), 51 (1910) 19–59.
42 H. Dale, Adventures in Physiology, The Wellcome Trust, London, 1953, quoted from a 1965 reprint.
43 J.H. Burn, J. Pharm. Pharmacol., 28 (1976) 342–347.
44 W.E. Dixon and P. Hamill, J. Physiol. (Lond.), 38 (1909) 314–336.
45 J.H. Burn and M.J. Rand, J. Physiol. (Lond.), 144 (1958) 314–366.
46 H. Blaschko and A.D. Smith, Phil. Trans. Roy. Soc. (Lond.) B., 261 (1971) 273–440.
47 F.H. Scott, Brain, 28 (1905) 506–526.
48 F.H. Scott, J. Physiol., 34 (1906) 145–162.
48a H. Blaschko, Notes Rec. R. Soc. (Lond.), 37 (1983) 235—247.
49 E. Scharrer and B. Scharrer, Biol. Rev., 12 (1937) 185–216.
50 E. Scharrer and B. Scharrer, Physiol. Rev., 25 (1945) 171–181.
51 W. Bargmann and E. Scharrer, Am. Scientist, 39 (1951) 255–259.
52 E. de Robertis, Histophysiology of Synapses and Neurosecretion, Pergamon, Oxford, 1964.
53 T.R. Elliott, J. Physiol. (Lond.), 32 (1905) 401–467.
54 H. Blaschko, Pharmacol. Rev., 4 (1952) 415–458.
55 M.C.L. Hare, Biochem. J., 22 (1928) 968–979.
56 M.C.L. Bernheim, J. Biol. Chem., 93 (1931) 299–309.
57 H. Blaschko, D. Richter and H. Schlossmann, J. Physiol. (Lond.), 90 (1937) 1–19.
58 H. Blaschko, D. Richter and H. Schlossmann, Biochem. J., 31 (1937) 2187–2196.
59 K. Bhagvat, H. Blaschko and D. Richter, Biochem. J., 33 (1939) 1338–1341.
60 H. Blaschko, Br. Med. Bull., 9 (1953) 146–149.
61 H. Blaschko, Rev. Physiol. Biochem. Pharmacol., 70 (1974) 84–148.
62 R. Kappeller-Adler, Amine Oxidases and Methods for Their Study, Wiley, New York, 1970.
63 E.B. Kearney, J.I. Salach, W.H. Walker, R.L. Sang, W. Kenney, E. Zeszotek and T.P. Singer, Eur. J. Biochem., 24 (1971) 321–327.
64 J. Hawkins, Biochem. J., 51 (1952) 399–404.
65 J.P. Johnston, Biochem. Pharmacol., 17 (1968) 1285–1297.
66 CIBA Foundation Symposium 39 (new series). Monoamine oxidase and its Inhibition, Elsevier, Amsterdam, 1976.
67 G.A. Alles and E.V. Heegaard, J. Biol. Chem., 147 (1943) 487–503.
68 H. Blaschko, Biological Inactivation of Adrenaline, Ph.D. Thesis, Cambridge, 1936 (unpublished).
69 O. Warburg, Pflüger's Arch. Ges. Physiol., 154 (1913) 599–617.
70 G.L. Cotzias and V.P. Dole, Proc. Soc. Exp. Biol., 78 (1951) 157–160.
71 J. Hawkins, Biochem. J., 50 (1952) 577–581.
72 H. Blaschko, Pharmacol. Rev., 6 (1954) 23–28.

73 L.L. Iversen, The Uptake and Storage of Noradrenaline in Sympathetic Nerves, Cambridge, University Press, Cambridge, 1967.
73a J. Axelrod, Science, 126 (1957) 400–401.
74 Z.M. Bacq, Pharmacol. Rev., 1 (1949) 1–26.
75 S.M. Partridge, D.F. Elsden and J. Thomas, Nature (Lond.), 197 (1963) 1297–1298.
76 C.E.M. Pugh and J.H Quastel, Biochem. J., 31 (1937) 286–291.
77 H. Blaschko, J. Physiol. (Lond.), 93 (1938) 7P.
78 P.J.G. Mann and J.H. Quastel, Biochem. J., 34 (1940) 414–431.
79 O. Neubauer, in A. Bethe, G. von Bergmann, G. Embden and A. Ellinger (Eds), Handbuch der normalen und pathologischen Physiologie, Vol. V, Springer, Berlin, 1928, pp. 670–989.
80 K.W. Rosenmund and H. Dornsaft, Ber. Dtsch. Chem. Ges., 52 (1919) 1734–1749.
81 F. Knoop, Ber. Dtsch. Chem. Ges., 52 (1919) 2266–2269.
82 K.W. Rosenmund and H. Dornsaft, Ber. Dtsch. Chem. Ges., 53 (1920) 317–318.
83 F. Knoop, Ber. Dtsch. Chem. Ges., 53 (1920) 716–718.
84 H. Blaschko, J. Physiol. (Lond.), 101 (1942) 337–349.
85 H. Blaschko, Biochem. J., 36 (1942) 571–574.
86 P. Holtz, R. Heise and K. Lüdtke, Arch. Exp. Pathol. Pharmacol., 191 (1938) 87–118.
87 H. Blaschko, J. Physiol. (Lond.), 96 (1939) 50P–51P.
87a W.B. Cannon and A. Rosenblueth, Autonomic Neuro-Effector Systems, MacMillan, New York, 1937.
88 Z.M. Bacq, Ann. Physiol., 10 (1934) 467–553.
89 R.L. Stehle and H.C. Ellsworth, J. Pharmacol. Exp. Ther., 59 (1937) 114–121.
90 K.I. Melville, J. Pharmacol. Exp. Ther., 59 (1937) 317–327.
91 C.M. Greer, J.O. Pinkston, J.H. Baxter and E.S. Brannon, J. Pharmacol. Exp. Ther., 62 (1938) 189–227.
92 H. Blaschko, Br. Med. Bull., 13 (1957) 162–165.
92a H. Langemann, Br. J. Pharmacol., 6 (1951) 318–324.
93 C.E. Dalgliesh and F.G. Mann, J. Chem. Soc., I (1947) 658–662.
94 H. Blaschko, P. Holton and G.H. Sloane Stanley, Br. J. Pharmacol., 3 (1948) 315–319.
95 K.H. Beyer, H. Blaschko, J.H. Burn and H. Langemann, Nature (Lond.), 165 (1950) 926.
96 E. Werle and J. Snell, Biochem. Z., 236 (1954) 110–122.
97 W.J. Hartman, R.S. Pogrund, W. Drill and W.G. Clark, J. Am. Chem. Soc., 77 (1955) 816–817.
98 H. Blaschko, Biochim. Biophys. Acta, 4 (1950) 130–137.
99 H. Blaschko, C.W. Carter, J.R.P. O'Brien and G.H. Sloane Stanley, J. Physiol. (Lond.), 107 (1948) 18P.
100 W.O James, Nature (Lond.), 161 (1948) 851–852.
101 U.S. von Euler, Pharmacol. Rev., 18 (1966) 29–38.
101a U.S. von Euler, J. Physiol. (Lond.), 78 (1933) 462–466.
102 S. Gurin and A.M.C. Delluva, J. Biol. Chem., 170 (1947) 545–550.
103 E. Annau, S. Huszák, J.L. Svirbely and A. Szent-Györgyi, J. Physiol. (Lond.), 76 (1932) 181–186.
104 H.O. Schild, J. Physiol. (Lond.), 79 (1933) 455–469.
105 E.R. Trethewie, Austral. J. Exp. Biol. Med. Sci., 16 (1938) 225–232.
106 W. Feldberg, J. Physiol. (Lond.), 99 (1940) 104–118.
107 W. Cramer, J. Physiol. (Lond.), 52 (1918) VIII–X.

108 H. Blaschko and A.D. Welch, Naunyn-Schmiedeberg's Arch. Exp. Pathol. Pharmakol., 219 (1953) 17–22.
109 H. Blaschko, P. Hagen and A.D. Welch, J. Physiol. (Lond.), 129 (1955) 27–49.
110 H. Blaschko, Adv. Biochem. Psychopharmacol., 25 (1980) 251–253.
111 N.-Å. Hillarp, S. Lagerstedt and B. Nilson, Acta Physiol Scand., 28 (1953) 257–263.
112 N.-Å. Hillarp and B. Nilson, Acta Physiol. Scand., 31 (1954) Suppl. 113, 79–107.
113 J.D. Lever, Endocrinology, 57 (1955) 621–635.
114 F.S. Sjöstrand and R. Wetzstein, Experientia, 12 (1956) 196–199.
115 H. Blaschko, J.M. Hagen and P. Hagen, J. Physiol. (Lond.), 139 (1957) 316–322.
116 A.D. Smith and H. Winkler, J. Physiol. (Lond.), 183 (1966) 179–188.
117 A.D. Smith, in P.N. Campbell (Ed.), The Interaction of Drugs and Subcellular Components in Animal Cells, Churchill, London, 1968, pp. 239–292.
118 N.R. Eade, J. Physiol., 132 (1956) 53P–54P.
119 H.J. Schümann, J. Physiol. (Lond.), 137 (1957) 318–326.
120 N.-Å. Hillarp, Acta Physiol. Scand., 43 (1958) 82–96.
121 H. Blaschko, G.V.R. Born, A. D'Iorio and N.R. Eade, J. Physiol. (Lond.), 133 (1956) 548–557.
122 K. Helle, Mol. Pharmacol., 2 (1966) 298–310.
123 A.G. Kirshner and N. Kirshner, Biochim. Biophys. Acta, 181 (1969) 219–225.
124 A.D. Smith and H. Winkler, Biochem. J., 103 (1966) 483–492.
125 P. Banks and K. Helle, Biochem. J., 97 (1965) 40C–41C.
126 J. Barcroft, The Respiratory Function of the Blood, Cambridge University Press, Cambridge, 1914, p. 291, Fig. 136.
127 G.H. Hogeboom, W.C. Schneider and G.E. Palade, J. Biol. Chem., 172 (1948) 619–635.
128 A.G.E. Pearse and T.T. Takor, Clin. Endocrinol., 5 (1976) Suppl. 2295–2445.
129 H. Blaschko, in G.A. Cottrell and P.N.R. Usherwood (Eds.), Synapses, Blackie, Glasgow, 1977, pp. 102–116.
130 J.H. Burn, Clin. Exp. Pharmacol. Physiol., 4 (1977) 59–100.
131 P.H. Patterson, L.F. Reichardt and L.L.Y. Chun, Cold Spring Harbor Symp., 40 (1976) 389–397.
132 W. Trendelenburg, Z. Psychol. Physiol. Sinnesorg., 37 (1904) 1–55.

G. Semenza (Ed.) Selected Topics in the History of Biochemistry: Personal Recollections (Comprehensive Biochemistry Vol. 35)

Chapter 8

The Svedberg and Arne Tiselius The early development of modern protein chemistry at Uppsala

KAI O. PEDERSEN

Institute of Physical Chemistry, University of Uppsala, Uppsala (Sweden)

Today the soluble proteins are considered as well defined macromolecules. The student of today takes it for granted that the proteins have a definite chemical and structural composition and a well defined molecular weight.

In the twenties, when I was studying chemistry at the University of Copenhagen, the situation was quite different. The soluble proteins were generally regarded as lyophilic colloids which formed micelles of varying size. Very little was known about the detailed chemical composition and structure of the proteins. From the work of Emil Fischer (1852–1919) and of Franz Hofmeister (1850–1922) at the beginning of this century it was known that the proteins contained various amino acids held together by peptide linkages. When dissolved, they gave colloidal solutions indicating a fairly high particle weight, at least several thousand; perhaps even higher than 10 000.

The colloidal particles were not considered as true chemical compounds, but rather as more or less random aggregates. The colloid chemists therefore thought that it was a waste of time to study the properties of the proteins as if these substances were well defined, genuine, chemical individuals. Even if the colloid point of view was quite general during the first decades of this century, it was never universal, and there were always some protein chemists who looked at the proteins in very much the same way as we do today [1].

At that time many different proteins had been prepared mainly by salting out or by other fractional precipitation methods. As, however, the starting materials usually contained a mixture of different proteins together with other chemical substances, the composition of the protein salted out varied with the experimental conditions. In order to improve the purity of the protein, it was often reprecipitated a few times to remove foreign material and in the hope of obtaining a pure protein. However, the great problem was that there hardly existed any criteria of purity for proteins. They were characterized mainly by their origin, solubility, optical rotation, and elementary composition. In a few cases, an approximate amino acid composition had been determined after many months' work.

Some proteins, e.g. haemoglobins and seed globulins could be crystallized and formed well defined crystals, the shape of which varied with the origin of the protein even among similar proteins from closely related species. At the beginning of the century, some American scientists had shown that for the specialist it was even possible to distinguish one species of a certain genus from another species of the same genus by comparing their haemoglobin crystals [2]. These results should have indicated strongly that the haemoglobins formed very well defined molecules. But the colloid chemists and most of the protein chemists at that time did not seem to have noticed these findings, or they have neglected them.

During the first World War, S.P.L. Sørensen and his collaborators at the Carlsberg Laboratory in Copenhagen studied the osmotic pressure of various solutions of egg albumin. They came to the conclusion that the most likely mean particle weight for this protein would be about 34 000. This result was published in 1917 [3]. Thirty two years later, Güntelberg and Linderstrøm-Lang [4], from the same laboratory, made a critical study of the old experimental data, and they found that a molecular weight of about 45 000 would fit the old data much better than the 34 000 chosen in 1917. At that time 34 000 was undoubtedly considered a very high mean particle weight, making it difficult for Sørensen and his collaborators to trust the still higher values indicated by a great number of their experiments.

Nowadays we know that proteins are macromolecules and that they are very well defined as to structure, composition, and size. They belong to the most complicated organic molecules. They are built from hundreds of α-amino acids linked together through the so-called peptide bonds into one or several long chains (primary structure). These chains are folded in a characteristic way, the α-helix, and kept in position by intra-chain bridges (secondary structure). The chains are packed in a special way by other intra-chain bridges (tertiary structure). These units may further be bound together to still larger units by means of inter-chain bridges.

Today we know the exact amino acid composition for many of the proteins; it may actually be determined overnight by means of the fully automated amino acid analyser. Likewise the sequence of the amino acids in the peptide chain is known for a great number of proteins. The X-ray crystallographers have succeeded also in determining the actual structure and conformation for many proteins. A number of developments in experimental techniques during the last sixty years have together made this progress possible. Some of the early developments were carried out at Uppsala by The Svedberg (1884–1971) and by Arne Tiselius (1902–1971), the basic thing being that these scientists introduced new separation techniques which made it possible to characterize and isolate the different proteins in a new way.

Svedberg introduced and developed the ultracentrifuge and demonstrated that the proteins formed well defined molecules in solution. Tiselius improved and developed the electrophoresis technique so that it could be used to characterize the proteins electrochemically, as well as for preparative purposes. He made important contributions to the theory and practice of chromatography.

The Svedberg

The (Theodor) Svedberg was born 30 August 1884 at Fleräng in the parish of Valbo near Gävle, Sweden. He was the only child of Elias Svedberg and Augusta Alstermark. His father was manager at different iron works in Sweden and Norway, and the family lived at various places in Scandinavia. Since his boyhood Svedberg had been deeply interested in chemistry, physics, and botany. He spent the years 1900–1903 at a well known grammar school in Örebro. Here he had some understanding teachers who allowed him to study by himself in the physical and chemical laboratories of the school in the afternoons after the ordinary lessons.

In January 1904 Svedberg matriculated at the University of Uppsala. Before he left home, he had been quite uncertain as to whether he would study biology, especially botany, which was his great interest all through his life, or chemistry, his other great interest. Finally he decided to study chemistry as he expected that many problems in biology would sooner or later get their explanation as chemical phenomena.

Svedberg studied intensely, and in record time he passed the necessary courses and examinations and got his Fil.kand (B.Sc.) in September 1905. He now felt ready to start doing research himself. In his spare time he had read the 1903 edition of Nernst's *Theoretische Chemie*, and had been

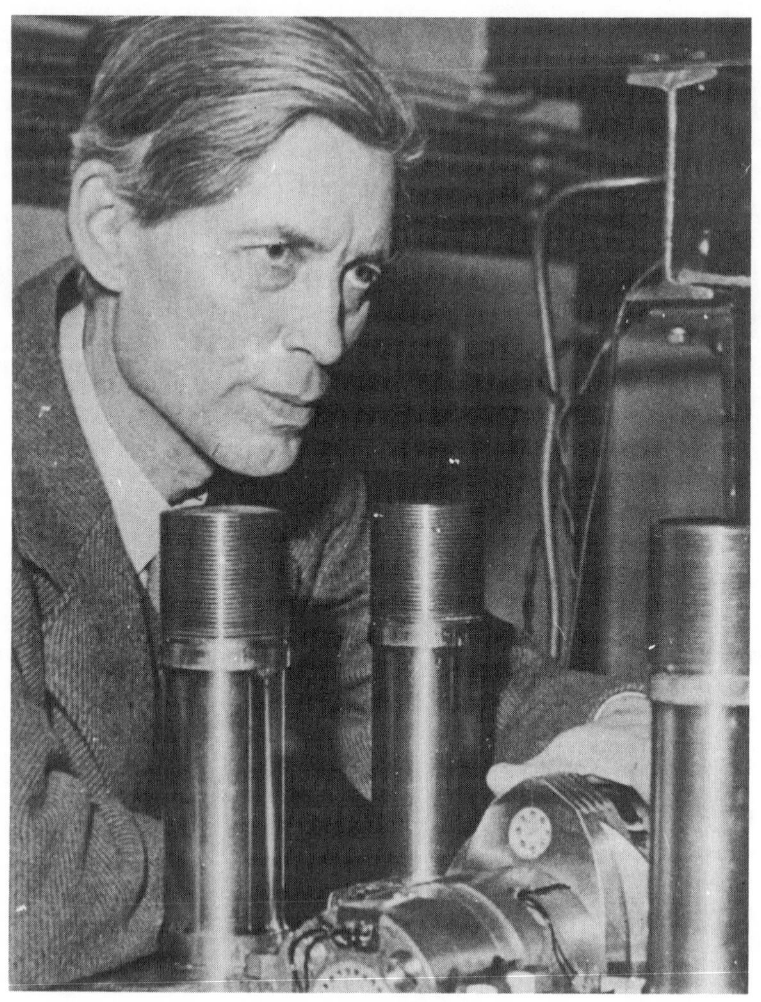

Plate 9. The Svedberg.

especially interested in the part dealing with colloids. In one of Uppsala's bookshops he found Zsigmondy's *Zur Erkenntnis der Kolloide.* After having read this book and Bredig's *Anorganische Fermente* he started to prepare organosols of various metals by a modification of Bredig's method. His first publication came as early as December 1905 [5]. In the following month he studied the preparation process in some detail, and in a second publication [6] he described organosols from more than 30 different metals, some of them very stable, others stable only for one day or less. Now he knew how to prepare a number of stable metal sols in a reproducible way, and he could start quantitative studies on the physicochemical properties of these metal sols, especially with regard to particle size, one of his main interests for the following fifteen years.

In 1904 the Uppsala chemists moved into a newly built chemical institute. However, the experimental facilities were very limited and primitive. Some rooms had been reserved for physical chemistry although the equipment was missing and had to be constructed and built by the students themselves. Together with Carl Benedicks (1875–1958), who was a new "docent" (assistant professor) in physical chemistry, Svedberg succeeded in obtaining the necessary parts to build a Zsigmondy-Siedentopf ultramicroscope. With this instrument Svedberg could study the Brownian movements of the particles in the metal sols. He was able to interpret his experimental results by the new theories of Einstein (1879–1955) [7,8] and of von Smoluchowski (1872–1917) [9].

In December 1907 Svedberg defended his thesis *Studien zur Lehre von den kolloiden Lösungen* for the degree of doctor of philosophy [10]. A few weeks later he became "docent" in physical chemistry at the university, obtained better laboratory facilities and could have his own students. He was now ready to start an extended study of various colloidal systems; he even got time to publish some books. In a monograph he described the different methods for preparing colloidal solutions of inorganic substances [11]. On the basis of his own experiments he published another monograph; *Die Existenz der Moleküle* [12]; he even found time to write a more popular book on the development of the concept of matter from ancient time up to the present time [13].

It was quite evident for the two chemistry professors at the faculty, Strömholm and Widman, that Svedberg had such an intellectual capacity that he had to be given a permanent position at the university. They therefore started an action for a personal chair for Svedberg. The proposal was strongly supported by Svante Arrhenius who gave his opinion on the matter.

On 29 June 1912 Svedberg got a royal appointment as Professor of

Physical Chemistry at Uppsala University, the first professorship in physical chemistry in Sweden. His inauguration lecture took place on 1 February 1913 and had the title: *Atomic research in modern physics and chemistry*.

In the following decennium Svedberg's work was mainly concentrated on the study of the physicochemical properties of colloidal systems. Many of the studies of his students resulted in dissertations on colloidal problems (for more details see [14]).

For a very long time Svedberg had been interested in particle sizes and particle size distribution in colloidal solutions. Toward the end of the decenium this interest had especially centered around the formation and the growth of the newly formed particles after the electrical synthesis of metal sols.

The particle size distribution was determined by means of sedimentation measurements in the gravitational field [15]; Herman Rinde (1889–1949) had developed a method to measure light absorption at different heights in a small sedimenting system. By determining the variation with time of the concentration at different heights in a small sedimenting column the particle size distribution could be calculated. For the larger particles in the metal sols the method worked fine, whereas for the smaller particles the sedimentation was much too slow to be measured accurately. It was evident that some kind of a centrifugal method had to be introduced, if the smallest particles should be studied.

In July 1922 Svedberg made a first draft of a system for following the sedimentation in a centrifuge. It was based on ultramicroscopic observation during centrifugation by means of darkfield illumination using a cardioid condenser. The idea remained on paper and so did other proposals drafted during the fall of 1922 (see Fig. 1 in [16]).

At this time colloid science had become quite popular, and at the Department of Chemistry, University of Wisconsin, they planned to start a division for colloid chemistry. For this reason they contacted one of the most prominent colloid scientists in Europe, viz. Svedberg. He was invited to come to Madison, WI, and act as a guest professor for the spring and summer terms of 1923. Svedberg accepted the invitation with enthusiasm and started to make plans for the research work to be carried out at Madison: sedimentation in centrifugal fields, electrophoresis, diffusion, electrical colloid synthesis, etc. He should lecture on colloid science and organize research in colloid chemistry.

On his way to Madison in January 1923 Svedberg presented his and Rinde's paper at a symposium in Rochester, NY [15]. After the lecture he asked one of his young listeners, Burton Nichols (b. 1902) what he thought about the problem, and his answer was: "I would like to tackle that problem!" And so he did later on at Madison together with Svedberg.

The optical centrifuge

In the spring of 1923 they constructed a centrifuge where the sedimentation could be followed optically while the centrifuge was running [17]. In this centrifuge they studied the sedimentation of various metal sols and some other colloidal solutions and compared the results with ultramicroscopic measurements on the same solutions. The experiments had been promising. As, however, the centrifuge had cylindrical and not sector-shaped cells, the particles were partly carried down by convection along the walls of the cell.

There are indications that Svedberg already at this time was interested in particle size determinations on proteins. In a lecture at the First Colloid Symposium in June 1923 at Madison he said:

> "It is possible to build up a method for recording distribution curves from observations of diffusion."

and a little later:

> "Experiments of this kind are planned in my laboratory. They are of importance because we are here dealing with one of the few possible means for studying the distribution of sizes in protein sols." [18]

As a colloid chemist he was convinced that the proteins were lyophilic colloids and that the particles in protein sols varied in size.

On the boat trip across the Atlantic he made sketches of rotors for future centrifuges. An essential thing was that the cells should be sector-shaped.

Back at Uppsala he and Rinde worked during the autumn on the construction of a new centrifuge. A sheet of paper with a rough sketch of a centrifuge, dated 23 November 1923, is marked "centrifugal analysis of proteins". However, their first aim was to construct a centrifuge that could be used for sedimentation studies on metal sols.

The first ultracentrifuge

The name ultra-centrifugen (the ultracentrifuge) appears for the first time in Svedberg's laboratory records for 15 February 1924. At the beginning Svedberg had a number of difficulties with convections that disturbed the regular sedimentation. By letting the rotor spin in a hydrogen atmosphere the convections disappeared. Before the summer holidays they could send a manuscript to the *Journal of the American Chemical Society: The ultracentrifuge, a new instrument for the determination of size and distribution of size of particle in amicroscopic colloids* [19].

With this first ultracentrifuge they could sediment the smallest particles in the gold sols, and Svedberg's dream was that it should also be possible to sediment smaller particles such as proteins. According to the textbooks in biochemistry, egg albumin should have an average molecular weight of 34 000. By means of a light absorption method using ultraviolet light it should be possible to follow any protein sedimentation. Svedberg was anxious to test this, and Nichols came over from Madison to study in Svedberg's laboratory. The first experiments were disappointing, no sedimentation of the egg albumin could be observed.

At about this time Svedberg got acquainted with Robin Fåhraeus (1888–1968), Mrs. Svedberg and Mrs. Fåhraeus being old friends. Fåhraeus was Assistant Professor of Pathology at the Caroline Institute (the Medical School of Stockholm). His dissertation on *The suspension stability of the blood* led to the introduction of the well known method of measuring the sedimentation rate of the red corpuscles. He was profoundly interested in proteins, and in the fall of 1924 he came to Uppsala to study some months with Svedberg.

After the unsuccessful experiments with egg albumin Svedberg and Fåhraeus started to run bovine milk in the ultracentrifuge on 12 September 1924. The light absorption method showed that the casein sedimented. From exposures taken at different times during the run they could calculate the particle size distribution. From a series of experiments they came to the conclusion that the casein particles had a very broad frequency distribution with coarse particles of the order of 10 to 70 nm diameter (unpublished experiments, see [20]). This was just what Svedberg had expected of a protein.

Svedberg has told that after these successful experiments Fåhraeus asked:

"Why not try to centrifuge a solution of haemoglobin?"

Svedberg:

"Haemoglobin should only have a molecular weight of about 17 000 according to its iron content, and egg albumin with the double molecular weight did not sediment in the ultracentrifuge."

Fåhraeus:

"I see, but one could in any case try."

And so they did! They started to prepare CO haemoglobin from horse blood. As soon as the preparation was ready late in the evening of 16 October 1924, the experiment was started. As, however, Svedberg did not expect to observe anything before the next day, he went home to bed and let Fåhraeus watch the centrifuge. In the middle of the night he was awakened by a telephone call from Fåhraeus who shouted:

"The, I see a dawn."

Svedberg rushed to the laboratory where he found a marked lightening of the red colour at the top of the cell. The haemoglobin had actually started to sediment. Shortly afterwards a crack developed at the bottom of the cell, and the solution leaked out!

Some weeks later the cell had been repaired and in the middle of November some successful sedimentation equilibrium experiments were made. It was found that the molecular weight was four times 16 700 and that it was constant from the meniscus to the bottom of the cell. Svedberg had here for the first time a monodisperse protein. Could it be that some proteins were monodisperse and not polydisperse as he had assumed? Svedberg was excited!

However, the sedimentation equilibrium method is not very sensitive to small differences in particle sizes. A more sensitive method would be the sedimentation velocity method. In this the shape of the sedimenting boundary in a dilute monodisperse solution would be determined solely by the diffusion coefficient of the sedimenting substance. A larger spreading of the boundary than that corresponding to the diffusion of the substance, or to the appearance of separate boundaries, would indicate inhomogeneity of the material. From the results obtained with haemoglobin it was evident that, in order to introduce the sedimentation velocity method, it would be necessary to dispose of a centrifugal field of 70 000 $\times$ g to 100 000 $\times$ g, corresponding to a 15–20-fold increase of the centrifugal field available in the first ultracentrifuge. An entirely new type of ultracentrifuge had to be constructed, and a number of new problems, concerning technique and safety, had to be discussed and solved.

The high-speed oil-turbine ultracentrifuge

Svedberg approached Mr. F. Ljungström (1875–1964) of the Ljungström Steamturbine Co., Stockholm, who proposed the use of oil turbines for driving the rotor; this would simplify the lubrication problem.

Before any detailed planning and construction of the new centrifuge could be started, Svedberg had to find the necessary money for financing the project. Money for research was very scarce in Sweden at that time, and it was almost impossible to finance such a comparatively big project. However, Fåhraeus advised him to send an application to a new private foundation for medical research. Svedberg applied for 25 000 Swedish crowns, a huge sum at that time. In spite of much opposition he got this grant as well as another one from the chemistry section of the Nobel Foundation.

Shortly afterwards Svedberg met one of his colleagues who had heard about the grants and said:

"How do you dare accept such a huge grant for a risky project? Just imagine that you fail!"

"I am not going to fail,"

was Svedbergs answer.

Early in 1925 the work on the new ultracentrifuge could be started. F. Ljungström had made further valuable suggestions and let one of his young men, A. Lysholm (1893–1973), assist Svedberg in the construction and testing of the ultracentrifuge, which was built at the workshop of the Ljungström Steamturbine Co., Stockholm, during the spring and summer of 1925. Auxiliary equipment, such as pumps and measuring instruments, were obtained partly from Sweden, partly from abroad. The installation of the new ultracentrifuge at Uppsala started in the fall, but not until January 1926 could the first test run be started. It was very disappointing; instead of the 40 000 to 42 000 rev./min aimed at, only 19 000 rev./min were reached. Many unexpected technical difficulties had to be solved. The turbines were drastically modified, the bearings reconstructed, the oil system improved. Vibrations made the building of a special balancing machine necessary. Heat convection current in the solution during sedimentation together with the very marked gas friction around the rotor at the high speed made it essential to let the rotor spin in a hydrogen atmosphere at low pressure (10 to 20 mm Hg). The introduction of these various modifications resulted in a gradual increase in the speed obtained.

The turbines presented a special problem, but this difficulty was eventually overcome, partly due to the skill and interest of the first instrument mechanician at the laboratory, Mr. Ivar Eriksson (1888–1967). He never believed in the design of the oil inlets and the turbine blades according to the engineering drawings, and he started to experiment on his own. One morning he told Svedberg about his new model (already finished and ready to be put into the centrifuge). Against the reluctant opinion of Svedberg he got it tested. As soon as the new parts were installed, the speed went up markedly, and finally, on 7 April 1926, a rotor speed of 40 000 rev./min was achieved, corresponding to a centrifugal field of about 100 000 $\times$ g in the cell. It was evident that one ought to follow Eriksson's ideas.

In the early 1920s chemical laboratories at the universities in Sweden rarely had mechanical workshops, nor mechanically skilled personnel; even smaller construction work usually had to be given to workshops in the city. In the work on the ultracentrifuge, however, it soon became necessary to have more easy access to mechanically skilled persons in addition to workshop facilities at the laboratory. Ivar Eriksson, who held a position at the workshop of a bicycle factory at Uppsala, used to help Svedberg in his spare hours. However, one day early in 1924, Svedberg was astonished to meet him at the laboratory already in the morning. Eriksson then told Svedberg that he had left his job at the factory to come and work full time on the centrifuge. He thought that Svedberg needed his help. When Svedberg, who had not been notified in advance, worried about the salary, Eriksson calmly replied that such matters would probably take care of themselves, which luckily turned out to be true.

Besides being occupied with the planning and installation of the new high-speed ultracentrifuge, Svedberg was also engaged in the more theoretical aspects of sedimentation velocity and sedimentation equilibrium methods. In the *Zsigmondy Festschrift* [21] he first published the well-known Svedberg formula $M=(RTs)/[D(1\text{-}V\rho)]$ for calculating the molecular weight by combining sedimentation and diffusion measurements. He also showed how the charge would influence the sedimentation of a colloidal electrolyte, and how the concentration distribution in the cell would vary in a polydisperse system.

In the first years, all measurements of concentration in the rotating cell were based upon light absorption. As substances like the carbohydrates, however, do not have sufficient light absorption, there was a need for another method of measuring concentration or concentration gradients in the rotating cell. Among Svedberg's notes from this time there are paper sheets dated 14–16 December 1925, where he refers to O. Wiener's paper [22] and calculates the deflection which a light pencil will undergo when passing through a concentration gradient in a centrifuge cell. He also indicated the equation for calculating the molecular weight from the deflection for certain distances from the axis of rotation at equilibrium.

The problem seems to have been put aside for some time until Tiselius proposed to photograph a scale through the rotating cell which made the experimental application of Wiener's method much more convenient. The whole problem of the introduction of the scale method and its theoretical background was handed over to Ole Lamm (1902–1964) when he came to work with Svedberg in 1927 (see [23–24a] and [25] pp. 253–273).

The year before Svedberg and Fåhraeus had sent their paper on the molecular weight of haemoglobin to the *Journal of the American Chemical Society,* G.S. Adair (1896–1979) at Cambridge had found an average molecular weight of 66 700 for haemoglobin from ten different species [26] by using a refined osmotic technique. Svedberg and Fåhraeus did not become aware of this work until some time after their paper had been published.

In the summer of 1926 Adair visited Uppsala to meet and discuss with Svedberg and to see his ultracentrifuges. On his way back from Uppsala Adair stopped for some days in Copenhagen. From there he sent a long letter to Svedberg. He starts by saying:

"I would like to thank you very much for all the trouble you took in showing me your method, the pleasure of seeing it is an ample reward for the long journey and the rough crossing of the North Sea!"

Then he evidently continues a discussion they have had at Uppsala on the "Donnan effect" and on the possibility of eliminating "acids and bases from a protein preparation so completely that the osmotic pressure against distilled water equalled that of the protein particles themselves". Later in the letter he writes:

"In the matter of future work, there are a good many points in which the ultracentrifuge would be of the greatest value in clearing up, points which the membrane work leaves doubtful, and it is possible that great advances could be made by carrying out experiments by both methods on the same protein preparations."

In the spring of 1939 Adair came to Uppsala to take up this problem. However, the work was discontinued when he had to leave Uppsala some weeks after the outbreak of World War II.

While Adair was in Copenhagen, Tiselius met him at the Carlsberg Laboratory. In a letter to Svedberg of 18 August 1926 Tiselius wrote among other things:

"At the Carlsberg Laboratory I met Linderstrøm-Lang, a very interesting gentleman who with the greatest amiability let me interview him about everything. I also met Adair who was passing through. Both these gentlemen expressed severe doubt whether the gas laws were valid for hydrophile colloids in the presence of electrolytes. They therefore did not have adequate respect for the possibilities of our ultracentrifuges. I strongly emphasized that in any case we could determine the thermodynamic properties of colloids from sedimentation equilibrium experiments. This was a thing that evidently had not appeared to them."

These were problems in which Tiselius was vividly interested at this time. In June and July 1926 he had been outlining the theory for the law of mass action in a centrifugal field. He had sent several letters to Svedberg about this matter, and he had even been to Svedberg's summer cottage at the west coast of Sweden to discuss this matter. In the fall he submitted a now classical paper on the calculation of thermodynamical properties of colloidal solutions from measurements in the ultracentrifuge [26a].

After it had been possible, in April 1926, to run the new high-speed ultracentrifuge at the speed aimed at, viz. 40 000 rev./min, a number of tests and adjustments had to be carried out. It was important to minimize the risk of getting convections in the centrifuge cell. Svedberg therefore began a systematic study of how the convection currents started and how they developed. Thermoelements were placed at various places in the casing to make it possible to follow the temperature variation near the rotor and at the bearings. Resistance thermometers were inserted in the pipes carrying the turbine oil and the lubricating oil for the bearings, as well as into the oil coolers and the pipes for the cooling water. The sedimentation was followed by studying the sedimentation of dilute solutions of haemoglobin. Svedberg has often emphasized how important it was that all the early experiments were carried out with coloured solutions, first with the gold sols in the first low-speed ultracentrifuge and later with coloured proteins in the high-speed ultracentrifuge. In this way convections could be discovered as soon as they started, and the development could be followed in detail and could be referred to the temperature distribution in the system and the cooling of the turbine oil. The effect of a change in temperature of the turbine oil could immediately be observed in an otherwise worthless experiment.

In June 1926 Svedberg made some melting-point observations with azoxybenzene in the cell in order to see if the cell temperature would be much different from the temperature on the rotor casing just outside the periphery of the rotor. He came to the conclusion that the cell temperature was 1.5°C higher than the inside temperature of the casing.

In September 1926 the ultracentrifuge was finally ready to be used for routine runs, and Svedberg could attack the main question about the uniformity of the protein molecules. Together with Nichols he started a new series of experiments on CO-haemoglobin with the sedimentation velocity method. The experimentally obtained curves were compared with the theoretical curves for the combined sedimentation and diffusion. In his laboratory record for an experiment made 17 September 1926 Svedberg wrote:

"Not the slightest indication of the presence of differently sized molecules could be noticed."

The uniformity of the CO-haemoglobin had thus been confirmed by the new sedimentation velocity method, and an important new criterion of protein purity had been introduced.

Even before this important confirmation of the uniformity of the CO-haemoglobin was clear, Svedberg had decided to concentrate completely on the study of a possible monodispersity of the proteins. This is quite evident from correspondence he had with some prospective foreign students during the summer of 1926. In a letter to Eugen Chirnoaga (b. 1891) he wrote:

"...Before you finally decide upon coming to us, I think you need some information about the kind of work we are doing in my lab just now. During the last two years we have specialized on the colloid chemistry of the proteins, and due to the success which we have had, we decided to use all the resources of the lab in the service of this study. At the present time, therefore, all the facilities of the lab are organized for the study of the proteins, and in case you want to come to us, you would have to join in this big work. As you will see from the enclosed reprint, our chief work, the study of the proteins by means of centrifuging, is rather interesting, and I think that a year spent in such work would not be wasted."

Several foreign students came to Uppsala to study with Svedberg, and they were all set to study various proteins, mainly chromoproteins, as their sedimentation in the ultracentrifuge so easily could be followed with the light absorption method.

In December 1926 three prominent colloid scientists received Nobel Prizes. Richard Zsigmondy got the postponed 1925 year's prize in chemistry "for his elucidation of the heterogeneous nature of colloid solutions and for the methods he had devised in this connection, which have since become of fundamental importance in modern colloid chemistry". Svedberg was awarded 1926 year's Prize in chemistry for his work on disperse systems. The physics Prize for 1926 went to Jean Perrin for similar work with special reference to his discovery of the sedimentation equilibrium.

For Svedberg's part this was in some way concluding his long period as an active colloid scientist. From now on, and for about the following 15–20 years, his interest and energy were centered on the proteins and the development of the ultracentrifuge.

The Nobel Prize was a great stimulation for Svedberg and made it much easier for him to obtain funds for his research activities. After his return from Madison in 1923 he had tried in vain to get an institute for physical chemistry at the University of Uppsala. Now the government realized the need for such an institute. This building became ready in 1931.

Svedberg's molecular weight hypothesis

Some of the chromoproteins studied in the centrifuge sedimented faster than CO-haemoglobin and were also found to be monodisperse. According to Svedberg's own remarks, the greatest sensation was the discovery of the giant haemocyanin molecules from the snail *Helix pomatia* in 1927. From the copper content of this blue respiratory protein a minimum particle weight of 15 000–17 000 could be calculated, which would mean slow sedimentation and large diffusion. Instead, it sedimented very fast with a knifesharp boundary. An estimate of the particle size showed that it had to be in the millions, and all the molecules had to have the same size. This was the first time such uniformly sized molecules had been observed. Later on it could also be demonstrated that this haemocyanin showed a constant molecular weight in buffer solutions over a wide pH range.

In the following years other proteins were studied in the ultracentrifuge. Most unexpected to Svedberg, as well as to most other chemists, all the proteins first studied were found to be monodisperse. In a few cases the proteins were found to be paucidisperse, that is, two or more distinctly different size classes were present. By suitable fractionation or sometimes simply by changing of the pH, such solutions could yield monodisperse proteins. In only one case, besides casein, a real polydisperse protein was found, namely gelatin, which at that time was often used as a "model protein". Nowadays one would hardly consider gelatin as a typical soluble protein.

After some years' study of the proteins with the ultracentrifuge, it appeared to Svedberg that the molecular weights for most proteins were multiples of that for egg albumin (M_r approx. 35 000) and this idea was put forward in some of his papers from that time. Thus in the summary of a paper with N.B. Lewis (b. 1902) on the molecular weight of phycoerythrin and phycocyan [27] the authors say:

"It was pointed out that the molecular weight of phycoerythrin and of phycocyan, as well as that of haemoglobin, approximately were multiples of that of egg albumin."

This idea is further developed in a paper together with A.J. Stamm (b. 1897) [28], where it is said:

"The molecular weight of normal edestin is practically the same as that of phycoerythrin. The specific sedimentation velocity and diffusion constant are likewise the same within the experimental error. Calculation of the radius of the particle by applying both Stoke's law and Einstein's law shows that the edestin molecules are practically spherical as in the case with phycoerythrin.

This research gives considerable further evidence in favour of the senior author's theory that the proteins all have molecular weights that are integral multiples of the molecular weights of egg albumin, namely 34500. Normal edestin has a molecular weight very close to six times this value. Everyone of the proteins thus far studied in this laboratory has a molecular weight of either 1, 2, 3 or 6 times that of egg albumin with the exception of the two haemocyanins which are such large multiples that it is impossible to determine whether they are integral multiples.

The nature of the dissociation products of edestin at pH 11.3, obtained by the new method of Lamm, gives perhaps the best single piece of evidence obtained to date of the validity of the theory. Not only are the molecular weight of the dissociation products even multiples, but they are the same multiples as other known proteins and have the same specific sedimentation velocities and diffusion constants."

Svedberg's ideas about the multiple system were further developed in a letter to *Nature (London)* published 8 June 1929 where he says:

"Our work has been rewarded by the discovery of a most unexpected and striking general relationship between the mass of the molecules of different proteins and the mass of the same protein at different acidities, as well as of a relationship concerning the size and shape of the protein molecules. [29]

It has been found that all stable native proteins so far studied can, with regard to molecular mass, be divided into two large groups: the haemocyanins with molecular weights of the order of millions, and all other proteins with molecular weights from about 35 000 to about 210 000."

Later on in the same letter to *Nature* Svedberg writes:

"When looking for an explanation of these unexpected regularities, it would be well to bear in mind the fact already brought out by so many biochemical experiences, namely that Nature in the production of organic substance within the living cell seems to work only along a very limited number of main lines. The great variety appears in the specialization of details. Thus it would seem that the numerous proteins are built up according to some general plan which secures for them only a very limited number of different molecular masses and sizes when present in aqueous solution. By varying the constituents of the different proteins (different percentage of different amino acids etc.) the chemical and electrochemical properties may be varied sufficiently to enable the cells to make use of them for their different purposes."

After having put forward the hypothesis of the multiple system for the proteins in 1929, Svedberg was of course anxious to test its validity. He wanted to study as many different proteins as possible. He now had an assistant, Bertil Sjögren (1905–1973), to help him in the protein work. In the following years Sjögren prepared several different types of proteins and studied them in the ultracentrifuge. Here may be mentioned Bence Jones protein, amandin, excelsin, serum albumin, and serum globulin, legumin, lactalbumin, cocosin, and insulin. Most of these proteins could be fitted into Svedberg's multiple system. After a few years Sjögren left Uppsala to get a position in one of the leading pharmaceutical companies in Sweden; eventually he became their director of research.

The first real polydisperse protein was found at this time when K. Krishnamurti from India studied the sedimentation behaviour of gelatin [30].

Svedberg thought that the amino acid composition of a protein from a certain species could vary to some extent, even for proteins from the same individual at different times, depending upon the availability of the various amino acids. He thought that the molecular weights should be regarded as average values. This may be a reflection from Svedberg's days as a young docent when he and Strömholm made an investigation on isomorphic coprecipitation of radioactive compounds. They were probably the first to attempt to fit a part of the disintegration series into the Periodic Table. They were also the first who indicated that what we now call isotopes is not limited to the radioactive elements, but is likely to be found all through the Periodic Table. This was several years before the conception of isotopes was introduced into chemistry. Maybe Svedberg thought that there existed some kind of a Periodic System for the proteins also.

At the beginning, quite a number of scientists would not believe that the proteins were so well defined in regard to size; they were by most scientists regarded as lyophilic colloids. Therefore, when Svedberg put forward his hypothesis about the molecular weights of the proteins as being distrubited over a limited number of weight classes, where the molecular weights of the higher classes could be expressed as simple multiples of these for the lower classes, scepticism became even more prominent.

I can illustrate this scepticism with some personal experiences from that time: before I left Copenhagen for Uppsala, in January 1930, many chemists expressed grave doubt about the findings of Svedberg, and some were wondering if what Svedberg and his coworkers were measuring could be artifacts of some kind, perhaps due to the ultracentrifugation method.

When, after my graduation in Copenhagen in 1929, I told Professor Brønsted that I should like to go to Uppsala to study with The Svedberg, he was rather sceptical. He admitted that Svedberg was an excellent experimental physical chemist and a very charming man, but his ultracentrifuge could hardly be of any great importance. Brønsted thought that it would be much better for me to go either to Germany to study with Herbert Freundlich at Berlin-Dahlem, or to England to study with Rideal at Cambridge or with Hinshelwood at Oxford. Furthermore, at those places I could learn a useful foreign language! However, I had become so fascinated by Svedberg's work and by his papers and books, that I went to Uppsala in January 1930 to study there for a couple of years – and I have stayed there ever since.

Before I went to Uppsala, I visited the grand old man in protein chemistry, Professor S.P.L. Sørensen at the Carlsberg Laboratory in Copenhagen. He expressed his doubt about Svedberg's multiple hypothesis. He did not expect it to be of a general nature, nor did he believe that proteins were substances so well defined in regard to mass and size as claimed by Svedberg. One can understand Sørensen, as at that time he was studying the solubility behaviour of serum globulin. He assumed that most proteins formed reversible, dissociable systems with no uniform molecular weight.

Arne Tiselius

Arne Wilhelm Kaurin Tiselius was born in Stockholm, 10 August 1902, the son of Hans Abraham J:son Tiselius and Rosa Kaurin, a daughter of the rector of a mountain parish in the center of Norway.

Most of Tiselius' ancestors on both sides were scholars and many had shown great interest in science, especially biology. His father had taken a degree in mathematics at Uppsala University.

His father died as early as in 1906, and his mother moved with Arne and his sister to Gothenburg where his grandparents lived and where the family had close friends.

Arne Tiselius' profound interest in science was already awakened in grammar school in Gothenburg, where he had an inspiring teacher of chemistry and biology who discovered Arne's ability in chemistry. He gave him a private key to the laboratory of the school to allow him to carry out his own chemical studies in the afternoon after the ordinary lessons. Gradually it became clear to him that he wanted to study at Uppsala with The Svedberg. In September 1921 he entered that university with which he remained associated for the rest of his life. In May 1924 he received the M.A. in chemistry, physics, and mathematics.

Svedberg had for several years been interested in studying the electrophoresis of proteins. The first publications were from 1923 and 1924. In the summer of 1925 Tiselius came to Svedberg as a research assistant, and after a year they published a paper on the use of the light absorption method for following the electrophoresis of proteins [31].

Svedberg was at this time so heavily engaged in the development of his high-speed oil-turbine ultracentrifuge and its use in the study of the proteins, that he turned the electrophoresis problems entirely over to Tiselius who was given a very free hand, or as said by Tiselius in a kind of autobiography [32]:

"...Electrophoresis appeared so much simpler from a technical standpoint, suitable for a young man to play around with more or less on his own. However, the excellent instrument workshop greatly facilitated the testing of new ideas."

Tiselius now made a very careful study of the various sources of errors in the electrophoresis technique and how to control them. The limiting current loading, consistent with convection-free migration, was established from experiments with alternating current. Silver – silverchloride reversible electrodes replaced the previous arrangements. The requisite volume of the electrode vessels was calculated from theoretical consideration. The optical and photographic conditions were refined, a concentration scale for light absorption was established with the aid of microphotometric registration, and a theoretical justification adduced for utilizing the 50% concentration point as defining the authentic boundary position. Various types of boundary anomalies were discussed. Finally the electrophoretic behaviour of several well defined proteins, such as egg albumin, serum albumin and phycoerythrin was examined. In mixtures of such uniform proteins the components were shown to migrate indepen-

Plate 10. Arne Tiselius presenting his Ph.D. thesis.

dently of one another. Preparations of salt-fractionated horse serum globulin displayed inhomogeneous migration, but no definite inflexion points were detectable on the photographic registration curves.

In 1930 Tiselius submitted a thesis for the degree of doctor of science at Uppsala University. In this he described the results of his comprehensive theoretical and experimental studies of the moving boundary method for investigating the electrophoresis of proteins [33]. It has since been the classical work on electrophoresis for many years.

As a consequence of the excellence of his thesis, Tiselius was appointed "docent", but commented himself:

"Although it was very well received by the Faculty and by Svedberg himself and led to my appointment as docent, I remember very vividly that I felt disappointed. The method was an improvement, no doubt, but it led me just to the point where I could see indications of very interesting results without being able to prove anything definite.... I decided to take up an entirely different problem, but a scar was left in my mind which some years later would prove to be significant." [32]

The years before his thesis, Tiselius read some biochemistry which at that time was not included in the chemistry curriculum at Uppsala, and he became fascinated by the variability and especially the specificity of biochemical substances. The fact that protein preparations, which appeared uniform in the ultracentrifuge, did not necessarily behave uniformly in electrophoresis, gradually convinced him that *definition*, *separation*, and *purification* were problems fundamental to the whole of biochemistry, demanding the application of diverse techniques in view of the wide range of substances involved. In his afore-mentioned autobiography he writes:

"Not only ultracentrifugation, not only electrophoresis, but other methods as well, would have to be explored, preferably those which depend on physicochemical phenomena, as these are likely to be more gentle... I remember speculating much about further development of chromatographic and adsorption methods..."

During his later research activities he made major contributions just into these fields.

Although Tiselius had been appointed "docent", his salary remained that of an assistant in Physical Chemistry for a further two years, until a "docent" fellowship in chemistry became vacant. Such departmental fellowships were held initially for three years and were usually prolonged for another three years with a possible seventh year's extension. In very exceptional cases a special research fellowship for another six years could

Plate 11. The Svedberg and Arne Tiselius in the old chemical laboratory (about 1926).

be awarded, but there were very few of these, and *all* faculties competed for them. The university had few permanent academic appointments, and at the expiry of their fellowships "docents" usually were forced to seek a living outside the university. At Uppsala University there was one chair in organic, one in general and inorganic chemistry, and in addition Svedberg's personal chair in physical chemistry. The situation at other Swedish universities was similar. Because of retirement the chair in general and inorganic chemistry was due to become vacant in 1936, but no other chair in chemistry in Sweden was likely to be vacated for a considerable time. As he had married in 1930, he felt it his duty to try to obtain a chair in chemistry and decided to direct his research interest into a field that would favour his candidature. Tiselius therefore left the electrophoresis problems for several years.

Through his extended reading Tiselius had learned about the unique capacity of certain zeolite minerals to exchange their water of crystallization for other substances, the crystal structure remaining intact even after the removal in vacuo of the water of crystallization. It was known that the optical properties changed, when the dry crystals were rehydrated, but until then no quantitative study of the phenomenon had been made. Tiselius saw the possibilities of this accidental observation, found the governing factors, and developed a very elegant and accurate optical method for the quantitative measurement of the diffusion of water vapour and other gases into zeolite crystals. The later part of the work was carried out at the Frick Chemical Laboratory at Princeton University in 1934–35 while Tiselius held a Rockefeller Foundation fellowship for study under Hugh S. Taylor.

Before leaving for Princeton, Tiselius admitted that, if he were completely independent, he would concentrate on biochemistry, where he thought that some of the most intriguing problems in modern science still awaited their solution. However, even if he could not concentrate on biochemistry during his stay in the USA, it proved to be a very stimulating year which decisively influenced his career. The atmosphere at the Frick Laboratory was inspiring, but of even greater importance for Tiselius' later work was his frequent contact with research carried on by the Rockefeller Institute at its laboratories at Princeton and in New York. This contact led to friendships with John Northrop, W.M. Stanley and M.L. Anson, and gave the opportunity of meeting K. Landsteiner, Michael Heidelberger, and L. Michaelis. From discussion with these scientists it became clear to Tiselius that to solve some of their problems they needed some new methods that had been in his mind for years, but which he had been unable to realize. Encouraged by the discussions with these friends, he was again convinced that the development of new and more efficient separation methods was a key problem in biochemistry, and he decided to concentrate on this problem. While still in the States, he began to work on a total reconstruction of the electrophoresis apparatus.

Further development of the high-speed ultracentrifuge

Svedberg was not quite content with his first high-speed oil-turbine ultracentrifuge, and soon after having got the Nobel Prize he started new discussions with Lysholm about improvements in the ultracentrifuge machinery. He wondered if he could get rid of some of the troubles with the oil system by introducing air-turbines instead of oil-turbines (see Fig.6 in [16]). He was also much interested in increasing the centrifugal field.

Lysholm thought they should try with some of the new types of steel that had become available, and they should also make an oval-shaped rotor with a larger diameter. However, before he could start with any calculations of a new rotor, he wanted to have more information about the gas friction around the spinning rotor and also about the bearings of the old ultracentrifuge.

In March 1928 Svedberg wrote to Lysholm to tell that he had got new funds for further development of the ultracentrifuge. In the meantime he had come to the conclusion that they should stick to the oil-turbine and not try the air-turbine drive. They should also stick to the cylindrical rotor with a diameter of 15 cm. By using the newer types of steel it should be possible to spin the rotor at a higher speed. Before getting started with the calculations of the new rotor, Lysholm became technical chief of the Ljungström Steamturbine Co., Stockholm, and he delegated the further work with the centrifuge to one of his younger engineers, Gustaf Boestad (b. 1899). Now a period of very intense collaboration between Svedberg and Boestad started. The brought new ideas and proposed effective improvements in the bearings of the old ultracentrifuge; the old rotor could then easily be run at 45 000 rev./min, and 48 000 rev./min was even realized once, instead of hitherto just a little over 40 000 rev./min.

At about this time the Swedish Parliament had granted funds for building an Institute of Physical Chemistry at the University of Uppsala. Svedberg could now start planning laboratories specially adapted for the different types of ultracentrifuges (high-speed oil-turbine centrifuges and low-speed electrically driven centrifuges for sedimentation equilibrium experiments). The experiences gained with the first ultracentrifuges should be utilized for a complete reconstruction of the whole equipment. Before this could be done, it was important to know what size the rotor should have for the high-speed equipment.

Boestad's first attempt showed that a cylindrical rotor could not stand any speed higher than 55 000 rev./min. He then tried Lysholm's proposal of an oval-shaped rotor, this too with a diameter of 15 cm. His calculations showed that such a rotor should be able to withstand a speed of up to 60 000 rev./min. With the equipment available at Uppsala it would probably be possible to reach a maximum of 55 000 rev./min. Boestad and Lysholm recommended that Svedberg try such an oval rotor in the old apparatus. After much reluctance Svedberg agreed. Finally, at the end of July 1929, the first oval rotor could be tested in the old equipment. In the first test run 55 000 rev./min. was reached corresponding to a centrifugal field of 154 000 $\times$ g in the center of the cell. In the following runs Svedberg had a number of difficulties with convections in the cell and vibrations of the rotor. All the time he had been very sceptical about this rotor type and had much argument with Boestad about the matter. Finally, at the end of October 1929 the oval rotor was put aside for a little more than two years.

Svedberg now thought it safer to make the calculations for a cylindrical rotor for the new ultracentrifuge, even if the centrifugal field should not be so strong as he had hoped. As a secondary matter calculations could of course be done for an oval rotor as well. At all events they had to build a new and convenient ultracentrifuge, more easy to handle than the old one.

At the beginning of May 1930 the first parts of equipment for the new ultracentrifuge laboratory were ready and could be installed in the basement of the new institute under construction. One month later the new rotor with 180 mm diameter – called Rotor I – was ready; however, it was not until January 1931 that an oil-pump, constructed especially for the new ultracentrifuge, was ready to be tested. Two months later the first test run could be made with Rotor I in the new ultracentrifuge laboratory. A number of difficulties turned up: it was not possible to get Rotor I up to the desired speed of 55 000 rev./min, the acceleration time was too long, and the rotor got too warm. The bearings and the turbines had to be modified. Finally, when a new driving oil with much lower viscosity had been introduced in September, a speed of 55 000 rev./min was attained (200 000 $\times$ g). More test runs and a few routine runs were made during the fall of 1931; but on November 4 Rotor I exploded at a speed of 50 000 rev./min, probably due to fatigue after having been run on a total of 120 times at varying speeds up to 56 000 rev./min.

Based on the experiences gained from Rotor I, Boestad introduced some modifications on a new rotor, and exactly three months after the explosion of Rotor I the new rotor could be tested. Rotor II worked satisfactorily and could almost from the beginning be used for routine runs ([25] p. 120), [34]).

Shortly before Rotor II could be tested, we discovered a tiny crack at the periphery of the old cylindrical rotor, just outside one of the cell holes, and this old rotor from 1926 had to be taken out of use immediately. Fortunately we had the old oval rotor which was put aside in the fall of 1929; it could now easily be run in the old ultracentrifuge equipment after a few minor adjustments. More information could also be gained about an oval rotor. It was successfully used for five months until the entire old ultracentrifuge with all its equipment was dismounted during the summer of 1932.

After the success with the old oval rotor, Svedberg agreed that Boestad started to calculate an oval rotor for the planned new second ultracentrifuge at our laboratory. The new Rotor III was ready to be tested at the beginning of April 1932. As, however, it could only be tested in the same ultracentrifuge equipment as that used for routine runs with Rotor II, and there was such a demand for "centrifuge time", the testing of Rotor III was postponed until the end of May 1932. It was then tested up to a maximum of 70 000 rev./min and could afterwards be run routinely up to 63 000 rev./min. From now on no cylindrical rotor was run in the high-speed ultracentrifuges and all new rotors were oval.

The old ultracentrifuge laboratory was completely reconstructed after the summer. The ultracentrifuge and all the machinery was now separated by a heavy wall from the control room with all the measuring instruments. This was a great advantage in the following years when so many new rotors were tested and many exploded.

In an attempt to develop the ultracentrifuges as far as possible, a number of different rotors were designed and tested in the following years. Very little was at that time known about the behaviour of metals at such high rotational speeds; it was therefore important to test the different types of steel as well as different rotor designs. Several of the rotors exploded, some due to faults in the steel, others to improper design.

More detailed accounts of the developmental work on the ultracentrifuge during the thirties have been given in [25] pp. 114–156, and by Pedersen [16].

Until 1937 all the new rotors were tested by Svedberg who personally directed the testing in every detail. The testing of the rotors put a heavy stress on Svedberg, and the years 1932–1937 were probably the most hectic ones in his life. After several explosions and after great difficulties with vibrations in 1934–1935, Svedberg was sometimes about to give up; it seemed so hopeless to find a satisfactory construction. He wondered if it was really worthwhile to do further work on the improvement of his ultracentrifuges, or if it might not be better to concentrate on other problems. His interest in the proteins and his anxiety to prove, or disprove, his hypothesis of the multiple system for their molecular weights made him continue. After his retirement he described the thirties as the happiest period in his scientific life.

Further protein studies

In the early summer of 1931 Svedberg moved over to the new Institute of Physical Chemistry where he got much more laboratory space. He also got more grants-in-aid, enabling him to appoint assistants and technicians. In the autumn of 1931 he got Inga-Britta Eriksson (1909–1962) as his private assistant; she had been studying chemistry for some years at the university. She was first engaged in a study of a number of different phycocyans and phycoerythrins prepared from materials collected by Svedberg at various places in Sweden and at the Stazione Zoologica at Naples, Italy.

Svedberg got a substantial grant for special equipment to his new institute from The Rockefeller Foundation. From January 1932 he started to get large yearly contributions for assitance etc. from the same foundation. At this time I (b. 1901) became research assistant at the institute. I had come to Uppsala two years earlier.

At the new institute the protein studies were intensified and we got more technicians to help with the running of the centrifuges, the photometer registration of the photographic plates from experiments with the light absorption method, and later on the measurements of the scales from exposures taken with the Lamm scale method. Before that time we had to do all that work ourselves. We also had to do all the routine calculations of the experimental results. A few of us had been fortunate to have our wives to assist with parts of these jobs.

Most of the work during the first period was concentrated on the respiratory proteins. An advantage of this was that these proteins usually are coloured and are the only distinctly coloured substances in the blood of the animals, which meant that they could often be run in the centrifuge without being isolated from the blood. This was just diluted with 1% NaCl to a suitable concentration before being run in the centrifuge. The sedimentation could be followed visually and with the more simple light absorption method. Svedberg also considered it a great advantage to study the proteins under conditions as nearly native as possible and without first having been precipitated with salts [35].

Svedberg wanted to get proteins from as many different groups of animals as possible. To assist him in the selection, the classification and the bleeding of the animals he got Astrid Hedenius (b. 1905) as assistant and secretary in the autumn of 1932; she had taken a B.Sc. in zoology.

Much of the material for the protein studies was obtained on trips to the woods and lakes in the surroundings of Uppsala or further away. All kinds of snails, worms, snakes, crayfishes, and tiny creatures were collected and brought to the laboratory. In the summer Svedberg gathered material near his cottage at the west coast of Sweden as well as at the nearby Marine Zoological Station at Kristineberg. Some material, considered important in the systematic study, could not be found in Sweden and was collected by Svedberg on his journeys abroad. Thus, in the autumn of 1931, he obtained blood from *Octopus vulgaris* when he visited Naples, Italy. In 1933 he got various samples of haemocyanins etc. from the Oceanographic Institution at Woods Hole, MA, while being in the USA.

Many of the proteins studied were from closely related species and had very similar sedimentation coefficients. Were they identical? (1). Were the values for s_{20} identical? (2). Were the chemical compositions of the proteins from related species identical? The first question was studied by means of the so-called "mixture test". The second question was determined by electrophoresis experiments.

In the mixture test two proteins were mixed in about equal proportions and centrifuged. If only one boundary appeared in the sedimentation diagram, the two s_{20} were taken to be identical [36].

In the electrophoresis experiments the mobility of the protein was measured at several pHs on both sides of the isoelectric point (IP) and the slope of the mobility-pH curve as well as the IP was determined [37].

During the years 1932–34 about one hundred different respiratory proteins were collected and examined at Uppsala. Their sedimentation coefficients were found to belong to a small number of groups, whereas their electrophoretic behaviour varied from species to species even between closely related ones, which means that they were chemically different ([37] and [25], UC Tables 40, 42, 44, and 48). Referring to the small number of s_{20} found, Svedberg says:

"It is obvious from these regularities that biological kinship is usually accompanied by identity in the sedimentation constant. On the other hand, the number of different sedimentation constants observed is so small that the same constant must with necessity occur in more than one animal class and in respiratory proteins containing different active groups. It seems that only a few molecular masses are stable, and it would depend upon the composition of the molecules with regard to various amino acids whether one or the other possibility is realized. The constancy of the molecular weight within a certain animal group would then be a measure of the similarity of certain chemical processes leading to the formation of the respiratory protein." [36]

Later in the same paper Svedberg says:

"One gets the impression that the protein molecules are built up by successive aggregation of definite units. It seems that the higher the molecular weight, the fewer are the possibilities of stable aggregation. The steps between the molecular weights, therefore, become larger and larger as the weight increases."

Svedberg also points to the fact that for example the IPs of the blood pigments of the invertebrates are all very low compared with the IP of haemoglobins, the respiratory proteins of the vertebrates. He says:

"The situation at the isoelectric point is therefore to a certain degree a measure of the kinship."

He summarizes the results as follows:

"The sedimentation constant and the molecular weight may be used as a group characteristic, the isoelectric points as a species characteristic."

Serum and serum proteins

In May 1930 Svedberg received a letter from Herbert Freundlich in Berlin asking if it would be possible to send one of his students, Paul von Mutzenbecher (1904–1940), to Uppsala to study some serum proteins prepared by electrodialysis of serum according to a method by G. Ettisch. At the "Kaiser Wilhelm Institut", Berlin-Dahlem, where the method had been developed, they believed it to be a very mild fractionation method and therefore thought that the proteins obtained were native and had a uniform molecular weight. They now wanted to have these proteins studied in the ultracentrifuge.

Svedberg answered that von Mutzenbecher was welcome and that he himself would be interested in the problem, as he and Sjögren [38] had already studied albumin and globulin prepared by salt fractionation of horse serum. Mutzenbecher came to Uppsala in the autumn of 1930, and he made a number of preparations according to Ettisch's method and found that the pseudoglobulin as well as the euglobulin obtained was polydisperse, just as Svedberg and Sjögren had found earlier [39].

During von Mutzenbecher's stay at Uppsala Svedberg discussed the possibility that he could return to Uppsala and study native serum, when the new high-speed ultracentrifuge was ready for routine runs, hopefully at 200 000 $\times$ g.

Mutzenbecher was granted a Rockefeller fellowship and came to Uppsala in October 1931. Unfortunately Rotor I in the new ultracentrifuge exploded on November 4, 1931, and Mutzenbecher had to start his experiments with serum in the old centrifuge. The first observations, made on undiluted horse serum with the light absorption method, showed one boundary only. Even after the introduction of the Lamm Scale method, the sedimenting boundary could not be resolved; the centrifugal field in the old centrifuge was too weak (100 000 $\times$ g).

Three months after the explosion of Rotor I the ultracentrifuge had been repaired and partly reconstructed. A new cylindrical Rotor II had been designed and tested successfully. At the beginning of March 1932 it was ready for routine runs. Now, with a centrifugal field more than twice as intense as in the old centrifuge, the sedimenting boundary could be resolved, when the scale method was used.

The problem of the independent existence of the albumin and the globulin in serum turned up to be much more complicated than anticipated. As long as the centrifuged solution was dilute, the concentration of the albumin component calculated from the sedimentation diagram corresponded fairly well with the value from chemical analysis. On the other hand, in the more concentrated solutions or in undiluted sera the concentration calculated from the sedimentation diagram was too high for the albumin. The effect has been explained later by Johnston and Ogston [40] as a "boundary anomaly" caused by changes of the concentration of the slower component due to difference in its rate of sedimentation in the presence or absence of the faster component on the two sides of the boundary (see also [41]).

In the case of human serum the situation was still more complicated than for the horse serum. Besides the apparent stronger increase in the concentration of the slower sedimenting albumin after increase in the total serum concentration, a "new component", sedimenting slower than the albumin, was observed in the sedimentation diagram from the more concentrated serum solutions. Mutzenbecher assumed this component to be some kind of a split product from the serum proteins. It was later shown by Pedersen [42, 43] to be caused by a low-density β-lipoprotein present in most human sera.

At the end of his stay in Uppsala, Mutzenbecher made some runs on dilute mixtures of serum albumin and globulin. The sedimentation diagrams from these runs were completely like those obtained from dilute solutions of the corresponding sera. Before Mutzenbecher left Uppsala in March 1933, he and Svedberg sent a note to *Naturwissenschaften* [44] where they stated that albumin and globulin in serum existed as free molecules and not combined to some kind of a serum molecule.

While Mutzenbecher was still working at Uppsala, Svedberg got a letter from A.S. McFarlane (1905–1978) asking for permission to come and work in Svedberg's laboratory for a period of six months to one year, supposing he got a fellowship he had applied for. McFarlane was in charge of the biochemical laboratories of the Glasgow Royal Cancer Hospital, and his main interest was in chemical pathology. He had been studying particle sizes of proteins and viruses by means of graded filtration through collodium membranes; now he wished to acquire a first-hand experience of the ultracentrifuge technique for measuring the size of protein molecules in the hope of applying it at a later date to the problems presented by pathogenic viruses.

Svedberg answered that he should be very glad to receive him at his laboratory and added that the problems mentioned in the letter were of great interest and were in line with the work going on at the laboratory. No doubt that Svedberg was anxious that Mutzenbecher's studies should be continued and his bold conclusions further tested.

After his arrival at Uppsala in September 1933, McFarlane made preparations to start directly on a study of normal sera. From Mutzenbecher's work it was quite clear that the sedimentation studies on serum had to be carried out by means of a refraction method and not by the light absorption method. On the other hand the scale method with a fixed scale, as used by Mutzenbecher, was too inflexible; an improvement in the method of observation was therefore a most urgent task. Hitherto it had been necessary to select a certain scale distance before the run was started and keep that scale distance fixed during the entire run.

The difficulties were overcome when Lamm proposed the use of a scale-projecting system whereby it became possible to change the scale distance during the run and at any moment select the most suitable one. It even became possible to project the scale image very close to the middle of the rotating cell so that even undiluted serum could be studied.

McFarlane made a very careful and comprehensive study of the problems connected with ultracentrifugal studies on normal sera and on a few pathological sera.

After his sojourn at Uppsala, McFarlane should move to London where he should continue his studies as a Beit Memorial Fellow at the Lister Institute of Preventive Medicine, where plans had been made to set up a department for biophysics. He had planned to start with electrophoretic studies on serum proteins as a supplement to his ultracentrifugal studies on these proteins. It was his dream, however, some day to work with an ultracentrifuge in England.

In the spring of 1934 Svedberg had mentioned the problem to some of the directors from The Rockefeller Foundation at one of their visits to Uppsala. There had also been some preliminary talks between professor Ledingham, the director of the Lister Institute, and the Rockefeller people.

When McFarlane in June, on his way back to England, visited Svedberg at his summer cottage at the west coast of Sweden, Svedberg was rather optimistic regarding a British ultracentrifuge, while McFarlane was afraid that it was just wishful thinking.

After McFarlane's first day at the Lister Institute he wrote to Svedberg and told him that there was a very great interest there for the prospect of getting an ultracentrifuge. He anticipated:

"That if they got the centrifuge, he would be asked to perform miracles with it!"

Some weeks later Ledingham wrote to Svedberg and told about the great interest the Lister Institute had in getting an ultracentrifuge. After discussions with the directors of the Rockefeller Foundation he hoped it should be possible to get one.

In his answer to Ledingham Svedberg wrote among other things:

"...It is with great satisfaction that I hear of the keen interest that you take in the project of installing an ultracentrifuge in your institute... Personally it would be very gratifying to know that the ultracentrifugal method on which I have been working now for about ten years were to become of real service to the medical science. It has always been my dream to be able to do something towards the development of methods for medical investigations."

In a letter to Svedberg in November Ledingham emphasized the importance of being able to demonstrate that the biological activity may be removed from the supernatant fluid after centrifugation of certain active solutions at high speed and for prolonged periods. Other types of semi-quantitative experiments would also be important for many of the scientists.

Svedberg got quite upset by this letter and answered Ledingham:

"...It is not easy to answer the questions in your letter of November 12th. The cells which are used in the ultracentrifuge at present are the results of extended series of trials. One could hardly tell offhand what difficulties might arise from cells of the type you indicate. On the other hand, using our accumulated experience, it does not seem unlikely to me that cells of the types you want could eventually be worked out.

I should like to point out, however, that my primary interest is to supply methods for accurate quantitative studies of high molecular compounds such as the proteins, and that I have always been a little suspicious of such qualitative and semi-quantitative methods as you suggest in your letter. For instance in cells holding large quantities of liquid convection currents might arise, and unless the process of sedimentation can be followed by optical measurements, you are entirely in the dark about what happens in the cell during centrifuging. All-metal cells, therefore, are rather hazardous unless you know from observations with similar, but transparent, cells that regular sedimentation actually does take place.

The types of ultracentrifuges which we have worked out here are not meant for the separation "in substance" but for optical observations. It has taken us many years of hard work to arrive at the present types, and it may very well take much time and labour to modify the machines for the purpose you indicate. We have often found that even a slight change may cause much unforeseen trouble.

All I can say, therefore, is this: I can offer to supply you with ultracentrifuges of the best types which we have worked out so far, but we cannot just now promise to undertake the experiments necessary for the modifications you indicate. On the other hand, if McFarlane wishes to come here for some time to work along these lines, I am quite willing to put our experiences and the resources of our work-shops at his disposal. For the present, though, I think it would be wise to concentrate upon installing in your institute machines of the types we have worked out and the function of which we know and to regard the modifications you suggest as possibilities for the future."

The separation cell

Nowadays it is taken for granted that enzymes and antibodies are proteins; fifty years ago the situation was quite different and at the beginning of the thirties no decisive proof had yet been made to show that for example enzymes and antibodies were proteins; many thought that they were some kind of an "agens", perhaps of rather low molecular weight, and that the accompanying protein just acted as a "carrier" for the active agens.

It was therefore natural that the scientists at the Lister Institute were anxious to be able to take samples from the cell after the conclusion of the run in order to test them for biological or chemical activity. The same was the case for some of their colleagues at Uppsala. However, we felt that the ultracentrifuge cells were not yet sufficiently developed to be adequate for this special purpose. So even if we felt that it was urgently needed, we had to try to be patient.

When, in a purification of an enzyme for example, one finds that the enzymatic activity increases in the same way as the relative concentration of one of the components in the sample increases, one cannot say definitely that this component is the enzyme, although it is likely to be so. Thus in this example the rate of sedimentation of the enzyme can only be determined by measuring the enzymatic activity in different well defined sections of the cell after the conclusion of the run and provided no mixing of the contents in the different sections has taken place. From the changes in activity in the different sections the rate of sedimentation may be computed and compared with the rate of sedimentation calculated from optical measurements of the sedimentation in the same run.

Tiselius and Pedersen had many discussions on these problems which are vital when studying biological active material in the ultracentrifuge, as well as in electrophoresis experiments. The problem in the ultracentrifuge is how to prevent the mixing of the material in the different sections when stopping the rotor and pipetting out the samples. It was not until 1937 that this problem was solved by the construction of a special separation cell [25] (pp. 152, 301); [45]. Here the content of the cell is divided into two parts separated by a thin wall of Bakelite with a large number of fine holes, supporting a piece of hardened filterpaper. The sieve plate is made in one piece with the Bakelite center-part of the cell. This arrangement effectively prevents mixing for a time sufficient to remove the cell from the rotor and to pipette out first the content of the upper compartment and then rinse it quantitatively; secondly, mix the content in the lower compartment and pipette it out.

The arrangement of such a highly permeable membrane does not obstruct free sedimentation, as Pedersen found in a series of experiments with different concentrations of haemoglobin in the cell. As the cell had the usual rock-crystal windows, the whole process could be followed visually from the start of the experiment until the solutions had been taken out of the two compartments. The effectivity of this arrangement is probably due to local convection at the membrane rapidly equalizing any small difference in density which would tend to occur as a consequence of an accumulation of material just above, or dilution just below the membrane. One must remember that density differences are multiplied enormously by the centrifugal force.

Ultracentrifuges to go abroad

In the autumn of 1934 we had had a lot of trouble with vibrations of the centrifuge rotors. It was then discovered that too little attention had been paid to the dynamic balancing of the rotors and cells, especially after the introduction of the oval rotors. After reconstruction of our balancing machine and insertion of a new type of bearings, the vibrations could be eliminated.

Plate 12. The Svedberg lecturing (about 1935).

The absence of any promise from Svedberg that it should soon be possible to measure the sedimentation of biological activity does not seem to have had any negative influence on the willingness of the Rockefeller Foundation to make a grant to the Lister Institute for ultracentrifuge equipment. In the middle of January 1935 it became quite clear that this grant would be made.

In a letter in January 1935 Svedberg wrote to McFarlane after his congratulations to "the centrifuge station master":

> "I presume you know that Peters has succeeded in getting money from the Royal Society for a high-speed ultracentrifuge. Isn't that surprising? I just received a letter from him yesterday. I suppose the two equipments will be ready for use at about the same time. Yours will be the more complete, though, with both a high-speed and a low-speed centrifuge and a cataforesis equipment."

It came as a great encouragement to Svedberg; two of his ultracentrifuges should now be constructed and sent abroad to be used for protein studies, one at the Lister Institute, the other at the University of Oxford. The year before, a high-speed ultracentrifuge had been made at Uppsala and shipped to the DuPont Experimental Station, Wilmington, DE, where it was used in the study of high polymers.

During the summer of 1936 the ultracentrifuge was installed at the Lister Institute. To celebrate this occasion Professor Ledingham had invited a number of prominent British scientists to meet Svedberg at a "housewarming" party at the "centrifuge station" on September 29th.

Five month later Svedberg again came to England, this time to be present at the opening of the ultracentrifuge laboratory at Oxford on February 27, 1937.

Just before going to Oxford, Svedberg had tested Rotor XIX successfully, the fourth rotor to go abroad, this time to the University of Wisconsin, Madison. After J.W. Williams' studies at Uppsala 1934–1935 he had applied for funds for ultracentrifuge equipment to his laboratory. In June 1936 the Rockefeller Foundation made a grant for purchase and installation of velocity and equilibrium ultracentrifuges of the Svedberg type. The centrifuges were installed and tested during the summer and autumn of 1937.

The later periods in Svedberg's life

At the end of the thirties, Svedberg had taken up some new problems.

Together with Sven Brohult (b. 1905) he had started investigations on the effect of ultraviolet light, α-particles and ultrasonics on solutions of proteins, particularly haemocyanins, as these substances were easy to study with the ultracentrifuge [46, 47].

At this time Svedberg began to extend his investigations to other macromolecules of biological interest, such as for instance the polysaccharides. He had been wondering if he could find similar well defined regularities for the molecular weight of carbohydrates in native solutions as he had found for the proteins. Together with Gralén he showed the existence of soluble carbohydrates of high molecular weight in the sap of the bulbs from different species in the family Liliaceae [48]. A more exhaustive investigation of about 75 different species in the Liliflorae revealed the presence of two classes of carbohydrates, distinguished by their sedimentation behaviour. It was found that there was in general a similarity among the species of the same genus with regard to the content of high molecular material which could elucidate problems in systematic botany [49].

An ultracentrifugal investigation of the brewing process was started in collaboration with a Swedish brewing industry [50]. Similarly a collaboration with Swedish cellulose industries was initiated by Svedberg who was interested in studying the cellulose molecule. A review of this work has been given by Svedberg [51]. During this period investigations of cellulose as well as cellulose derivatives were started, and some doctoral theses dealing with cellulose problems were published.

We may now say that the climax in Svedberg's work on the ultracentrifuge and with the proteins was attained just before and after New Year 1939.

On November 17, 1938, Svedberg was invited to give the "Opening address" at a discussion on "The protein molecule" held by the Royal Society, London [52]. He started by saying:

"The proposal of the subject for this discussion is in itself a remarkable thing and the symbol of the spirit of this meeting. A few years ago, the proposal would have looked preposterous. Proteins were known as a mysterious sort of colloids, the molecules of which eluded our search. What is it then that has happened in these years? Why is the most distinguished scientific society of this country inviting a discussion on the protein molecule?"

Later on Svedberg continued:

"Investigations along different lines have given the result that the proteins are built up of particles possessing the hall-mark of individuality and therefore are in reality giant molecules. We have reason to believe that the particles in protein solutions and protein crystals are built up according to a plan which makes every atom indispensable for this completion of the structure."

The first rotor of the final shape, Rotor XXI, was successfully tested on January 5, 1939.

At the beginning of February 1939, the manuscripts for the monograph *The Ultracentrifuge* and *Die Ultrazentrifuge* were sent to Oxford and Leipzig, respectively.

After the outbreak of World War II Svedberg became heavily engaged in work for the government and for other State Agencies. One activity, which took much of his time, was the development of a Swedish production of synthetic rubber (polychloroprene). Owing to the blockade during the war, no oil-resistant rubber could be obtained from abroad. For the armed forces in Sweden it was essential to have this material available. This work was successful and led to a small production plant in the north of Sweden. However, it meant that during several years more than half of the research facilities of his institute were used for pilot plant work and development.

Svedberg had also to advise and direct experimental work in his own laboratory as well as in outside institutes and industries closely related to the war situation. This gave him less opportunity than before for experimental work with his own hands and forced him to devote more time to planning and directing all the various activities.

Svedberg was at this time one of the most influential scientists in Sweden, and he was in the position of playing a very important role in the creation in 1942 of the Research Council for Technology. This was the first Research Council in Sweden and was soon followed by several others. He was a member of the Research Council for Technology from 1942 to 1957. Closely connected to this activity were his relations with the Atomic Research Council (founded 1946) of which he was a member from 1946 to 1959. From 1947 to 1956 he was also a member of the Board of AB Atomenergi which was partly owned by the government.

Svedberg continued to direct some of the work which had his particular interest at the Institute of Physical Chemistry. This included research on the properties of the cellulose molecule, both in solution and by means of electronmicroscopy in the fibrous state, as well as osmotic work on some other chain molecules [51].

Svedberg's old interest in radioactivity was also extended to include the neutron. After having studied the effect of α-particles on proteins, it was natural for him to ask what the action of an uncharged particle would be. It is a typical example of his enthusiasm and openness towards new ideas and methods that he was able to build a small neutron generator and have it working in his laboratory only ten years after the discovery of that particle in 1932. This apparatus was used for irradiation purposes and also for production of some radio isotopes for tracer work. Though only actively used for a short period, the results obtained were of very great importance to Svedberg personally. He now became interested in building a cyclotron to be used in the production of radioactive isotopes and in the field of radiation chemistry.

Svedberg had been collaborating in the use of radioactive tracers produced by the neutron generator with one of his old friends in the Medical Faculty, Professor John Naeslund. The latter, who had personal contact with a public spirited industrialist in Gothenburg, Gustaf Werner, suggested that they should approach Gustaf Werner about the possibility of getting financial help to build a large cyclotron. This would enormously increase their research potential in the field of radiation chemistry and radioactive isotopes. Svedberg was most enthusiastic about this idea, and the response from Gustaf Werner was equally positive, partly because of his general interest in the possible medical applications of such research.

The Gustaf Werner Institute with its big synchrocyclotron (185 MeV) in its huge underground hall was inaugurated in December 1949. This was soon after Svedberg's 65th birthday.

Svedberg retired from his chair in physical chemistry at the mandatory age of 65. However, by a ruling unique in Swedish administration, he was allowed to remain head of the Gustaf Werner Institute as long as he desired.

Svedberg now decided to devote the last years of his time in Uppsala to the creation of a new research institute around the cyclotron. The work here should be focused on biological and medical applications, but should of course also include radiation effects on macromolecules as well as radiochemistry and radiation physics.

He resigned as head of the Werner Institute in 1967 and then left Uppsala for good. Some years earlier he had moved together with his fourth wife and their three young children to an old mining inspector's farm, Sundet, Kopparberg, situated not far from the place where he spent most of his boyhood.

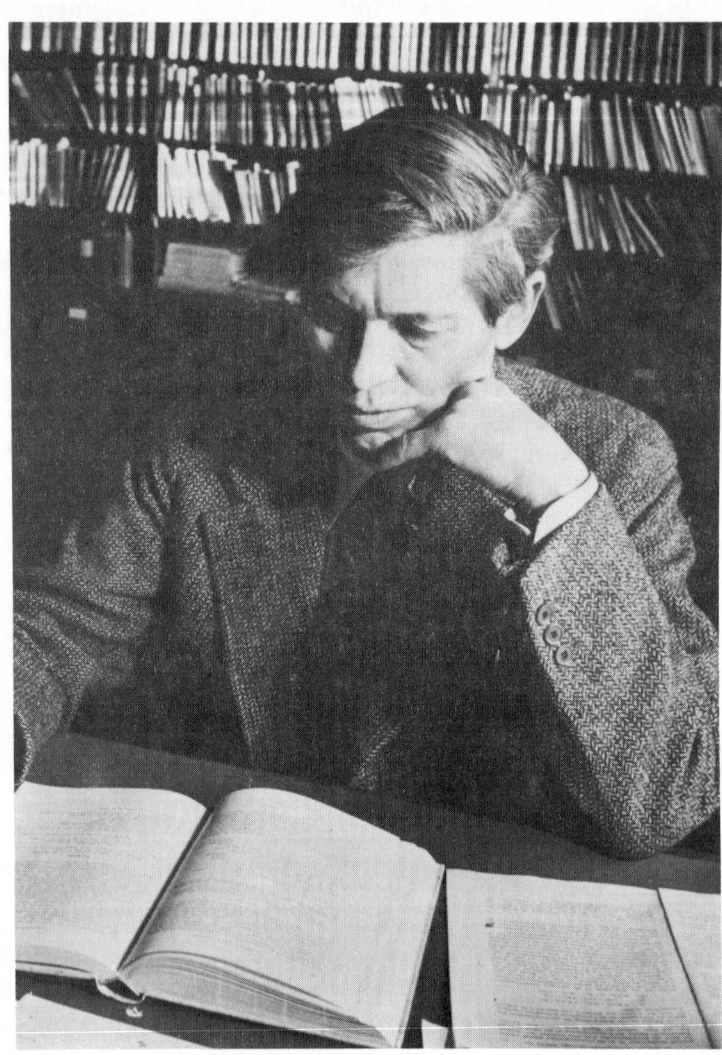

Plate 13. The Svedberg in his office.

Chromatography of colourless substances

Soon after his appointment as professor in biochemistry, Tiselius started to widen the range of his research interests. He had hoped that his new electrophoresis technique might also be useful in elucidating a problem of great interest to him, viz., the isolation and identification of the large fragments and polypeptides obtained by a mild breakdown of protein molecules. In this respect the new electrophoresis technique was a disappointment; he felt that it was hardly specific enough for separating the multitude of substances occurring in materials of biological origin. He then became interested in adsorption methods which had been used to some extent in organic and biochemical preparations. The separation had hitherto been studied mainly *on* the column after the conclusion of the experiment. Tiselius saw the possibilities of developing a new quantitative *analytical* method in which the separation in the eluate emerging from the column could be observed by refractometric methods similar to those used in electrophoresis. He also gave a theoretical treatment that related the retardation volume of an adsorbed substance to its adsorption coefficient and the mass of adsorbent in the column. He considered the modification of adsorption behaviour arising from the presence of a second more strongly adsorbed solute [53–55].

All the early experiments were carried out by frontal analyses that alowed determination of the concentration of the components in a mixture, but did not result in their separation. The latter could be done by using an elution method. The eluted components, however, showed a very marked "tailing". In 1943 Tiselius showed that this could be prevented by adding to the eluting solution a substance with higher adsorption affinity than any of the components in the mixture. The method has since been called displacement analysis [56].

Specific retardation volumes were determined for a number of amino acids and peptides, and it was found that the length of the carbon chain had a decisive influence, each additional CH_2 group producing a marked increase. The retardation volumes of neutral amino acids remained unaffected over a wide range of pH, while those of the acidic and basic amino acids showed a large pH dependence.

A very important technical improvement was made by Tiselius and Claesson [57] when they introduced interferometric methods to measure the concentration of the eluate. The object of this development was to overcome the instability arising from the very slight density difference between neighbouring layers of eluate by restricting convective mixing to very small volumes. The volume of the interferometric channel was only 0.13 ml, and this apparatus was operated thermostatically. In addition, a primitive hand-operated fraction collector was provided to collect effluent fractions. The equipment can be seen as the forerunner of the complex modern fully automated chromatographic analyser-fractionator assemblies. This experimental arrangement was exceptionally well suited to the detailed study of the fundamental processes underlying chromatographic analysis and its use led to some important theoretical advances and to their experimental verification.

The distinctions were pointed out between frontal analysis, elution analysis, and displacement analysis. In frontal analysis a solution is applied continuously to a column of adsorbent, and because of differing retardation volumes a series of concentration gradients corresponding to the various solutes develops sequentially resembling the situation in moving boundary electrophoresis. With this procedure the complete separation of components in a mixture is not possible, but only their identification and estimation.

In elution analysis a sample of solution is applied to a column followed by solvent alone, and under these circumstances a series of gradually separating zones develops, the procedure permitting preparative purification.

In the following decade Tiselius and his co-workers made several modifications and improvements in the chromatographic technique. In most of the work active carbon had been used as adsorbent and many attempts were made to modify its adsorptive properties by various pretreatments. Tiselius [58] also tried to use calcium phosphate in the hydroxylapatite form as an adsorbent for proteins in conjunction with phosphate buffers as eluting agents, with some degree of success, but the definitive solution to the problem for protein chromatography was to come with the development of the cellulose ion-exchangers by Peterson and Sober [59]. The decisive contribution of Tiselius and his co-workers in the field of chromatography lay in the elucidation of the fundamental processes involved.

Until the middle of the forties, Tiselius did a large part of the experimental work himself, sometimes with the assistance of a technician. After that time great demands were made on him for committee work in Stockholm and abroad, and to his regret he could only spend little time in his laboratory. Studies on paper- and zone-electrophoresis were continued by his co-workers and students under his direction, while much work on other separation problems was delegated to his collaborators.

Here should just be mentioned two important new separation methods originating from Tiselius' laboratory. The dramatic separation of particles and of macromolecules obtained by P.Å. Albertsson (b. 1930), making use of partition in aqueous polymer two-phase systems of for instance dextran and polyethylene glycol [60]. The other is the so-called gel-filtration method with Sephadex by J. Porath (b. 1921) and P. Flodin (b. 1924), whereby fractionation is obtained according to size and shape of the dissolved molecules [61].

Tiselius was awarded the 1948 Nobel Prize in chemistry "for his work on electrophoresis and adsorption analysis and especially for his discovery of the complex nature of the proteins occurring in blood serum". He was given honorary doctor's degrees from 12 universities. He was elected member or honorary member of more than 30 learned societies all over the world, including The Royal Society, London, and The National Academy of Science, Washington, DC. He was awarded a number of scientific medals.

In the summer of 1944 Tiselius became a member of a small governmental committee that should recommend measures for improving conditions for research in the sciences, especially basic research. Most of the proposals made were approved by the Swedish Parliament, and a number of improvements were introduced in the science departments. Furthermore the Swedish Natural Science Research Council was established, Tiselius being appointed by the government as its chairman from 1946 for the first four years. At this time great demands were made on him also from other governmental councils and committees.

In 1947 he became member of The Nobel Committee for Chemistry and Vice-President of the Nobel Foundation. In 1960–64 he was President of The Nobel Foundation.

At the First International Congress of Chemistry after the war, held in London in 1947, he was elected vice-president in charge of the section for biological chemistry of the International Union of Pure and Applied Chemistry. Four years later, at the conference in New York, 1951, he was elected president of the union for a four year period.

In the sixties he was very active in the creation of the Science Advisory Council to the Swedish Government which, under the chairmanship of the prime minister, is concerned with Swedish research policy.

In the last decade of his life Tiselius was quite concerned about the problems created by the evolution of science and he was eager that society should benefit from the advances, being aware that scientific developments may involve a severe threat against mankind. He was much concerned about the opportunity for the Nobel Foundation to use its unique position and status in some way that could be complementary to the prize-awarding role. He took the initiative by starting Nobel Symposia in each of the five Nobel Prize fields. The small number of participants in each Nobel symposium would not only discuss the latest developments, but also try to assess the social, ethical, and other implications of such developments. He firmly believed that the Nobel Foundation could and should play an important role in bringing science to bear in the solution of the most pertinent problems of mankind.

Tiselius retired from his chair in biochemistry in July 1968. He now devoted most of his time to the planning of Nobel symposia. The first cross-cultural Nobel Symposium was held in September 1969 just outside Stockholm. Two lines of development emerged from this symposium. They are of interest to mention, because both indicate a beginning of historical processes in conformity with the hopes and ideals expressed by Tiselius. One of them concerns the creation of some kind of world intellectual or scientific community and the other the continuation of cross-cultural Nobel symposia. It is typical of Tiselius' persistence and practical leadership that immediately after this symposium he started to think about the steps which might next have the greatest prospect of success.

At the beginning of October 1971 the Nobel and Rockefeller Foundations, following suggestions by Tiselius, jointly sponsored a workshop meeting of a small number of prominent scientists and humanists. Tiselius was the convener of this meeting which had the general title: "Contacts, co-operation and collaboration". It was held at the Rockefeller Foundation's Villa Serbelloni in Bellagio, Italy. This small group recommended the two foundations to sponsor jointly a new workshop as early as March 1972, to include the directors of some leading research institutes in different disciplines and from different parts of the world, in order to test the feasibility of forming a federation of institutes for advanced study. Tiselius was asked to act as chairman. In the midst of the preparations for the March workshop and after having had an important meeting in Stockholm during the morning of 28 October 1971, Arne Tiselius suffered a severe heart attack in the afternoon when on the way to visit his daughter, and he died at the hospital the next morning.

The two personalities

THE SVEDBERG was a versatile personality with interests in many different fields. From his student days and throughout his life botany had his greatest, indeed dominating, interest. He had a large herbarium, containing all Swedish flowers. Very early he started to photograph plants in their natural surroundings, and he had an impressive collection of such photographs. When he left Uppsala in 1967, both collections were presented to the Institute of Plant-Biology at the University of Uppsala.

For a long time Svedberg had been particularly interested in the migration of plants in Scandinavia; he studied rare species in the north and along the west coast of Scandinavia. He even extended his studies to Svalbard and to the east coast of America. When his friends asked him what he would like to get as a present for his 80th birthday, he said: "A trip to Greenland". The trip was realized in the summer of 1965, and it became his last long botanical journey.

Besides his scientific engagements Svedberg was extensively interested in all kinds of cultural activities. He enjoyed following the modern literature and liked to read poetry, especially French in the original language. He had a fine collection of flora from different countries and a fine selection of modern and old Swedish and foreign literature. All kinds of art had his great interest. During his extensive travelling he never missed an opportunity to visit the large museums. In the summertime he also did some painting himself, particularly landscapes, in watercolours. He even designed some draperies. They were commercially produced and sold under the names: Atomics and Genetics.

The only thing Svedberg strongly disliked was music which to him just meant annoying noise.

Svedberg was an extraordinary man in many respects and he was a genius in finding new and wide open areas for his various activities. His immense capacity for work and his passion for research were infectious. He created a very stimulating atmosphere for all of us who worked at his institute.

The Svedberg died on February 25, 1971, and is buried at Ljusnarsberg's cemetery, Kopparberg.

A biography of Svedberg's life and a bibliography of his various publications is to be found in *Biographical Memoirs of Fellows of the Royal Society* [14].

ARNE TISELIUS was a most modest, quiet and warmhearted person, possessing a very fine sense of humour. Whenever he rose to speak at a meeting, everybody would listen attentively and confident to what he would say. He presented his point of view in a modest way and in a low voice explaining what he considered the crucial facts in the problem under discussion.

It was characteristic for Tiselius that in his experimental work he would take up well recognized experimental phenomena, analyze them critically and establish their theoretical basis. As a consequence he was able to introduce essential improvements in experimental technique. His contributions to the development of new methods for analysis and separation of biological systems mark an era in the study of macromolecules and have vitally contributed to the enormous development in biochemistry since the end of the thirties.

Like many Swedes Tiselius was deeply interested in natural history, and he had a wide knowledge of botany and ornithology. In the thirties and forties he would often go hiking in the woods around Uppsala together with some of his friends, especially in the autumn and in springtime. In the winter he enjoyed cross-country skiing.

Gradually Tiselius became more and more interested in bird watching and in photographing birds in their natural environment. In the spring he made long excursions by night together with his son, Per, to places some distance from Uppsala, where it would be possible to observe the courting behaviour of the capercailzie and hear the crooning of the blackcock (P. Tiselius [63]). At other times of the year Tiselius and his friends made many excursions for bird watching and for photographing birds. This interest of Tiselius led in June 1961 to the formation of a small private "academy", the Bäckhammar Academy of Sciences, which consisted of a group of his friends who had a common interest in ornithology. The academy included five fellows and their wives. Tiselius was the president; the vice-president was Dr. Victor Hasselblad, the constructor of the well-known Hasselblad camera. The academy assembled for about ten days every year, usually during springtime at Jonsbol manor house, belonging to Dr. Hasselblad and once part of the Bäckhammar estate. From Jonsbol they made excursions to various parts of Värmland to study and photograph the very rich bird life in this part of Sweden.

The finest impression of Tiselius' versatile personality one gets by studying his article from 1968 "Reflections from both sides of the counter" [32].

Arne Tiselius died on October 29, 1971, and is buried at the cemetery of Uppsala, not far from his home.

A biography of Tiselius' life and a bibliography of his various publications is to be found in *Biographical Memoirs of Fellows of the Royal Society* [62].

In retrospect

Looking back on the development of protein chemistry during the last 50–60 years, one finds that much progress was gained through the use of physico-chemical methods. Svedberg's introduction of the ultracentrifuge led to the discovery that the proteins are well defined chemical substances, having definite molecular weights. The sedimentation velocity ultracentrifugation, and later also the Tiselius electrophoresis technique, made it possible in many cases to visualize in a much more direct way how far the isolation and purification of an individual protein had been successful. It led Svedberg to advance his hypothesis of the multiple system for the molecular weights of the proteins. And although this theory was not at all of such a general nature as Svedberg at first assumed, it has meant very much for the development of protein chemistry, especially in the thirties and at the beginning of the forties. It initiated new and greater interest in this group of substances. Several chemists and physicists discovered that the proteins no longer had to be considered ill-defined lyophilic colloids, but were well defined, exceedingly interesting and important substances, well worth studying.

One of the first steps had been taken into the new science of molecular biology. However, several more new steps had to be taken before the problems connected with the detailed chemical composition and structure of the proteins could be tackled.

Such new steps were introduced in the forties and the following decades by Tiselius and his co-workers. They developed and introduced a number of new and more specific separation and fractionation methods into protein chemistry.

REFERENCES

1 J.T. Edsall, Arch. Biochem. Biophys., Suppl. 1 (1962) 12–20.
2 E.T. Reichert and A.P. Brown, The Crystallography of Hemoglobins, Carnegie Inst. Washington Publ., No. 116, Washington DC, 1909.
3 S.P.L. Sørensen, C.R. Trav. Lab. Carlsberg, 12 (1917) 255–361.
4 A.V. Güntelberg and K. Linderstrøm-Lang, C.R. Trav. Lab. Carlsberg, Ser. Chim. 27 (1949) 1–25.
5 T. Svedberg, Ber. Deut. Chem. Ges., 38 (1905) 3616–3620.
6 T. Svedberg, Ber. Deut. Chem. Ges., 39 (1906) 1705–1714.
7 A. Einstein, Ann. Phys. (4), 17 (1905) 549–560.
8 A. Einstein, Ann. Phys. (4), 19 (1906) 371–381.
9 M. von Smoluchowski, Ann. Phys. (4), 21 (1906) 756–780.
10 T. Svedberg, Nova Acta R. Soc. Scient. Upsal. (4), 2 (1907) No. 1, 1–160.
11 T. Svedberg, Die Methoden zur Herstellung Kolloider Lösungen anorganischer Stoffe, Steinkopff, Dresden and Leipzig, 1909 (Dritte unveränderte Auflage, 1922).
12 T. Svedberg, Die Existenz der Moleküle, Akademische Verlagsgesellschaft, Leipzig, 1912, 244 pp.
13 T. Svedberg, Die Materie. Ein Forschungsproblem in Vergangenheit und Gegenwart, Akademische Verlagsgesellschaft, Leipzig, 1914.
14 S. Claesson and K.O. Pedersen, Biographical Memoirs of Fellows of the Royal Society, 18 (1972) 595–627.
15 T. Svedberg and H. Rinde, J. Am. Chem. Soc., 45 (1923) 943–954.
16 K.O. Pedersen, Biophys. Chem., 5 (1976) 3–18.
17 T. Svedberg and J.B. Nichols, J. Am. Chem. Soc., 45 (1923) 2910–2917.
18 T. Svedberg, Colloid Symp. Monogr., 1923, pp. 75–96.
19 T. Svedberg and H. Rinde, J. Am. Chem. Soc., 46 (1924) 2677–2693.
20 T. Svedberg, Kolloidzeitschrift, 51 (1930) 10–24.
21 T. Svedberg, Kolloidzeitschrift, Zsigmondy-Festschrift (1925) 53–64.
22 O. Wiener, Ann. Physik Chemie N.F., 49 (1893) 105.
23 O. Lamm, Z. Phys. Chem., A 138 (1928) 313–331.
24 O. Lamm, Z. Phys. Chem., A 143 (1929) 177–190.
24a O. Lamm, Nova Acta R. Soc. Scient. Upsal. (4), 10 (1937) No. 6, 1–115.
25 T. Svedberg and K.O. Pedersen, The Ultracentrifuge (UC), Clarendon, Oxford, 1940. Reprinted by The Johnson Reprint Corporation, New York, 1959; Die Ultrazentrifuge. Theorie, Konstruktion und Ergebnisse, Handbuch der Kolloidwissenschaft in Einzeldarstellung, Band VII, Steinkopff, Dresden, 1940.
26 G.S. Adair, Proc. Phil. Soc. (Biol.), 1 (1924) 75–78.
26a A. Tiselius, Z. Phys. Chem., 124 (1926) 449–463.
27 T. Svedberg and N.B. Lewis, J. Am. Chem. Soc., 50 (1928) 525–536.
28 T. Svedberg and A.J. Stamm, J. Am. Chem. Soc., 51 (1929) 2170–2185.
29 T. Svedberg, Nature (Lond.), 123 (1929) 871.
30 K. Krishnamurti and T. Svedberg, J. Am. Chem. Soc., 52 (1930) 2897–2906.
31 T. Svedberg and A. Tiselius, J. Am. Chem. Soc., 48 (1926) 2272–2278.
32 A. Tiselius, Annu. Rev. Biochem., 37 (1968) 1–24.
33 A. Tiselius, Nova Acta R. Soc. Scient. Upsal. (4), 7 (1930) No. 4, 1–107.
34 T. Svedberg and I.-B. Eriksson, J. Am. Chem. Soc., 54 (1932) 3990–4010.

35 T. Svedberg, Nature (Lond.), 128 (1931) 999.
36 T. Svedberg, J. Biol. Chem., 103 (1933) 311–325.
37 K.O. Pedersen, Kolloidzeitschrift, 63 (1933) 268–277.
38 T. Svedberg and B. Sjögren, J. Am. Chem. Soc., 50 (1928) 3318–3332.
39 P. von Mutzenbecher, Biochem. Z., 235 (1931) 425–437.
40 J.P. Johnston and A.G. Ogston, Trans Faraday Soc., 42 (1946) 789–799.
41 B. Enoksson, Nature (Lond.), 161 (1948) 934.
42 K.O. Pedersen, Ultracentrifugal Studies on Serum and Serum Fractions. Dissertation, Uppsala, 1945, also in University Microfilms International, 178 pp.
43 K.O. Pedersen, J. Phys. Colloid Chem., Ithaca, 51 (1946) 156–163.
44 P. von Mutzenbecher and T. Svedberg, Naturwissenschaften, 21 (1933) 331.
45 A. Tiselius, K.O. Pedersen and T. Svedberg, Nature (Lond.), 140 (1937) 848–849.
46 T. Svedberg and S. Brohult, Nature (Lond.), 142 (1938) 830.
47 T. Svedberg and S. Brohult, Nature (Lond.), 143 (1939) 938.
48 T. Svedberg and N. Gralén, Nature (Lond.), 142 (1938) 938.
49 T. Svedberg and N. Gralén, Biochem. J., 34 (1940) 234–248.
50 O. Quensel and T. Svedberg, Meddr. Carlsberg Lab., 22 (1939) 441–448.
51 T. Svedberg, J. Phys. Colloid Chem., Ithaca, 51 (1947) 1–18.
52 T. Svedberg, Proc. Roy. Soc. Lond., A 170 (1939) 40–56.
53 A. Tiselius, Ark. Kemi Miner. Geol., 14B (1940) No. 22, 1–5.
54 A. Tiselius, Ark. Kemi Miner. Geol., 15B (1941) No. 6, 1–5.
55 A. Tiselius, Science, 94 (1941) 145–146.
56 A. Tiselius, Ark. Kemi Miner. Geol., 16A (1943) No. 18, 1–11.
57 A. Tiselius and S. Claesson, Ark. Kemi Miner. Geol., 15B (1942) No. 18, 1–6.
58 A. Tiselius, Ark. Kemi, 7 (1954) 443–449.
59 E.A Peterson and H.A. Sober, J. Am. Chem. Soc., 78 (1956) 751–758.
60 P.Å. Albertsson, Partition of Cell Particles and Macromolecules. Dissertation, Uppsala, 1960, also published by Wiley, New York, 1960, 232 pp.
61 J. Porath and P. Flodin, Nature (Lond.), 183 (1959) 1657–1659.
62 R.A. Kekwick and K.O. Pedersen, Biographical Memoirs of Fellows of the Royal Society, 20 (1974) 401–428.
63 P. Tiselius, Phys. Chem. Biol. (Japan) 17 (1973) 147–148.

G. Semenza (Ed.) Selected Topics in the History of Biochemistry: Personal Recollections (Comprehensive Biochemistry Vol. 35)

Chapter 9

Survey of a French biochemist's life

PIERRE DESNUELLE

Centre National de la Recherche Scientifique, Centre de Biochimie et de Biologie Moléculaire, 31, Chemin Joseph–Aiguier, B.P.71, F 13277 Marseille–Cedex 9 (France)

De tout temps, il s'est glissé parmi les hommes de belles imaginations que nous venons à croire parce qu'elles nous flattent et qu'il serait à souhaiter qu'elles fûssent véritables.
Molière
(Le Malade Imaginaire)

My grandparents and my parents

I have had the good fortune to have known three of my grandparents and to have often been with them over long periods. Although he was scarcely over fifty-five when I myself reached the age of reason, my maternal grandfather already seemed to me an old man; he was tall and had an open and intelligent face, adorned by a handsome moustache. He was the son of a farmer from the Jura Mountains in France, near the Swiss border. In the family the story was told of how after his father had gone through considerable effort to have him sent to a religious college, isolated in the middle of the high plateaus, he escaped in the heart of winter and walked through the snow for dozens of kilometers to return home. There he quite naturally received an unpleasant welcome from his father, who without further discussion sent him back to the college where he then stayed until the end of his studies. At the time I knew him, he was Head of Administration at the Town Hall in Clermont-Ferrand. The capital of Auvergne, which was later greatly transformed in large part due to the Michelin factory's influence on extensive urban development, was then a small quiet town; the streets, bordered with houses built out of the region's black lava, were often swept by a salubrious wind descending from the nearby volcanos. These volcanos, inactive for many millennia, resembled huge mole-hills, rising unexpectedly from the deserted, heather-covered moor. At that time, neither fences nor forbidding hampered the indefatigable step of my grandfather, who sometimes brought along his grandson to hunt or to hike towards the mineral water springs, endowed, it was said, with invaluable properties. In spite of my instinctive horror at seeing animals, which were so beautiful in movement and flight, fall miserably to the earth, by evening I was proud to see the filled game-bag on my grandfather's shoulder. The victims were carefully prepared by my grandmother, who was an excellent housewife. Her floors were always flawlessly waxed and she further protected them by insisting that the family and visitors move about gliding on scraps of wool. It was, in short, tantamount to the Japanese custom of taking off one's shoes before entering the house. My grandmother did not love her floors only. She also dearly loved her only grandson; in spite of limited financial means she covered him with costly gifts. It was without pleasure that she saw that I preferred to play with an old board equipped by my grandfather with four small wheels.

My maternal grandfather and grandmother lived in a large house whose principal attraction was to give onto a vast courtyard where two large hunting dogs frolicked; white-haired and spotted with black, they were invariably called Cybèle and Diane. Next came the vegetable garden at the back of which my grandfather had built a small shed where he kept his tools and made hunting cartridges. My favorite places were that shed and the dog kennel where, with their welcome agreement, I often spent much of my time.

Towards the end of his life, my grandfather replaced the joys of open-air activities with those equally vivid inspired by the works of Greek and Latin authors, over which he had maintained a perfect command. Young schoolboy that I was, as little gifted for the dead languages as for the modern, I marveled at this old man who fluently read Horace and Tacitus in the original, while the translations of Julius Caesar's works, although easier, caused him such torment.

My paternal grandmother, whose husband died in the year of my birth, came of good country stock from the high plateaus in the Jura region. When I knew her, her three sons had already left home to become a finance officer, a military officer, and an engineer. She lived with her daughter in St. Claude, a small, extremely picturesque, town, situated between two steeply banked rivers, and renowned for its non-polluting tobacco-pipe and diamond industries. She hid a warm heart under the severe appearance of her sombre dresses fitted with long sleeves and high collars.

Her smooth face was framed by two white head-bands and crowned by a tightly twisted chignon.

In fact, I lived very little at St. Claude, but spent much time on my grandmother's agricultural estate half-way down the Bienne Valley on the sunny side of the slope. The property included a small house, a large farm, and numerous acres of green fields, bordered with black fir-trees and shadowed by vertical, grey limestone cliffs. There, thanks to the devotion of my grandmother, mother and aunt who agreed to live in the discomfort of a small house, I led for many summers the rustic lifestyle of that era's modest French farmer; helping to tie sheaves of wheat, get in the hay, harvest potatoes, care for the horse (out main energy source), guard the cows in the fields, and carry on my back the milk-filled containers that were brought down to the valley twice a day in order to make cheese. Doubtlessly it was in the Jura and Auvergne that the strong ties, which have always attached me to what my Parisian friends so willingly call "la province", were formed.

My father, the oldest of three sons was a finance officer of middle rank. He had met the girl who was to become my mother during his first assignment in Lons-le-Saunier, the chief town in the Jura. Lons-le-Saunier is well known in France because of one of its citizens, Rouget de Lisle, author of our national anthem. My birth was not accompanied by any particular manifestation except perhaps the fact that my mother's friend, who had come to help at that noteworthy event, overturned a paraffin lamp amidst the commotion, almost setting the house on fire. My mother still talked about it forty years later.

Other than that, I recall very little about Lons-le-Saunier which I left at the age of one to follow my parents to Roanne, another small town situated on the banks of the Loire, where the beautiful river descends from the Massif Central. The high school in Roanne was an old college built by Jesuits and requisitioned by the French government, as were many others at the beginning of the century, in the hope of conciliating its political beliefs, noble aspirations towards better education for young French students, and chronic lack of funds for things which were really useful. Half of the school building had been converted into a hospital during and just after World War I. In the other half the professors took great care of us, with the possible exception of one English professor who thought, sincerely or not, that we could learn that language by reading the French translations of Conan Doyle's best novels.

I had the choice between two routes to return from school to my parents' house. One was direct, passing through a very crowded street. The other required several extra minutes, carefully counted by my mother. But, it had the considerable advantage of being partly shared with the route taken by my best friend; and of passing by a small store full of fascinating objects, which, by the way, we never bought because of lack of money.

It was in Roanne that my father died when I was twelve years old. The treatment for the pulmonary infection which he caught had not at that time been discovered. My mother often told me that he had fallen ill during one of the bicycle trips he was required to take into the surrounding mountains twice a year in order to meet the mayors of the small isolated villages and help them put their books in order. At that time, the French administration was not generous towards modest civil servants. My father either had to use his own bicycle, or go on foot. Neither did he have an office. To my mother's despair many taxpayers in financial straits were received at home. I remember him as a tall man, simple, good and above all endlessly patient with his rather boisterous son. He was entirely devoted to his work and possessed a pleasing talent as a flutist, playing the instrument during his idle moments.

Several months later, my mother who henceforth had the full responsibility for my education, decided to move to Lyon, thereby accentuating the family's continual drift towards larger and larger towns. At the age of seventeen my mother had left the Clermont-Ferrand High School to live in Paris with her uncle, Professor of Mathematics at Chaptal College, and to prepare for the entrance examinations to the Ecole Normale Supérieure de Jeunes Filles; then simply called "Sèvres", due to its location which was at that time distinguishable from the Parisian conglomeration where it was implanted. Many years later she still kept a dazzling memory of the school. In those far-off days at the beginning of the twentieth century, young girls were rarely permitted to leave the school. The higher echelons of the administration seemed to believe that to have them mix with the boys in the Sorbonne amphitheatres would disturb their thinking, and also perhaps their hearts, which needed to be protected at all costs. Therefore, the Sorbonne professors travelled about in hackney-carriages in order to give the girls their school courses. My mother often recounted her excitement at watching such illustrious characters as Paul Langevin and Jean Perrin descend from their coach as she consulted the school's chefs regarding the nature of the meal to be served after classes which would alleviate their travel fatigue. After studying for three years, she passed the Competitive Aggregation Exam in Physics and was assigned to a high school in Lons-le-Saunier, where, as I have already said, she met my father.

After his death, my mother's two main centers of interest were her teaching and her son's education. Until the end of her life, she was a remarkable and demanding teacher, arousing strong feelings among the students whom she enthusiastically prepared for exams and competitions. Her only weakness, a small one at that, was to highly value her vacations; she would leave on a trip the first day and would not return until the last. It is true that in that area she was also only a forerunner. Vacations in France are now a veritable institution protected by the State.

In the beginning, we vacationed in the country, in the Jura or in Auvergne. Then, as the family finances improved, we developed the agreeable habit of embarking on still farther expeditions to Italy, Greece, Morocco, Norway and Spain. The latter trip was taken by car as I had just turned eighteen, the required age in France to have a motor vehicle license. We went as far as southern Andalusia in the company of two ladies who had long ago replaced their fleeting youth by a great dignity. My mother had insisted on bringing them along in order to share the expenses, and I was generously entrusted with the driving as well as with the group's safety. At times, we had to ford rivers, a procedure not only demanding delicate navigation between more or less stable stones, but also intricate manoeuvres through rude downgrades followed by difficult ascents quite incompatible with the car's mechanical possibilities. At eighteen, everything seems simple; I confronted the problems without any difficulty and there were no serious incidents. When I recount this anecdote, certainly a rather anodyne one, to my Spanish friends, they eagerly ask me to go with them and verify on the spot that excellent bridges now exist in Andalusia.

My mother also strove to be a good instructor, even though the results were not always satisfactory. She did, however, put all her efforts into encouraging me to complete suitable secondary studies, in spite of a certain indolence on my part; and teach me how to conduct myself appropriately in life.

High school and university studies

On arrival in Lyon I enrolled at Ampère High School, named after a famous physicist born in a small nearby village. The time-honored building, constructed by Jesuits like the one in Roanne, was located on what is known in Lyon as "the peninsula", the part of town between the Saône and Rhône Rivers, on the banks of the latter. A small suspended footbridge served as entrance and exit. I had to cross that bridge four times a day, except when it was very windy and the bridge was closed as a safety precaution. When that happened I had to take a long detour which was not at all compatible with the high school's strict regulations regarding punctuality. The school's caretaker was an old soldier who had been wounded during World War I. He regulated everything in the school, the professors' movements as well as those of the students, by beating his drum. The stone thresholds giving access to the classrooms were so worn by the passing of so many heavily soled shoes that their centers had become several centimeters lower than the bottoms of the doors. Owing to this fortunate disposition, the coal stoves which provided our winter warmth were allowed to be heated so much that our cheeks were permanently red-hot and our feet constantly frozen. The discomfort, which we really barely even noticed, did not stop us from working diligently under the command of our demanding professors. The excellent reputation of secondary school professors in France is well-known. Our professors were no exception. Their knowledge was more or less widespread. Their pedagogical skills were more or less noteworthy. But, they all did their best to instill into us an appreciation of reason and well-executed work.

The most demanding of my professors in Lyon was undoubtedly the one who undertook to teach me the last course of the Elementary Mathematics Program seven hours a week. That man lived for and by the solid geometry with which he inundated us. Unfortunately, few of my fellow students knew how to solve the cosmography problem we had on the final exam. I myself was saved by a philosophical essay on the death penalty, having probably been lucky enough to happen upon an examiner who shared my ideas on the controversial subject. Another incident which was almost, but only almost, fatal was one that had already occurred the previous year when a geography examiner had asked me what I knew about Provence. I knew very little indeed about that beautiful region east of the Rhône River on the Mediterranean coast. I was interrupted soon enough by the examiner who told me that he could easily see that I had never set foot there, which was, by the way, perfectly true. The low mark which accompanied his remark would have posed a problem if I had not had the good luck to be saved by the examiner in French. This latter very politely asked me to comment upon Alphonse Daudet's story about sheep transhumance in Provence. After several apparently well-chosen sentences, the dear man said to me with a big smile: "To have such a good understanding of the text, you must have been born in Provence." Thus, I learned for the first time that imagination is a precious tool given to man so that he may avoid reality's pitfalls and make it tolerable.

At that time, as is still true today, the Baccalauréat Diploma opened the alleged magical doors of the university and the "Grandes Ecoles" to young French men and women. Theoretically, it is the time when many should decide upon the nature of their careers. This decision, in appearance formidable, is happily tempered in many cases by the intervention of luck. In my case, my health, which had bothered me during my last year in high school, would not have allowed me, even if I had been capable of doing it, to study for the competitive entrance examinations to the "Grandes Ecoles". My secret wish, however, would have been to go to the Navy School, because I was enraptured by both the promise of travel and the handsome uniforms. My mother, who was anxious to keep me near her for as long as possible, skillfully dissuaded me by having the family doctor tell me that my kidneys were at the time a bit fragile and would suffer enormously in a salt air environment. Therefore, I entered the Industrial Chemistry Institute in Lyon, directed by Victor Grignard, a Nobel Prize Laureate in Chemistry as a result of his famous work on organomagnesians. Victor Grignard's Organic Chemistry courses were descriptive and very detailed. They were often very difficult, and sometimes dull. I willingly admit that I rarely reviewed my notes; they made an impressive pile on the closet floor at home. Doubtlessly it was at that moment that I resolved that if I myself were ever to become a professor, I would not overload the students' minds with too many details or numerical results; on the contrary, I would have them discover a limited number of facts, and would give them both the desire and the opportunity to go back over their work later in a more concentrated fashion. This pedagogical method has moreover now been adopted by most of my young colleagues. At any rate the fact remains that Grignard guided us through the vast domains of organic chemistry with dexterity and ease. The respect for that facet of chemistry which he instilled in me was later very useful in biochemistry.

The little that I know about organic synthesis I owe to one of his students, Jean Collonges, who was then Assistant Professor at the school. His associates supplied us with the compounds that were necessary for a reaction, and they carefully controlled the weight and the purity of the final product which we brought to them some days later. Abnormally low yields and melting points were ruthlessly sanctioned. Many years later, even though the laboratory I directed was frequented by brilliant engineers, it was still I who, as much out of necessity as out of pleasure, realized the synthesis of the non-commercial products needed for our research.

On the other hand, things did not go as smoothly in inorganic chemistry where we spent many months identifying ions in complex mixtures of mineral salts with the help of a large-sized H_2S generator. I must admit to my great shame that I was disheartened by both the uninteresting operations and the smell of the principal reagent. A certain number of us secretly bribed the laboratory aid to reveal the compositions of the mixtures. We shamelessly referred to our actions as "Synthesis", which is as everyone knows a useful complement to "Analysis".

A more respectable episode in my career was when I was the first, or one of the first, to receive authorization to prepare the Licence Degree in Physics and Chemistry at the university. The generous authorization, so difficult to wring out of an administration covetous of its prerogatives, allowed me to take advantage of First Year Mathematics, First and Second Year Physics, and last but not least to take Third Year Biochemistry. Indeed, it offered the possibility of the career in teaching and research that I consequently chose.

All things considered, I was very fortunate to benefit from both the theoretical teaching at the university and the practical teaching at the school long before it became a standard practice.

In spite of the rapid increase of knowledge in all the sciences, the French university at that time still maintained the pretence of teaching the quasi-totality of physics in only two years to young students freshly out of high school; and it thus incited regrettable prejudices and even total rejection in certain students. Fortunately for me, the youngest physics professor, Paul Déjardin, had decided even before the regulations permitted him to do so, to put fire and sword to the archaic organization. He presented to us several judiciously chosen subjects only, such as the quantum theory, the emission spectra, and all kinds of spectroscopies which familiarized us with the dynamics of electrons gravitating around the atomic nuclei. In addition, a new course in nuclear physics, also taught by a young and very talented professor, was created at Lyon around that time. I was then preparing my thesis and I followed the new course with such enthusiasm that my Thesis Director, fearing no doubt an abrupt change in my orientation, did me the honor of worrying about it. He should have on the contrary rejoiced as the course revealed to me, not only the interior life of the nuclei, but also the existence of heavy and radioactive isotopes, and the techniques used for their determination.

The discipline which at that time presided over the school would seem harsh to many of us today. Victor Grignard was the most benevolent man in the world except concerning his precious organic chemistry, and the discipline was placed under the responsibility of the Assistant Director Paul Meunier, who was also our industrial chemistry professor. Paul Meunier was a very tall man with an impassive face framed by white hair. Constantly dressed in a dark blue suit, he made a great impression on us by the clarity of his lectures and diagrams. But, his prestige was especially due to the absolutely fantastic punctuality which marked his entry into the amphitheatre at exactly eight o'clock in the morning. Having since then experienced the worst difficulties in beginning my own courses on time – a fact with which one of my former students sharply reproached me at the time of my reception by the Académie des Sciences – I have begun to spitefully suspect that Paul Meunier waited behind the door for the large school clock to strike eight. At any rate this little story is dedicated to those of my colleagues, if any, who want to gain the goodwill of a recalcitrant audience.

The school laboratories, where we spent all our available time from eight until eleven in the morning and from one-thirty until six in the evening, were well-lighted and spacious. Surely, there was little expensive apparatus, but on the whole we had what was needed to work satisfactorily. The benches were made of oak and we had to wax them vigorously every Saturday at eleven o'clock. I remember our supervisor's stupefaction when one day he saw a student letting the liquid escaping from a distillation apparatus run onto one of those precious benches. The guilty party naïvely explained that it was of no importance because the product concerned would not show up until later. We laughed so much that the supervisor couldn't deal harshly with the situation. But the wax consumption was greatly increased due to that unorthodox manoeuvre.

My time at the school provided me with good laboratory experience. It also permitted me to make numerous friendships, some of which have lasted through half a century of busy activity. In the laboratory we worked in groups of two. Each group member called the other "mon nègre", an expression stemming from a deplorable racism, which at that time in France applied to any person who could be persuaded to do your work. In fact, each was "le nègre" of the other; the groups were on good terms with each other, save a few exceptions. We have not let our present activities entirely separate us. Thanks to a few devoted classmates, an enjoyable dinner reunites us every five years in Lyon, capital of French gastronomy. We have, however, recently decided to reduce the interval to two years and we will probably be obliged to reduce it again if we want to keep a good chance of still seeing each other several times.

First contact with biochemistry and research

When I left school I followed the advice of one of my professors and did not immediately enter industry as most of my fellow students did. Instead, I stayed at the university to do a doctorate. In so doing I hoped to give myself more time to think and realize what I was really capable of doing.

During our last semester at the school we were supposed to specialize by choosing one of the four big sections set up for that purpose. On my part I opted for the Bacteriology section which the directors had created with remarkable foresight in the early thirties to satisfy the needs of bioindustry. In this section the chemical and biochemical aspects of various industrial fermentations were developed by a young professor from Paris, Claude Fromageot, who had just been appointed to Lyon after having worked a year at Carl Neuberg's laboratory in Berlin-Dahlem.

Although Neuberg's concept concerning the possible involvement of methylglyoxal in alcoholic fermentation was soon abandoned in favor of the Embden–Mayerhof theory, the stay had given considerable incentive to Claude Fromageot, permitting him to convince his technician to paint his laboratory, and helping him to obtain a little money from France and from the United States. Thanks to those funds, he was able to hire several young scientists to work in the stimulating atmosphere with which pioneers surround themselves.

Claude Fromageot was a man of great talent. The combination of a quick intellect and a warm heart with exceptional dynamism and natural distinction rapidly made him one of the most remarkable figures in biochemistry in France and in the world. He gave ample proof of these gifts at first in Lyon where his work was continued by Danièle Gautheron, and later in Paris where he rapidly transformed the dilapidated building which was allocated to him into a world-renowned laboratory attracting such researchers as Julie and Bernard Labouesse, Roger Acher, Hubert Clauser, Pierre Jollès, and Roger Monier, who now occupy top positions. With several other men of his own generation and of the preceding one, he shared the honor of having given a decisive impetus to French biochemistry whose results can still be felt today.

Although my physicochemical training instilled in me a certain distrust of the so-called "Natural Sciences", Claude Fromageot's personality so fascinated me that I enthusiastically accepted his generous offer to become part of his laboratory team. Thus my orientation did not result from a deliberate choice in favor of biochemistry, which was still seeking its place between organic chemistry and physiology, and where my knowledge was not thorough. It resulted rather from the influence of one man's spiritual and intellectual qualities. I can only further admire those who at the right moment feel called upon by a vocation.

In the thirties, the undertaking of a science thesis in France was both risky and relatively simple. An Administration for Research Development had just been created thanks to the reputation and commitment of Jean Perrin. But its means were limited and moreover the university, badly neglected by the Republic even though many of its leaders were professors, had few openings, especially in "the provinces". Thus it was necessary to have a certain amount of courage, or as in my case a good deal of unconsciousness, to try to make it a career, even if it was only for a few years. My integration was possible thanks to a very ingenious arrangement whereby the assistant positions were divided in half, each beneficiary of a part-time position only receiving a half-salary, while he was expected to do the work, which was not really very exacting, of a full-time assistant.

Although the material aspects of French university life were uncertain, the choice of a research subject was, in contrast, very easy, for the simple reason that almost everything was yet to be discovered.

At the request of Claude Fromageot, I explored the possibility of yeast-synthesizing proteins from carbohydrates in the presence of ammonium hydroxide as a nitrogen source. The key step in the process was the conversion of α-keto acids into the corresponding α-amino acids. It was an exciting day in the laboratory when I isolated a small sample of pure alanine with a correct elemental composition and rotatory power from a complex mixture containing a large excess of pyruvic acid.

My fellow students and I would often drop by Claude Fromageot's office to discuss our work with him. His favorite saying was: "It's a small discovery, or a big mistake." Nevertheless, my work progressed from mistake to mistake, and the time came when I could foresee writing my thesis. It was a very laborious task which Claude Fromageot, with his usual generosity, helped me to accomplish. To tell the truth, at that time it was very convenient to have professors who were incontestable storehouses of wisdom and knowledge, we were then left with more time and freedom of thought for various other more agreeable activities, which we were frivolous enough to find.

The ink on my thesis had not dried before I set off to do my military service. During my university studies I had set aside part of every weekend to learn the rudiments of the art of war at a barracks on the outskirts of Lyon. Thus I had the privilege of being sent directly to a school in Poitiers where brilliant officers did their best to transform disobedient students into perfect artillerymen in six months. One of my most striking memories of the school is the overpowering lethargy which overtook me when just after a long ride on horseback through the surrounding countryside, I found myself shut into an amphitheatre to hear, and if need be to understand, the theoretical aspects of the art. Another memory which is just as clear is that the school was built on a plateau separated from the city by a deep valley. Twice a month we had authorizations to go out at night, provided of course that we were back by midnight at the school gate where old, ill-natured, non-commissioned officers kept a vigilant watch. With our boots, spurs, and long coats, we stampeded like frightened elephants and dashed up the two hundred and eighty steps to be back on time. In spite of these minor incidents, I have a fond memory of the time spent at military school thanks to the spirit of good fellowship which was always present, the equal balance between intellectual and physical activities, and the absolutely extraordinary zeal with which the instructors taught us the difficult skills which we were soon expected to use.

Not even a few weeks had passed after my arrival when Claude Fromageot wrote an official letter to my Colonel requesting authorization for me to go to Lyon for my thesis defense. The Colonel called me into his office and told me in a voice touched with emotion that in his eyes the duty of every young Frenchman to his country had priority over the needs of Science, however important they might be. However, being a very distinguished man, he gave me three days: two for the train trip and one for the actual defense. The defense was not brilliant as several weeks of military life had seriously dampened my meager scientific qualifications.

Claude Fromageot's haste to have me defend my thesis arose from his wish to have me receive a Rockefeller Postdoctoral Fellowship allowing me to work for a certain period in a foreign laboratory, as he himself had done several years earlier. After long debate, we agreed on Germany where Gustav Embden, Richard Kuhn, Otto Mayerhof, Otto Warburg and Otto Wieland rivaled England's Sir Frederic Hopkins, David Keilin, and J.B.S. Haldane for the world title in biochemistry. We finally decided on the laboratory directed by Richard Kuhn at the Kaiser Wilhelm Institut in Heidelberg where the renowned research on several vitamins, the "old" yellow enzyme, and the relationship between vitamins and coenzymes was in strong competition with the work of Paul Karrer in Zürich and that of Otto Warburg.

I had had two years of German at school, but still could neither read nor speak it readily. I hoped to remedy that distressing situation by having myself transferred to Metz near the French-German border for the last part of my military service. I had not counted on the supercilious patriotism of the inhabitants who, for the most part fluent in German, refused to speak it or even to admit that they were capable of doing so. On my arrival in Heidelberg, I was spared the biggest problems thanks to the kindness of Richard Kuhn's personal secretary. A very distinguished old lady with whom I was put in contact agreed to rent me a room in her lovely house on the Neckar, very near the Institute.

At that time Heidelberg was much as it still is today, a quiet, clean town, well-known not only for its university and institutes, but also for its surrounding vineyards, hilltop castle, and old bridge over the Neckar. However, these last two edifices had been halfway demolished by Louvois, Louis XIV's minister, an action which, as I often told my German friends, had at least the merit of having bestowed them with the charm and romance of ruins. Be that as it may, marble slabs bore the inscription "Durch den Franzosen zerstört" (destroyed by the French), in memory of the unhappy event. The Kaiser Wilhelm Institute was, and still is, a large H-shaped building, the cross-bar was occupied by the library, and the sides housed, respectively, the Departments of Biochemistry, Physiology, Experimental Medicine and Physics. These laboratories and the vast basement were full of modern material that seemed marvelous to the eyes of a young Frenchman little accustomed to that sort of display. The precursory signs of a fast-growing Nazism were unfortunately beginning to show. People in uniform marching in step behind loud brass bands were a common sight in the streets. A much more serious affair, an increasing number of academicians and researchers, including Otto Mayerhof who was obliged to leave Germany the following year, were put under more and more intolerable pressure. In that constraining atmosphere, my means of action were limited. I had to content myself with several ironic comments on the caricatures of the regime's high dignitaries published by the few rare newspapers which were still authorized. There I found some pleasure and the more prosaic opportunity of assimilating colloquialisms. Some interlocutors laughed. Others bore expressions of disapproval. But none of them created any real difficulties for me, doubtlessly attributing what they thought to be my inconsequence to the well-known mocking humor of the French. Thus, on the whole my year in Heidelberg passed agreeably in spite of the heavy thunderclouds we felt hovering over our heads. I formed lasting friendships with Theodor Wieland, Kurt Wallenfels and F. Weigand who were at the time also working under Richard

Kuhn's direction.

Richard Kuhn was endowed with an exceptional intelligence which allowed him to instantly grasp the most complicated problems. If during a conversation in the laboratory he felt that results could be published, he would lead me to his office where he would dictate a text to his secretary and at the same time continue to question me. Several minor corrections were enough to give the text its definitive form. In one year, we repeated that procedure for the Berichte and the Hoppe-Seyler's Zeitschrift für Physiologische Chemie six times, thus showing Claude Fromageot and the Rockefeller Foundation Administrators that they had not been entirely wrong in choosing me. I greatly admired Richard Kuhn for that ability, which unfortunately he was not able to pass onto me.

Upon my arrival in the laboratory, Richard Kuhn asked me to try under F. Weigand's supervision to synthesize a lactoflavin analog in which the ribose was replaced by arabinose. At first I obtained an unfavorable looking oil, which began, nonetheless, to crystallize after having been left on the corner of a bench for two weeks. This analog was later found to be about as active as the natural isomer. Another, probably more significant, line of research was to determine the amino acid composition of the protein moiety of the "old" yellow enzyme. The conditions seemed to lend themselves well to such an enterprise as, thanks to German organization, several kilograms of pressed yeast yielding about half a gram of crystalline enzyme were processed each week in the basement. In contrast, the finding that proteins could be obtained in a chemically homogeneous form, and not merely as poorly defined "colloids", was still recent. The yellow enzyme was known to be split upon acid dialysis into a low-molecular-weight, yellow compound (the lactoflavin pyrophosphate) and a colorless compound which remained in the bag, thus giving a good experimental confirmation of the Von Euler concepts of coenzymes and apoenzymes. But the protein nature of the apoenzymes was not yet firmly established. We set out to hydrolyse gram amounts of the apo yellow enzyme and to apply to the hydrolysates various colorimetric techniques known at the time to be roughly specific for amino acids extracted from other sources or obtained by synthesis. Of course, positive responses given by colorimetric techniques could hardly prove the existence of amino acids in the apoenzymes. However, their existence was confirmed by crystallizing various derivatives of arginine, lysine, histidine, aspartic and glutamic acids, from the hydrolysates, and by checking their properties with those of authentic samples. This was a very exciting period of my life in Heidelberg. In spite of serious difficulties which later opposed Richard

Kuhn and my country during World War II, I made a point to attend the formal ceremony held in his honor by our German colleagues a year after his premature death.

My return to Lyon and World War II

With the alanine and the "old" yellow enzyme from yeast, I had acquired a specialty which would not alter in later years. During my stay in Germany, Claude Fromageot had oriented part of his laboratory's activity towards the metabolism of amino acids containing sulfur. A young Iranian student was studying the production of hydrogen sulfide from cysteine by a variety of microorganisms under anaerobic conditions. We soon found that the rate at which hydrogen sulfide was liberated by *Escherichia coli* increased exponentially with time and that the initial rate of the reaction was considerably higher for cells previously incubated with cysteine. The enzyme responsible for this desulfuration, which we incorrectly called cysteinase, was one of the first known examples of an adaptative enzyme. We were content to draw attention to this phenomenon's importance regarding the life of prokaryotic cells, and we did not try to actually interpret it, which was done a short while later by Jacques Monod.

That agreeable life in which my time was equitably divided between biochemistry and skiing and climbing in the Alps with a group of good friends was suddenly interrupted by the advent of World War II in 1939. While most of my friends who were engineers remained assigned to their factories, I, on the contrary, due to my qualifications as an artilleryman and as an academician, was given the responsibility of a detachment composed of thirty men and twenty horses. Our mission was to establish an ammunition depot in a valley situated north of Nice near the Italian border. I still shiver when I think of the night we loaded twenty horses freshly out of pasture into the train cars. With some men pulling and others pushing the recalcitrant animals, we finally managed to leave before daybreak. As soon as we arrived, we set to work on an embankment where we dug holes to be filled with various types of ammunition. After having royally ignored us for over a month, we were tartly informed that headquarters found our camouflage insufficient due to a report filed by reconnaissance aircraft. The work could not have been completed on time anyway due to both lack of material and the torrential rains which sometimes swept over the meridional regions and half filled the holes.

After a few months the army realized that I was not only an

artilleryman, but also a chemist. I was thus snatched from the last beauties of autumn in Southern France and sent to a fort on the edge of the German border. There I was charged with the supervision of the masks and detectors intended to protect our personnel from war gases in the event the enemy decided to use them. I had many squadrons at my disposal, posted at the entrances to the fort; they were to palliate any eventual breakdowns in the automatic machinery and to detect the odor of any toxic gases which might arrive. I was not really aware of my usefulness until the day when one of our soldiers was wounded as he left the fort by an exploding grenade which had remained hanging from the branches of a tree. The poor fellow, who was badly hurt, was anesthetized before being carried back inside the fort on a stretcher. One of my men did not grasp the connection between the wounded man in front of him and the odor of the drug which had been used to put him to sleep. Discerning an odor which was apparently chemical, he immediately set off the general alarm. Calm was reinstated when the garrison saw that the expert they took me to be was walking the length of the fort's dark underground tunnels without a mask. Another episode in my military life, and a very funny one at that, concerns the day when an officer from headquarters decided to pay us a visit. I thought it would be a good idea to give him a description of a machine which I had conceived and which I thought would permit rapid identification of any known gas. The V.I.P. warmly congratulated me and declared that as my stay of several months thirty meters underground seemed to have done so much for my creative faculties, it could be arranged to prolong it for as long as possible.

Several weeks before the Germans attacked our territory, I made a second request to the military authorities for an exceptional three-day leave. This time it was to get married. The ceremony was hardly over when my wife accompanied me to the train platform at the Gare de l'Est in Paris which had its sad war-day appearance. Shortly afterwards I was taken prisoner and transported to a camp north of Berlin. I thus began my second stay in Germany, which was quite involuntary and under very different conditions. That stay lasted for a year and a half; my principal occupations were to serve as an interpreter and to give biochemistry classes to audiences who were very incredulous as they were ill-prepared to accept the idea that living beings could be composed of molecules and that vital processes could be explained in terms of reactions and interactions between these molecules. One of my colleagues, a humanities professor from Paris, gave us classes on Victor Hugo. At that time in France the answer to the question "Who is the greatest French poet?" was inevitably "Victor Hugo, alas". But our professor read the verse and commented upon it with such enthusiasm and so many subtle interpretations the author may never even have thought of, that we were impressed and inspired and at the end of the lecture we gave him a standing ovation.

The Germans had decided to progressively liberate those of us who were ill. I allowed those who were actually sick to leave. Then I announced to the camp's French doctor that I was going to use my knowledge of biochemistry to give myself the symptoms of acute nephritis. The strategy succeeded beyond our wildest hopes and the German doctor, convinced or not of the seriousness of my condition, announced that in principle I would be liberated after a week or two. In fact Richard Kuhn who had been alerted by my family, had written several letters to the Central Headquarters of the German Army. One of them was effective and an hour later I was authorized to leave the camp. It was a very painful moment when I saw my companions watching my departure through the barbed-wire fence. I would have felt still worse had I known that most of them were to remain prisoners for five years under increasingly difficult material conditions. After some difficulty in crossing the actual border supervised by the occupation troops and in entering the "free zone", I arrived in Lyon. There I found not only my mother and my wife, but also a little girl named Marie-Pierre who began to howl at the sight of the unknown man who claimed a right to some of her mother's attention.

At the beginning of the forties, the conditions in France were frankly awful. Nevertheless, thanks to the devotion of my family who took my place in the food-lines in front of the stores, I was able to work with a young pharmacy student, Louis Grand, who was actually my first pupil. We noted that serine and cysteine were deaminated by *E. coli* under anaerobic conditions while alanine was not. That observation suggested that deamination was coupled with another reaction in which the β-sulfhydryl group of cysteine, or the β-hydroxyl group of serine was involved. Our hypothesis was that two hydrogen atoms were transferred from the amino group to the β-group, thus leading on one side to the liberation of hydrogen sulfide or water, and on the other to an α-imino acid converted later into an α-keto acid and ammonia. Although that formulation was later to be altered, it was a landmark because for the first time I realized that I was capable of having my own ideas and of instructing students.

Life in Marseilles (Act I)

Installation

I still ask myself today how I could have made such a quick decision to leave Lyon and take my family three hundred kilometers south to live in Marseilles. My wife was born in Lyon and both of us had done our studies there, she in law and I in science. We had our parents, our friends, and our everyday world there. I belonged to a laboratory whose international reputation was continually growing. Lyon had, and still has, many excellent schools which would have been capable of supplying me with competent collaborators. The working habits and the creativity of the people from Lyon cannot be denied. In contrast, we knew little about the people or the city of Marseilles. Our friends in Lyon, who desperately tried to dissuade us from such a rash action, assured us that there were still epidemics of plague and cholera in Marseilles and that gangsters were everywhere.

In fact, things had begun in an insidious fashion. At the end of my stay in Heidelberg, a large pharmaceutical company had made me an offer which after much hesitation I had finally refused. I was made to understand in Lyon that due to the bitter lesson learned from the years preceding the War, certain French industrialists manifested a desire to develop research in every large sector of their profession, and that to achieve that end they were ready to finance institutes with a collective goal. An Institute for Fat Research had already been created and the directors wanted to reanimate a moribund laboratory located in Marseilles, the uncontested capital of oils and soaps. The Marseilles industrialists assured me that they were ready to approach the university and take the steps necessary for the creation of a biochemistry teaching and research center. Thus the offer gave me the opportunity to direct a laboratory for fat research and to create a new department at the university in a discipline which I did not want to abandon at any cost.

The perspective of doing work which was both fundamental and practical was not disagreeable. In France it is often said that the university should collaborate with industry, which clearly means that the problem has not been entirely solved. Some claim that researchers should remain "pure", others that they should choose their subjects in view of their possible applications. For myself, I choose my subjects according to my own ideas and those of the people who work with me. But I have never refused to help those who thought that the results obtained could be applied in industry or medicine.

Another important argument which pushed me to accept the proposal was that it would allow me to have working premises, salaries to remunerate collaborators, and a little money at my immediate disposal; they were all things which were badly lacking at the university. After all, a U.V. spectrophotometer acquired for the determination of conjugated double bonds in heated oils can also be used for measuring the protein content in any biological system. To be perfectly honest, however, I should add that because of the War and all the material difficulties which I had been up against for so long, I had the feeling of having lost many years which should have counted among the most productive of my life, and that therefore I would have to make a substantial effort. I realized that I would be able to make such an effort only if I confronted certain difficulties and confronted them alone. My dislike for simple solutions, which revealed itself then for the first time, was to create serious problems in the future.

Thus I rather abruptly left Claude Fromageot, who, with his usual generosity, did not hold it against me. We continued to keep in frequent contact in Lyon, and then in Paris. He often invited me to share the modest meal that one of the cleaning women, promoted to the rank of cook, prepared for him at the laboratory. We spent a month together in Japan, and it was after he returned from that trip that he felt the first attacks of an illness which was to carry him off in less than a week at the age of fifty-six.

In the meantime my family had grown, thanks to the birth of little Claire-Françoise. Thus all four of us left Lyon, capital of Gallo-Roman France and installed ourselves in Marseilles founded by Protis the Greek and by Gyptis the Ligurian. For a long time the city had resisted the authority of the French kings in order to keep its port open to African and Middle and Far Eastern countries.

The bewilderment of new surroundings which would have been overwhelming was fortunately attenuated by the existence of an excellent biochemistry laboratory at the Medical School in Marseilles. The laboratory was directed by Jean Roche who was soon afterwards to become a professor at the Collège de France and Rector of the Académie de Paris. Jean Roche offered me some very wise advice which was of great help in overcoming my first problems. When my laboratory was partially destroyed in an air-raid, he generously offered the use of his own laboratory, located in an old convent surrounded by fig-trees. He and his wife Elsie introduced us to their large circle of friends, thus easing our installation into an unknown city.

The laboratories which the university assigned to me were very large, but they were only frequented by a young Russian girl who smoked a pipe. A more serious situation was that the drains in the sinks were blocked because, as I was told, the former occupants had tried to perfect a formula for a new wax. Moreover, the water supplied by the city arrived in large zinc containers and was placed just below the roof and then distributed to the different floors by force of gravity. That original arrangement, whose justification I have forgotten, had the disadvantage of creating a very low water pressure on our floor, and was thus incompatible with the use of water ejectors. The means of action for those who wanted to work, although they were not numerous, were thus seriously limited.

The first signs of organization began nonetheless to materialize and young students full of enthusiasm were following the path leading to the laboratory when our efforts were brutally interrupted by the terrible air-raid mentioned above. Not only part of the university, but also many of the houses surrounding the one where we lived were destroyed. Therefore, we decided that my wife and our two little girls would leave for Lyon the next day and I would stay on for a few days to see how the situation evolved. In fact, the train they took to leave was one of the last to depart from Marseilles for several very long weeks. The air-raid was a prelude to the disembarkation of the allied troops east of Marseilles. Communications with Lyon were thus cut off until the German Army was pushed back further north. As soon as I heard the news, I jumped on a military jeep where the soldiers let me keep my place despite regulations, and I arrived in Lyon to find that my wife had left herself for Marseilles on a bicycle in the hope of joining me. We must have passed each other without realizing it somewhere along the road which was very crowded and scattered with debris from the recent fighting.

Our final problem was to get back to Marseilles while all the means of transportation, the train, bus, and of course private cars, were still completely disorganized. So we put 4-year-old Marie-Pierre and one-year-old Claire-Françoise into a military truck. Even though it had not been designed to carry children, in our eyes the truck still had the immense advantage of going in the right direction. We arrived safe and sound the following day. Our hasty return trip allowed us to find without much difficulty an agreeable apartment with a large square entrance hall, and a sunny terrace where the children could play. From that moment onwards, our family life progressively took on more normal forms. We had very good friends with whom we spent our Sundays in the calanques, the small fjords found a few kilometers from Marseilles. The calanques are an extraordinary playground for those who like open air activities. They are narrow valleys, entered by the sea and bordered by high limestone walls falling sheer into the blue water. We liked to go climbing there without suffering the drawbacks of cold, bad weather, and high altitudes. When the family finances allowed it, we bought a small sailboat, which was replaced by a larger one several years later. But our natural tendencies, doubtlessly inherited from former generations, carried us rather to the mountains and we were never really good sailors. We generally spent the month of August in the Southern Alps where we camped on land covered with larch-trees which we had bought from a farmer for the price of a cow. Our third child, a boy this time, was born several years after the War. We named him Claude, and asked Claude Fromageot to be his godfather.

Chemistry and physicochemistry of lipids

A number of my friends, especially those who are American, have reproached me for the deplorable habit I have had during my entire career of studying several subjects at the same time. In order to avoid overlapping and repeating myself, I am thus obliged to disregard the chronological order of the subjects I have studied and present them one after another while giving the names of the young scientists who participated in the work. It seems logical to begin with the lipids as theoretically it was because of them that I came to Marseilles.

Breastfed on amino acids, proteins and enzymes, I had ignored almost everything about lipids; fortunately for me, little was known about them at the time and we set ourselves courageously to work. With Maurice Naudet who is now Professor of Fat Chemistry at the university, we had the idea that the various fatty acids which compose the triglyceride molecules in natural oils were perhaps not distributed at random on the three-carbon glycerol backbone. If this assumption was correct, randomization by chemical interesterification could be expected to modify the structure of the glycerides and consequently the physical properties of the oils. Actual randomization was demonstrated with the relatively poor techniques available at that time and some properties were found to be significantly changed. But we were not able to benefit from the later important industrial development of the technique, because we failed to find the most potent catalyst and to realize that in continually eliminating one of the products of the reaction, interesterification could be "directed".

Another work on minor oil components called "mucilages" was done a little later with Jean Molines, Jean-Paul Helme and Robert Massoni. We showed that these substances were not merely complex mixtures of sugars or sugar esters, but mostly amphipathic phospholipids which swell in the oil phase under addition of water. The resulting myelin-like figures were shown to be formed of phospholipid-triglyceride mixed layers, thus explaining the often substantial oil losses observed when non-washed oils were refined. The triglycerides included in the swollen phase constituted an "unavoidable" loss to which should be added an "avoidable" loss due to poor phase separation and other technical difficulties. A short while later, with our limited knowledge of physics, Maurice Naudet, Elie Sambuc, and I were able to perfect a trichromatic technique for the determination of the color of crude and refined oils. The technique, for which we built a special colorimeter equipped with three interferential filters, produced good results. However, on an international level it had great difficulty in replacing the Lovibond colorimeter whose principle seemed very doubtful to us, but which was readily used by the British Oil Industry and, consequently, by many others.

With Maurice Burnet we also worked out a reliable technique for the determination of the oxidized fatty acids present in thermally or oxidatively altered oils. The technique was based at first on reverse chromatography. But we found later that better results could be obtained using silica or deactivated alumina columns. After more than twenty years, the technique has naturally evolved, but the principle remains the same.

The use of pancreatic lipase for elucidating the glyceridic structure of animal and vegetable oils, will be described later in a chapter devoted to this enzyme. In contrast, I can state here that I helped the Institute for Fat Research to create a Biochemistry Section in Marseilles which is now headed by one of my former students, Bernard Entressangles. Two main subjects both having to do with fat nutrition are studied there: (a) the passage across the intestinal brush border membrane of abnormal fatty acids (oxidized, cyclic, etc.) generated during oil processing and storage; (b) the incorporation into membranes, especially the outer and inner membranes of heart mitochondria, of the trans isomers formed during the hydrogenation of unsaturated oils.

The pancreatic zymogens (first period). The revival of biochemistry in France after the harsh war years was launched by a conference organized in Paris honoring Louis Pasteur, and a French-Belgian meeting held in

Liège at Marcel Florkin's initiative. Belgium was back on its feet sooner than France after the War, and I remember the emotion we felt at seeing not only our Belgian friends whom we had not met for so long and who greeted us with extraordinary warmth, but also the store windows full of merchandise which we had almost forgotten. The meeting in Liège ended as usual with a banquet. That year it was a truly memorable one as it was held in a room in the city's museum and we dined amidst paintings from Picasso's blue period. Professionally, those two international conventions were the first that I had the privilege to attend. It was very enjoyable and also immensely stimulating because I realized the long and painful distance we would have to cover before regaining the ground lost because of the War and its aftermath.

My activities regarding lipids led me to read a paper on a novel colorimetric method for the determination of glycerol after oxidation of the α-glycol structure with periodic acid. The same reaction could be applied to α-amino alcohols and therefore to serine and threonine. It could also serve for the characterization of the unusual amino acid hydroxylysine in collagen hydrolysates after separation of the basic amino acid fraction by electrophoresis.

In the years following the War, amino acid separation and analysis underwent extensive development, especially in England thanks to Martin's and Synge's work on paper chromatography. I obtained some money to spend three months at the Biochemical School in Cambridge. In spite of the delightful springtime allure of the Backs and the Cam, I spent most of my time in the laboratory where I met Fred Sanger who was working in the basement on the primary structure of insulin. K. Bailey, who unfortunately passed away a few years later, R.R. Porter, already quite enthusiastic about immunology, and S. Perry worked in the same laboratory. Fred Sanger generously welcomed me in spite of the restricted amount of equipment, which was limited to a few home-made cabinets for paper chromatography and one colorimeter. I often cite Fred Sanger's example to certain of my young colleagues who disdainfully declare that they do not want to start work until their laboratory is fully equipped. It seems to me that the most important factor in scientific research is still the brain and that great discoveries, like those made by Fred Sanger, are still possible with limited equipment. I always admired Fred Sanger's talent for interpreting paper chromatography patterns without ever making a mistake. We never possessed such a talent which requires a real artistic gift and allows one to work very rapidly while using small quantities of material.

Time spent in underground rooms has been decidedly beneficial for me; it was in that laboratory that I tried to discover if Fred Sanger's clever technique for the determination of the N-terminal amino acids in proteins and peptides could also be used to follow the so-called limited proteolysis during which a protein is converted into another protein by chemical or enzymatic cleavage of only a few bonds. Before leaving France, I had become interested in the nitrogen-oxygen transposition of acyl chains in α-amino alcohols under the influence of strongly acidic media. Preliminary assays with various N- and O-acyl serine derivatives had shown that stable amide bonds involving this amino acid could be transformed into more labile ester bonds, thus presenting perhaps the possibility of specific cleavages of the protein. As a matter of fact, very high amounts of the dinitrophenyl derivatives of both amino acids could be characterized after acidic treatment of several proteins. But other bonds were also cleaved, although at a much lower rate, and the technique was never used in laboratory practice. At the end of my stay in Cambridge, my wife came to join me and, considering the abominable reputation of the French for such things, we had to have a long discussion with my old landlady at the bottom of the stairs before she would allow us to climb them together.

On my return to France, I happened to buy J.H. Northrop, M. Kunitz and R. Herriott's beautiful book *Crystalline Enzymes* and I discovered in it that pancreatic enzymes were very abundant, relatively simple, molecules which could be fully purified in a high yield. And also that they were devoid of coenzymes so that the protein should be entirely responsible for activity and that they were important in gastroenterology. A fascinating problem was posed by the zymogens synthesized by the gland in an inactive state and later activated in the duodenum under the influence of trypsin. As they were triggered by trypsin, the conversions could be expected to be proteolytic in nature and to shed some light on the formation of an active site in an initially inactive protein. With Mireille Rovery, Colette Fabre and Mireille Poilroux, we thus decided to use Sanger's N-terminal technique to compare residues of chymotrypsinogen and trypsinogen with those of chymotrypsin and trypsin.

Meanwhile, the Rockefeller Foundation awarded a grant to the laboratory after one of its directors, Georges Pomerat, who was of French descent, came to Marseilles for two days to discuss the nature of my projects. I was later able to compare that system, in which someone is sent to seek on-the-spot information and to grant funding accordingly over a period of several years, with the French system where any person requesting money for work is considered a potential malefactor and should therefore produce progress reports to justify himself several times a year.

The Foundation also granted me a fellowship to spend three months in the United States. Naturally, I began my periplus at Rockefeller University in New York which was still called Rockefeller Institute at the time. Although the atmosphere there was certainly very different from that of the Biochemical School in Cambridge, it was just as agreeable and stimulating. My first visit was to Stanford Moore and William Stein who lavished upon me, and continued to lavish upon me the tokens of their high esteem and warm friendship. The two of them had just perfected their well-known techniques for the separation of amino acids, peptides, and proteins by chromatography on ion-exchange resins, and they had begun to apply them to ribonuclease. Many years later I asked Stanford Moore to attend the ceremony given for my reception into the Académie des Sciences. He did me the great honor of accepting and delivering a speech in French in which he recalled our numerous meetings in various places across the globe. I also met Georges Palade who was in the midst of working on his famous study of the intracellular transfer of pancreatic secretory proteins, and who made me realize the beneficial aspects and the need of close collaboration between biochemists and cell biologists. Unfortunately, it was only much later when we were studying the intestinal enzymes that I was able to profit from the idea.

Another encounter which was to have important consequences in the immediate future was that with Moses Kunitz. In spite of his advanced age, he still worked in a laboratory where an open bible was placed on a varnished wooden pulpit in the middle of the room. After having listened attentively to my projects, he opened the door of a large cold room virtually filled with crystalline enzymes and zymogens and he very kindly gave me several grams of the invaluable products.

From New York I went to the opposite coast of the United States to meet Hans Neurath in Seattle. I discovered that he and his colleagues were investigating the other side of chymotrypsin (the C-terminus) by the carboxypeptidase method. In principle, our aims were separated by the whole length of the enzyme chain. However, probably because of a special folding of the molecule, the two sides were not very far from each other. For some years this created a certain competition which always remained fair, moderate and stimulating.

Upon my return to the laboratory, we soon found that the samples so generously given by Moses Kunitz contained some autolyzed products which could not be entirely removed by repeated crystallization as more were formed at each step. However, the successful inhibition of chymotrypsin by diisopropylfluorophosphate had just been published, and we set out to prepare the reagent without knowing that it was one of the most potent war gases. During the preparation, sealed glass ampoules containing several grams of product had to be heated for several hours in an electric oven. None exploded and we survived without any sign of severe intoxication, except for contracted pupils which lasted for over a month.

When the inhibited enzymes or pretreated zymogens were crystallized, everything became quite clear. No N-terminal residues were detected in chymotrypsinogen, due, as was later shown by others, to the presence of a half cystine at the N-terminus. In contrast, α-chymotrypsin prepared according to Kunitz by "slow" activation of the zymogen, contained two N-terminal amino acids: isoleucine and tyrosine. With trypsinogen the situation was somewhat different since an N-terminal valine was replaced by an isoleucine, suggesting that the scissile bond was in that case located outside the disulfide bridge network. A little later, we showed that chymotrypsinogen activation also resulted from the cleavage of a single bond. The primary enzyme (π-chymotrypsin) formed by "fast" activation was found to contain only isoleucine, while the tyrosine of α-chymotrypsin was due to autolysis or initial attack on chymotrypsinogen by its own enzyme, chymotrypsin. Fig. 9 gives a simplified version of the events leading from chymotrypsinogen to the last member of the series, α-chymotrypsin.

Structural work on chymotrypsinogen and chymotrypsin was not continued in our laboratory, except for a novel technique for separation of the B- and C-chains of chymotrypsin by ion-exchange chromatography in concentrated urea. It was probably the first or one of the first examples of long peptides being separated in a dissociating medium. The brilliant sequence analysis work performed by Brian Hartley's group in Cambridge and Hans Neurath's group, as well as the crystallographic investigations by David Blow, then in Cambridge, J. Kraut and W.N. Lipscomb showed all the details of how the cleavage of a single peptide bond in a protein may induce a transconformation converting a latent site into one that is fully functional and active. Limited proteolysis first studied with pancreatic zymogens, later turned out to play a very important role in a number of other biological processes such as blood clotting, activation of pre-hormones, the co-translational processing of secretory pre-proteins, etc.

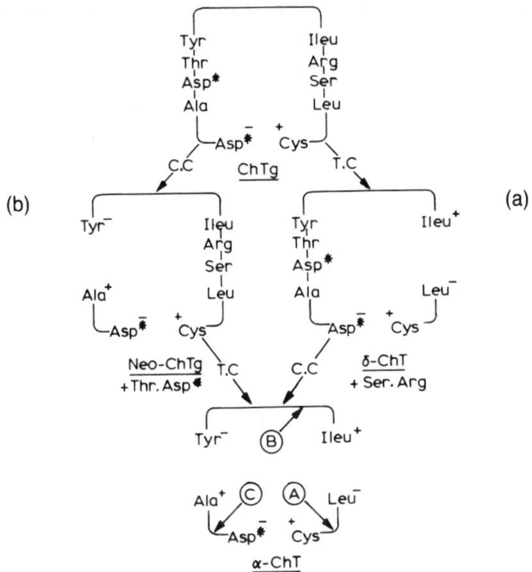

Fig. 9. "Fast" and "slow" activation of bovine chymotrypsinogen A showing the two routes leading to α-chymotrypsin: (a) Activating cleavage of the Arg 15-Ile 16 bond by trypsin followed by autolysis of the resulting δ-chymotrypsin. (b) Non-activating chymotryptic degradations of chymotrypsinogen during the "slow" degradation. These degradations lead to intermediary, still activatable neochymotrypsinogens. (From P. Desnuelle, The Enzymes, 2nd ed. (B.D. Boyer, H. Lardy and K. Myrbäck, Eds.) Academic Press, New York, 1960, pp. 93–118.

Pancreatic lipase (first period). Working on both the lipids and the enzymes of the pancreas, it was inevitable that I would eventually run into the pancreatic lipase which is excreted into the duodenum for the hydrolysis of dietary triglycerides. On thinking it over, that enzyme seemed to call for special interest as, opposed to the other enzymes operating in a homogeneous aqueous phase, it apparently acted very fast on insoluble substrates which formed emulsified particles separated from water by an interface. In that case, it seemed logical to think that the site at which lipase should exert its activity was the interface where both substrate and water were available, thus creating the conditions which we later called "heterogeneous biocatalysis". With Marie-Jeanne Constantin, we quickly noted that very little investigation had been done in that area, and that the first thing to do in order to make things a little clearer was to perfect an accurate technique for the dosage of the enzyme. Indeed, most of our predecessors had avoided employing emulsified substrates which they thought would contradict the laws of classical enzymology. We soon demonstrated that lipase only recognizes emulsified and micellar substrates, and that it was thus necessary whether desirable or not, to use them for the dosage. In fact, conditions which insured zero-order kinetics during an appreciable period of time were easily found; and the technique, now in current use in a number of laboratories, was automatized when reliable recording pH-stats became available. We were then able to begin purifying our enzyme, which for one reason or another took us a long time. I must admit that our studies on the properties of lipase were already well advanced when Robert Verger and Gerhart de Haas succeeded in appropriately purifying it. In our preparations there still remained a little colipase which slightly altered its behavior.

Quite unexpectedly, hyperbolic curves similar to those of Michaelis-Menten were obtained with lipase when the amount of emulsified substrate was increased for a fixed amount of enzyme. Therefore, it could be concluded that when more interface was offered, more enzyme was adsorbed and active, thus increasing the rate until complete enzyme adsorption was achieved. With Gilbert Benzonanna, we spent a long time demonstrating that the essential parameter controlling lipase action was not as usual the substrate concentration, but what we called the "interface concentration" expressed in square meters per volume unit of emulsion. An important characteristic of heterogeneous biocatalysis is that for the same weight of substrate and enzyme in the same volume of emulsion, the hydrolysis rate strongly depends on the interface area and consequently on the mean diameter of the emulsified particles.

A second characteristic of lipase action was discovered by Louis Sarda when he explored the behavior of the enzyme towards soluble substrates. The enzyme activity remained very low as long as the substrate concentration was below saturation; but it sharply increased when emulsified particles began to form in a saturated solution. In contrast, the activity of a true esterase extracted from horse liver was high in solution with no detectable increase beyond saturation. This finding clearly showed that lipase was not only able to act very fast on insoluble substrates, but also that it could not act, or acted very poorly, on isolated molecules dispersed in water. It was extended some years later to micelles by Bernard Entressangles thus showing that the enzyme required a certain degree of aggregation from its substrate; and it is therefore fully adapted to its function.

When we began our work on lipase, it was generally accepted that it completely hydrolysed triglycerides into glycerol and three fatty acids. A one-step mechanism which would have required quaternary collisions between one triglyceride and three molecules of water was highly unlikely, so the real problem was to know if intermediary di- and monoglyceride accumulated during hydrolysis because they were generated faster than degraded. In fact, Marie-Jeanne Constantin easily identified these products in mixtures resulting from lipase action in vitro, and she found that the reaction virtually stopped at the monoglycerides stage. She also identified high amounts of monoglycerides in the intestine of rats which had received triglycerides by gastric intubation 3h before killing. But there were groups other than our own, in particular that of Bengt Borgström in Sweden and that of F. Mattson in the United States, who discovered that pancreatic lipase possessed a "positional" specificity directed towards the two triglyceride outer chains which were hydrolysed much faster than the inner chain, thus leading successively to 1,2-diglycerides and practically stable 2-monoglycerides. This finding influenced our ideas concerning the mechanism of the passage of digested fats across the brush-border membrane and the resynthesis of triglycerides from 2-monoglycerides and fatty acids which takes place inside the enterocytes.

In contrast, Bernard Entressangles, Pierre Savary and I realized that the "positional" specificity of lipase could be used for investigating the structure of triglycerides in natural oils. If, as already pointed out, this structure was not random, the liberated fatty acids (outer chains) should be different from the inner chains remaining in the monoglyceride fraction. In this way, we found that the chain distribution was very peculiar. For example, the inner positions in triglycerides from oil seeds and fruits were shown to be almost exclusively occupied by unsaturated chains. Even though afterwards the method needed numerous improvements, it was sufficiently appreciated by the industry and laboratories responsible for suppressing fraud, and I was awarded the Médaille Chevreul by the Association Française des Techniciens de Corps Gras, and a short while later Professor A. Seher, then President of the Deutsche Gesellschaft für Fettforschung, presented me with the Médaille Normann in Strasbourg during a very agreeable ceremony.

Adaptation of pancreatic enzymes. I tried with J.P. Reboud, A. Ben Abdeljlil and J.C. Palla to find out whether mammals share with microorganisms the capacity of adapting their enzymatic equipment to external stimuli. Any adaptation to the diet similar to that resulting in microorganisms from the composition of the culture medium, can be expected to show up primarily in the pancreatic secretion which is involved in the digestion of all types of nutrients. In other words, our working hypothesis was that the normal balance existing between the different enzymes composing this secretion is altered by ingestion of an excess of starch, proteins or lipids.

To test this hypothesis, we set out to determine the amylase, chymotrypsin, and trypsin "levels" in pancreas homogenates originating from rats fed a starch-rich or a protein-rich diet. We actually found large differences in the expected direction. Adaptation starts immediately after diet is changed, it takes about five days to reach new steady-state enzyme levels and it is fully reversible. Parallel variations were observed in pancreatic juice, thus demonstrating that the output of individual enzymes is really affected by the diet.

Since enzyme levels are complex parameters including biosynthesis, degradation and possible effects on enzyme activity, the next step was to determine the relative and absolute rates at which the enzymes were synthesized in the two groups of animals. As judged by radioactive amino acid incorporation into purified material, the variation was five-fold for amylase and four-fold for chymotrypsinogen. A significant finding was that adaptation was triggered by glucose or free amino acids as well as by starch and casein. Therefore, it was concluded that the observed modulation of pancreatic enzyme biosynthesis is due to a still unknown effect of the digested products, either directly on the intestinal wall, or/and after their passage into blood.

In the latter respect, a stimulating hypothesis is that a higher glucose discharge into blood caused by the ingestion of a starch excess induces a higher insulin discharge which enhances amylase biosynthesis. In fact, as shown in Fig. 10, amylase biosynthesis is almost stopped in alloxan-diabetic rats and it returns to normal upon insulin injection. Although very large (11-fold), this effect is still unexplained.

Construction of our first institute at the university

When Claude Fromageot was appointed Full Professor of Biochemistry at the Sorbonne, he asked me to come to Paris and take over the position he had just vacated. The Director of University Studies, who was then

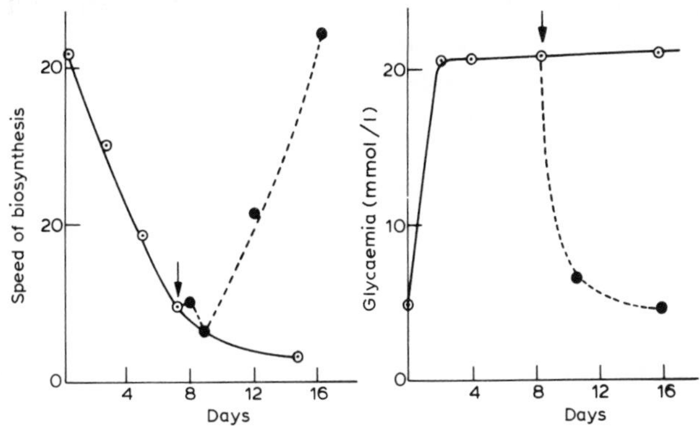

Fig. 10. Effects of insulin on the biosynthesis of pancreatic amylase. (Left) Amylase biosynthesis rate in alloxan-diabetic rats (solid line). After 7 days (arrow), some animals received daily injections of insulin (interrupted line). (Right) Blood glucose levels in same animals. (From J.C. Palla, A. Ben Abdeljlil and P. Desnuelle, Biochim. Biophys. Acta, 158 (1968) 25.

Gaston Berger, was quite worried and summoned me to his office. Gaston Berger was originally from Marseilles, which may explain part of his desire to have me stay there. He was above all a famous philosopher, who, in spite of that apparent handicap, understood better than many scientists the needs in France for graduate education in the sciences, and took official action in that direction. He was very agitated when I found him in his office because a colleague from another university had just asked him for money to build a private stairway leading directly from his laboratory to the street, so that he could avoid meeting a person with whom he had had a dispute. Gaston Berger calmed himself to tell me that if I stayed in Marseilles, he would finance the construction of a biochemistry institute and I would be given the directorship. The offer was so tempting, that I accepted it immediately.

The institute's construction was a very interesting adventure as the administration in Paris had not yet completely put on the iron gloves which it was afterwards to use to crush the initiative capacities and inventive spirits of its employees. I was given a free hand with the sole obligation of keeping a budget within the allocated funds. Under those conditions, our first problem was to find out where the institute could be built. My wife had discovered an attractive lot of land covered with pines and overlooking the sea a little way outside Marseilles. However, I preferred to stay closer to the university in the hope of attracting more students to biochemistry. Thus I laid claim to the last available building site on campus which had been until then occupied by a hangar. Old, useless things being practically indestructible in France, I would never have succeeded in my annexation enterprise without the constant and friendly backing of our dean, Pierre Choux, who well understood the benefits it would give our university.

Another problem which was just as important was to decide what would be built. Few architects in Marseilles knew exactly what a research institute was and how it would be used. Thus, I had to draw up the plans for the interior, calculate the pipe diameters, devise the laboratory fixtures, and see that the space reserved for offices was as small as possible. However, I did not regret the trouble I took as it was subsequently so rewarding to watch the gradual construction of the tridimensional structure whose blue-print plane representation had occupied my thoughts for so long. As usual, the longest delays were those caused by the administrative formalities. I was forced to go and request the return of "my" file from the Office of the Finance Controller who had not serious objections concerning the way the money would be spent, but did not like the plans for the central heating. I reassured him on the subject, and the file was able to continue its long journey through the offices. Once the construction began, things progressed normally and we were able to inaugurate the institute on a beautiful spring day with Gaston Berger present. We moved in shortly afterwards, not only my own group which was relatively large at the time, but also that of Roger Acher, a former student of Claude Fromageot and an excellent specialist on the structure of neurohypophyseal hormones. Roger Acher's scientific motivations were so strong that one summer day when the temperature reached at least 32° C, we received huge plastic bags containing chicken heads out of which he was supposed to extract the hypophyses. We called the students to our rescue, and many of them showed great commitment on that occasion. It was also an excellent test which allowed us to make a selection which was doubtlessly much more accurate than a simple biochemistry exam. The premises we occupied at the university were allotted to Maurice Naudet who could thus develop his own service of fat chemistry. The institute was well equipped in large part thanks to the grants given by the National Institute of Health, the Rockefeller Foundation, always well disposed towards me, and of course our Centre National de la Recherche Scientifique (CNRS). For convenience, the University's Institut de Chimie

Biologique will from now on be called "Institut St. Charles", named after the neighborhood in Marseilles where it is located.

Life in Marseilles (Act II)

Interlude in Paris

After Claude Fromageot's death, my closest friends, Jean Roche, René Wurmser and Georges Champetier, asked me to continue his work in Paris. The offer was tempting as it concerned trying to replace one of the best biochemists of the time in what was still called a "grande chaire", an important professorship. Moreover, the more or less avowed desire of most French civil servants, including the academicians, was and still is to finish their career in Paris, where better working conditions and more interesting people are found. And lastly, the laboratory created by Claude Fromageot was located in the heart of Paris and excellent groups were working there.

However, the offer put me in a cruel state of confusion because the construction of the institute which I described above had just been completed, and the research there was off to a good start thanks in particular to the generous American grants. Leaving would mean separating myself from my loyal colleagues who were already well aware of the problems we would be attempting to solve. So I was not really sure that I would be able to accomplish in Paris what I had already begun to achieve in Marseilles after so much difficulty. Moreover, many of my friends in the States and elsewhere had expressed their surprise at the excessive centralization in Paris and they voiced the opinion that it should be fought against. It was only later that I realized that that fight, like Don Quixote's, was lost in advance.

I stayed in Paris for three months before I decided to return to Marseilles. That difficult decision left me with few regrets, except that those who had so zealously supported my candidature were inconvenienced by my hesitation. However, a happier consequence of my leaving Paris was that a second biochemistry professorship was created. One was given to Jacques Monod and the other to Edgar Lederer. The future of biochemistry in Paris could not have been left in better hands. As for myself, I thereafter devoted part of my energy to developing biochemistry in "province"; and I must admit that I have had the great satisfaction of seeing excellent groups form and prosper in many of our big cities. In Marseilles itself, the arrival of Roger Monier, another of Claude Fromageot's students, allowed us to found a flourishing molecular genetics group, soon followed by a biophysics group directed by Michel Lazdunski. Thereafter we were well equipped to play a respectable role and take advantage of the possibilities which presented themselves.

Creation of a second research center (C.N.R.S.)

About ten years after the inauguration of the biochemistry institute at St. Charles, Jacques Monod asked me to become part of the molecular biology committee that had just been formed within a new organism, the General Delegation of Scientific and Technical Research. Thanks to the brilliant discoveries made by Jacques Monod, François Jacob and André Lwoff, the committee had been handsomely endowed by General de Gaulle. In spite of being little liked by most of the researchers, de Gaulle had a great admiration for research, and allotted it with considerable sums for the first time in our country's history. We were thus in a position not only to aid those groups which were already formed, but also to finance new operations and the construction of new biology institutes. After Paris and the nearby surrounding regions had nicely profited as usual, I took the initiative of presenting a project in favor of Marseilles, arguing that with the groups of Roger Monier, Mireille Rovery, Guy Marchis-Mouren, Michel Lazdunski and Louis Sarda, the biochemistry institute at St. Charles had reached the maximum compression threshold after which researchers' dispositions tended to darken. I had the good luck to be heard. The delegation allotted the C.N.R.S. with the money necessary to build a biochemistry and molecular biology center on land next to the campus it already possessed in Ste. Marguérite, a southern suburb of Marseilles.

Naturally, that new venture incited many difficulties which Roger Monier and I tried our best to resolve. The first problem was brought up by one of our colleagues who was a specialist in acoustics. He decided that the new ground should be dotted with small buildings reserved for a maximum of one or two researchers. He said that in his field, noise made by some might bother others. Thus it would be advisable to isolate each building by large deserted spaces, as is done in nuclear centers and ammunition depots. The problem took on alarming proportions over the next month, and was only resolved after the arrival in Marseilles of an important administrator from the C.N.R.S. main office. I was unable to completely stop the C.N.R.S. from applying what I called "cabanon" politics on their Marseilles terrain. "Cabanons" are small houses built on the seaside by Marseilles fishermen and the less wealthy summer residents. But at any rate, in the end a suitable site was allocated to us, and it was possible to construct a building of reasonable dimensions there.

Like the first one, this new construction progressed rapidly. This time the administration was well organized and took care of everything. We only had to furnish the plans for the laboratories and to sign the building agreements whereby we took on the responsibility without having the capacity to understand them. In the end all went reasonably well, with the exception of some of the architect's fantasies which took several years to be remedied. One of the most bothersome was that, following the gas company's recommendation, the architect had allowed for large openings in the walls to permit future ventilation in the spaces between the actual ceiling and the plaster. He had inadvertently placed them facing the Mistral, the sometimes furious north wind which blows over Marseilles more than a hundred days a year. The resulting turbulences naturally worked havoc. Roger Monier, who wanted to develop a microbiology section, had to have a sort of autonomous shelter made in his laboratory. Fitted with a roof, it kept the glass-wool debris and other rubbish from falling into the petri dishes.

It would have been naïve and even imprudent to think that by multiplying the available surface by two, we could double our scientific output. However, the construction of the biochemistry and molecular biology center at Ste. Marguérite did have the advantage of allowing more people to work in better conditions. On several occasions the center played the role of an interplanetary rocket by launching brilliant biochemists into orbit. Roger Monier was the first to leave in order to direct a large cancer institute in a Parisian suburb. Michel Lazdunski left for Nice to create ex nihilo an extremely active biochemistry service. And lastly, Michel Fougereau took charge of the direction of an important immunology institute at Marseilles-Luminy. Although those departures were not due to any dissatisfaction, but rather to the legitimate desire of competent men to fully exercise their responsibilities in a larger sphere, they did momentarily weaken the center. But each time the center was able to raise itself up again and maintain its activity at a satisfactory level. I do not doubt for an instant that it will still be capable of doing so today, after the serious shocks which followed the announcement of my retirement.

Proteolytic enzymes and zymogens (second period)

Enteropeptidase. As previously pointed out, the pancreas does not directly produce active enzymes which may destroy the cells in charge of their biosynthesis. Inactive zymogens are secreted and later activated upon their arrival into the duodenum by cleavage of a single peptide bond. It is noteworthy that all proteolytic zymogens and also pancreatic prophospholipase A are activated by trypsin. Therefore, the conversion of trypsinogen into trypsin should be regarded as a key step which controls all other activations by a cascade process.

J.H. Northrop and Moses Kunitz were the first to show that pure trypsinogen was activated in vitro by pure trypsin. However, the reaction was found to follow an autocatalytic course, thus implying a comparatively low initial rate. Side processes leading to substantial amounts of "inert" proteins were also noted. This meant that the trypsin-catalysed activation of trypsinogen in vivo is neither fast nor specific, contradicting the urgent need for active trypsin in the duodenum. Hence, the activation must be triggered by a more efficient enzyme.

Suzanne Maroux and Jacques Baratti thought that a good candidate for this important role was enterokinase (now called enteropeptidase), which had already been characterized in the duodenum by physiologists, but never studied chemically. Enteropeptidase was fully purified from porcine duodenum and found to be a large glycoprotein (molecular mass about 300 kDa; more than 40% sugars). It was also found to be a serine enzyme composed of two peptide chains linked by disulfide bridges; the active serine and histidine being located in the "light" chain. Moreover, it was shown that in contrast with earlier claims, enteropeptidase is not secreted into the intestinal lumen, but loosely bound to the duodenal brush-border membrane.

However, the most significant finding concerning enteropeptidase was that it is a "super trypsin" which does not merely recognize basic bonds in peptide chains, but exclusively distinguishes the sequence $(Asp)_4$-Lys present in all trypsinogens known so far. This unusual sequence is actually a written message which is read by enteropeptidase and insures a fast and fully specific activation. The crucial role of enteropeptidase in protein digestion was independently proved by others, who found that genetically induced enteropeptidase deficiency in children leads to the same symptoms as those observed after prolonged ingestion of a protein-deficient diet.

Bovine procarboxypeptidase A. Another proteolytic zymogen, pancreatic procarboxypeptidase A, is a trimer in bovine juice. One of the subunits was identified long ago by Neurath's group as being the real zymogen of the enzyme carboxypeptidase A. In contrast, the nature of the other two remained controversial due to the lack of a suitable technique for dissociating the complex under non-denaturing conditions. Antoine Puigserver thought that in such cases dimethylmaleyl anhydride might be a good dissociating agent, since dimethylmaleyl groups were known to be easily removed near neutrality. Indeed, the procedure worked very well and the three subunits of bovine procarboxypeptidase A were obtained in an apparently native state sufficient to allow reassociation. The second subunit was found to be a chymotrypsinogen of the C-type, which differs from the more classical A and B lines by a marked preference for leucine bonds after activation and a slightly higher molecular weight.

Another aspect of the work showed that the third subunit was an unactivatable zymogen, i.e. a zymogen possessing a latent active site which would not become fully functional upon trypsin action. So far two possible explanations for this curious property have been suggested by structural work. (A) The first seventeen amino acids of the N-terminal sequence are lacking in the subunit, including the two hydrophobic amino acids essential for activation. However, chemical grafting of these amino acids at the N-terminus of the protein has no effect. (B) One disulfide bridge is displaced as compared to those in normal chymotrypsinogens, and the new one may perhaps rigidify an essential region of the molecule which should be slightly distorted during activation. Sequence and crystallographic investigations are presently being continued in our laboratory to try to clear up this point. Pending further information, one may wonder why bovine pancreas still synthesizes an apparently inactive protein at a very high rate. The characterization of this protein, "killed" by one (or several) lethal modifications, was possible because of its still unexplained ability to form a complex with real zymogens.

Topology and mode of insertion of the hydrolases bound to the intestinal and renal brush-border membrane

The study on enteropeptidase drew our attention to a new family of enzymes that were very different from the pancreatic enzymes which we were used to examining. Instead of being extruded from a cell and dissolved into secretion fluid, the intestinal enzymes are bound to a membrane and therefore immobilized exactly where they are needed. Such enzymes, called membrane enzymes, present many important problems. What is their structure and their position with respect to the membrane core? How are they transferred from their biosynthesis site to the cell membranes? How are they assembled with the lipid bilayer? Several of these questions are still unanswered.

The last impurity removed during the purification of enteropeptidase was an aminopeptidase which was quite obviously one of the several hydrolases bound to the brush-border membrane of intestinal absorbing cells (enterocytes) as established by the investigations of R.K. Crane in the United States and G. Semenza in Switzerland. The brush border, which is the strongly invaginated part of the enterocyte plasma membrane facing the lumen, had already been isolated by Crane's group. With Suzanne Maroux and Daniel Louvard, we obtained the brush border in the form of sealed, right-side-out vesicles. The critical step of the procedure was to increase the density of the microsomes in the presence of calcium ions. Although our vesicles were not functional for transport, they turned out to be an excellent model for investigating the so-called surface constituents of the intestinal and renal brush border. We also purified two aminopeptidases (N and A) from pig and rabbit intestine and an aminopeptidase N from pig kidney. All were shown to be high-M_r glycoproteins composed of a large, enzymatically active, hydrophilic domain and of a short hydrophobic segment. Two molecular forms could be obtained in each case: a freely soluble "protease form" liberated from the vesicles by incubation with papain, and a strongly amphipathic "detergent form" extractable by neutral detergents. Suzanne Maroux and Daniel Louvard worked out a totally original technique to show that the number of antigenic determinants accessible to a monospecific antiaminopeptidase antibody in the papain solubilized form of pig aminopeptidase N was only slightly higher than in the bound form. They thus demonstrated that the major part of the enzyme surface protrudes from the external side of the membrane, and that therefore the enzyme like many others of the same group is a typical surface constituent of the intestinal and renal absorbing cells. As previously mentioned, the hydrophilic domain is external, thus permitting easy interactions with the substrate migrating along the intestine. The hydrophobic segment plunges into the lipid bilayer and anchors the catalytic moiety at the surface. The latter is split off by papain, probably because of the lability of the interdomain junction, while the whole molecule including both domains is extracted by detergents. The hydrophobic fragments obtained by limited proteolysis of several detergent-extracted brush-border hydrolases were purified and at first considered to have an apparent molecular mass of approx. 8 kDa. This value was later reduced to 3.5–4.5 kDa (35–45 amino acids only) using a more reliable isotopic dilution method. It is quite remarkable that the interactions of the hydrophobic segments with the membrane are strong enough to permit tight binding at the surface of a large, strongly hydrophilic structure. Semenza's group subsequently observed a similar mode of insertion for sucrase-isomaltase and Kenny's group in England did the same for dipeptidylpeptidase, thus suggesting that it may be general for this class of enzymes. The importance of the hydrophobic segment for enzyme binding has been very convincingly confirmed by Robert Verger who showed that the detergent form – but not the protease form – of aminopeptidase N can be reincorporated into liposomes.

Another aspect of this work demonstrated that all, or almost all, brush-border hydrolases are bound to the membrane by their N-terminus, a fact which is inconsistent with the scheme for secretory and membrane proteins proposed by G. Blobel in New York. The assembly of brush-border enzymes is better explained by one of the several "hairpin loop" mechanisms proposed in recent years. Sequence studies on the detergent form of rabbit aminopeptidase N, done in this laboratory by Jacques Bonicel, have shown that two positively charged amino acids are present among the first five residues of the chain: the N-terminal residue proper bearing the NH_3^- group and a lysine (Lys-4) which precedes an uninterrupted stretch of hydrophobic amino acids. This typical structure strongly suggests that the amino acids which emerge first from the ribosome and move ahead of the growing chain remain bound to the cytoplasmic face of the membrane by electrostatic forces, while the hydrophobic stretch penetrates freely into the membrane core. This model allows continued growth of the chain be the "hairpin loop" mechanism and its final attachment by the N-terminus. It is also a good example of cooperation between electrostatic and hydrophobic forces to unite a protein with a charged lipid phase, and it shows how a protein can span the membrane and be a surface constituent at the same time.

Amongst other studies done by Maroux's group in which I have been less involved, I would like to mention briefly those concerning the immunolabeling of aminopeptidase with the aim of following its intracellular transfer from biosynthesis to membrane assembly. Using ultra-thin frozen sections of intestinal mucosa, Suzanne Maroux, Alain Bernadac and Jean-Pierre Gorvel found that antibodies raised in rabbits by chemically pure pig aminopeptidase led to unspecific labeling of the goblet cells and of the enterocyte basolateral membrane. They realized later that the poor specificity of the enzyme located in the brush border was due to the fact that immunoglobulins generated by glycoproteins have a composite nature. Some are directed against the protein moiety of the antigen, while others recognize surface glycan structures which may be similar in several glycoproteins or complex sugars, thus leading to unexpected cross reactions. When the antibodies were depleted with intestinal mucus or group A erythrocytes, specific labeling of the brush border was obtained. With antibodies raised against group A erythrocytes, vesicles between the Golgi apparatus and the brush border were labeled. These vesicles might contain nascent aminopeptidase (and perhaps other hydrolases) en route towards the membrane. The complex nature of the antibodies generated by glycoproteins may have been the reason for serious errors made in the past.

Enterocytes are an excellent model of highly polarized cells where the essential proteins are unevenly distributed among distinct specialized regions of the plasma membrane. It is not yet known how this segregation occurs during cell differentiation, nor how it is stabilized in mature cells. As shown by Suzanne Maroux, Guy Fayet, Alain Bernadac and Hélène Ferraci, in the case of thyroid cells, these processes can be elegantly visualized by immunolabeling. In situ, these cells are assembled in follicles and they contain an aminopeptidase only in the region of the plasma membrane lining the secretory ducts. Upon cell isolation, the enzyme may be observed to diffuse laterally very quickly until complete occupancy of the plasma membrane. In contrast, the immunolabeling technique shows that cell repolarization and follicle structure are restored concurrently by cell incubation with TSH.

As already shown by Georges Palade's pioneer work on pancreas secretory cells, the complete unraveling of intracellular processes requires the simultaneous use of advanced techniques of biochemistry and cell biology.

The lipase-colipase system (second period)

From 1955 to 1981, during a quarter of a century, the laboratory continually studied lipase and began to investigate its cognate, colipase. Naturally, the studies were characterized by highs and lows for divers reasons. It is not easy to give an orderly account of the mass of data thus collected. Some of the information, provided that it was correctly interpreted, may have helped shed light on the heterogeneous biocatalysis process. But, as we will see later on, the problem is far from being entirely solved. It will only be possible to do so when the structure of the proteins implicated in the processes and that of the lipid phase on which they act, are better understood.

The monomolecular film technique. As previously pointed out, the lipase activity curve versus the substrate interface area is of a Michaelian type, provided of course that the interface properties remain the same in all assays. It was to the great credit of Robert Verger that to this purely quantitative concept, he added the very fruitful notion of interface "quality". The first demonstration showing that for a constant area the activity may be strongly affected by the properties of the interface, was obtained by the so-called monomolecular film technique. Films have the distinct advantage over more conventional emulsified or micellar systems, that in principle all substrate molecules are available to the enzyme, that the interface can be kept clean if the split products are soluble in the aqueous subphase, and that certain important parameters such as the pressure applied to the film can be varied at will. Professor D.D. Dervichian of the Pasteur Institute had already shown some years before that the lipase activity towards a film of insoluble substrate is strongly pressure-dependent. During a stay in Professors L.L.M. van Deenen's and Gerhard de Haas's laboratories in Utrecht, Robert Verger built a new type of "barostat". The principle of the apparatus is similar to that of a pH stat, except that the film pressure instead of the pH is automatically controlled. In this way, he was able to definitely show that the curve relating lipase activity to film pressure is bell-shaped. In the lower pressure range, corresponding to high surface energy, lipase is inactivated very quickly by denaturation. At high pressure, the kinetic curves show a lag period indicative of slow interfacial enzyme adsorption and this lag markedly increases beyond a certain pressure threshold. Therefore, the maximal activity observed at the top of the bell-shaped curve must be considered as corresponding to the best compromise between two fluxes with opposite effects: an adsorption flux which carries enzyme molecules to the interface where they become active, and a denaturation flux which inactivates already adsorbed molecules. This mode of regulation is typical of heterogeneous biocatalysis. The adverse effect exerted on lipase activity by high film pressures may indirectly support the view that lipolytic enzymes must "penetrate" into substrate aggregates, since "penetration" can be assumed to be hindered by exaggerated packing of the molecules at the interface. The "penetration" concept has been fully developed by Gerhard de Haas and Robert Verger for phospholipase A2. It has not yet been proved in the case of lipase for which the more general concept of interfacial adsorption is still used.

Colipase, a cofactor for lipase. A more physiological example of the role played by the "quality" of the interface in heterogeneous biocatalysis is given by the inhibitory effect of bile salts on the lipase-catalysed hydrolysis of triglycerides. Like other amphipaths, bile salts adsorb to triglyceride interfaces and modify their properties. Using the very convenient siliconized glass-bead system first proposed by Howard Brockman in the United States, Catherine Chapus and Michel Sémériva showed that the reason why bile salts hinder the interfacial adsorption of lipase is probably because the binding site of the enzyme does not recognize mixed triglyceride-bile salt interfaces. This inhibition brought about by a modification of the substrate is also typical of heterogeneous biocatalysis.

To counteract this inhibition which would normally block the intra-duodenal fat digestion, the pancreatic secretion contains a cofactor first discovered by B. Baskys in 1963 and studied later in several laboratories in Sweden, Belgium, the United States, and France. On our side, Marie-France Maylié purified the cofactor from porcine pancreas and found it to be a small protein that she called colipase. The sequence of this protein was fully elucidated the next year by Charlotte Erlanson, Mireille Rovery and Maurice Charles. Recently, the sequence of the more easily crystallizable horse colipase B has also been determined by Catherine Chapus and Jacques Bonicel. Both proteins are composed of about 100 amino acids forming a "core" strongly reticulated by disulfide bridges and two "tails" corresponding to the N- and C-termini of the chain.

Colipase plays a key role in fat digestion. It is also a quite interesting model for investigating protein–lipid interactions and protein–protein interactions in the presence of an organized lipid phase. Again using siliconized glass beads, Catherine Chapus and Michel Sémériva obtained evidence that colipase adsorbs first to a bile salt saturated interface and then serves as an anchor for lipase. A controversy arose for a couple of years concerning whether the mechanism was really sequential as postulated by Catherine Chapus, or initially required the formation of a lipase-colipase complex in the aqueous phase. The first scheme is now generally accepted. This implies the existence in colipase of two sites, one for lipid recognition and the other for lipase binding. Moreover, the conformation of the lipase binding site in colipase should be modified upon adsorption since the association constant between the two proteins is increased three orders of magnitude by the presence of a lipid phase.

The lipid recognizing site of colipase has been extensively investigated in the laboratory using isotropic micellar systems which are more amenable to the use of physical techniques than conventional turbid emulsions. Without going into too many details, it may be said that all assays done by analytical ultracentrifugation (Paul Sauve), UV-spectrophotometry (Hélène Sari), spectrofluorimetry (Simone Granon), proton nuclear magnetic resonance (Patrick Cozzone), and small angle neutron scattering (Michel Sémériva and Marc Chabre) were consistent with the presence in the colipase molecule of a single, topographically well defined lipid site in which two adjacent tyrosines located in the "core" (Tyr-56, Tyr-57) are involved. Other assays done in Sweden by Charlotte Erlanson and Bengt Borgström indicated that a lysine was also probably involved in the lipid site, thus again suggesting the participation of electrostatic forces in the protein–lipid association. Hopefully additional information about this important problem will be obtained soon by crystallography.

Structure, chemical labeling and mechanism of action of lipase. The amino acid sequence and disulfide bridges of porcine lipase have just been completed in the laboratory by Mireille Rovery's group. The enzyme is composed of a single chain with 449 residues. The disulfide bridges are arranged in such a way as to form small loops, thus insuring the chain with good flexibility. A remarkable exception in this respect is a disulfide bridge "knot" similar to that existing in colipase, which might be assumed in both cases to stabilize an essential region at the surface.

Unlike ordinary esterases, lipase is not inhibited by dissolved organophosphates. As shown by Marie-France Maylié, Maurice Rouard and Bernard Entressangles, it readily reacts with emulsified diethyl *p*-nitrophenyl phosphate or after inclusion of the reagent into bile salt micelles in the presence of colipase. Thus, the requirements of the enzyme binding site regarding substrates and organophosphate inhibitors are, as could be expected, the same. Moreover, inhibition runs parallel with the reaction of a single serine in the molecule. An histidine and a carboxylate are also essential so that lipase may appear at first sight to be a classical serine enzyme. However, as found with Catherine Chapus and Michel Sémériva, the serine is involved in interfacial binding and the carboxylate in stabilization of the superactive enzyme form at the interface. Only the histidine is probably part of the catalytic site proper. Thus, the original character of lipase compared to ordinary esterases resides in both binding and catalytic sites.

The most intriguing problem concerning lipase is still to find out how it is activated at the interface. Up to now, the lipase activity towards dissolved substrates, and consequently the extent of the activation brought about by interfaces are difficult to evaluate because even in the absence of added interface, the enzyme may be activated by adsorption onto the walls of the containers used for the assays. However, a fair estimate is that lipase is activated at least 1500-fold at the interface compared to its action in water.

It is noteworthy in this latter respect that Catherine Chapus succeeded in isolating sizeable amounts of a monoacetyl lipase derivative after incubation of the enzyme with *p*-nitrophenyl acetate. The point of attachment of the acetyl group is unknown due to its extreme lability. But the derivative has been observed to deacylate more than 100-fold faster in the presence of siliconized glass beads than in water. Therefore, we do not yet know whether the interfacial activability of lipase is due to a transconformation, a better fit of the binding site with substrate aggregates than with isolated molecules, or to any other reasons. But we do at least know which step of the catalytic cycle is accelerated most upon interfacial adsorption of the enzyme.

Miscellaneous

Like many scientists in France and elsewhere, I was often kept away from the laboratory for various reasons. The first and doubtlessly the most important reason was the teaching which I have already briefly mentioned. On my arrival in Marseilles, I was the sole professor responsible for teaching biochemistry at the science faculty.

Jumping nimbly from carbohydrate metabolism to the laws of thermodynamics and enzymology had the advantage of giving a certain homogeneity to the classes. Many of those who were my first students still tell me that they remember it. At that time we attracted some dynamic elements from the medical school for whom, extraordinary as it may seem today, we organized special Saturday afternoon workshop sessions thanks to the commitment of Marie-Jeanne Constantin and Maurice Naudet. But the system of having a single professor for each discipline could not be maintained for long due to the rapid succession of great discoveries in biochemistry, and to the growing specialization which they determined for each of us. Therefore, several young professors were appointed and we shared the responsibility which had become too heavy for the shoulders of one person. I reserved the structure of proteins and enzymes for myself, and later the structure–function relationships of membranes at the third-cycle level. In addition, after a difficult battle we succeeded in helping biochemistry emerge from a fragile auxiliary of chemistry and biology, and in giving it the right to be cited among the fundamental disciplines taught at science faculties. The first manifestation of that evolution was the creation of third-level biochemistry; one of the first such programs was developed in Marseilles. A short while later it was followed by first level and then second level where one could obtain a Master's in biochemistry at a middle point between chemistry and biology. The role of middle point is rather agreeable as, like a hinge, it gives the impression of offering a certain flexibility to a reputedly rigid ensemble, and it constitutes a bond between unjustly separated domains. But it involves a very serious risk of pinching one's fingers, which did not fail to happen to us, although it never succeeded in discouraging us. Our Master's Degree met with great success, perhaps too great, as, given the sluggishness of the university system, the number of professors could never keep up with the vertiginously rising curve of the number of students. One year we had as many as fifty-six students in the third level for whom we had to find laboratories able to receive them. Fortunately, the final employment placement of these students raised fewer problems than in other scientific disciplines, thanks especially to the growing interest accorded to biochemistry by scientific research, by medical and hospital research, and by industry.

It is said that certain French scientists spend so much time organizing others' work, that very little time remains for their own work. That criticism is partly justified. But many only agree to sit on multiple commissions and committees because of a laudable desire to defend the research section which they represent, and to obtain more money for their laboratories. It still remains that, taking into account the distance to Paris where all the important decisions come up and are taken, I submitted myself to incessant trips engendering both fatigue and loss of time for almost forty years. Nevertheless, the situation did offer the advantage of allowing me to have a first-hand observation of the formidable development of biochemistry and biology research ever since 1960. To give to certain men the impression that they participate in a great design is undoubtedly the best guarantee to their happiness and their commitment.

My father-in-law, who was a businessman, readily said that it is easier to make money than to keep it. That remark can also be applied to scientists, if "money" is replaced by "creative enthusiasm". We have been carried along by a current. Our young colleagues should on the contrary fight against the three-fold feeling that scientific research loses its attraction in a society where material goods are abundant, that they are part of a much larger body and are consequently more anonymous, and that their publications, where all the terms have been carefully weighed, are borne away as soon as they appear by a tumultuous stream where they emerge with great difficulty for a limited time. Admittedly, it was a very agreeable epoch when one or two names sufficed to symbolize the progress accomplished in an entire section of biochemistry, and when the fact of knowing and frequenting twenty people in the world was enough to give you the feeling of full participation. It is therefore just to admit that the biochemists of my generation were favored in comparison with their predecessors whose means of work were limited, and with their successors who are obliged to submit to the law of numbers. An analogous situation is found in the automobile world where traffic congestion takes away part of the pleasure one expects to feel when driving cars which are more and more sophisticated.

Given the universal character of science, the essential privilege of a scientist is that he meets people with a similar cultural background and way of thinking wherever he goes. I have greatly profited from such experiences during the course of numerous trips, and while participating in various international organizations. The first that comes to mind is the very famous Enzyme Commission of the International Union of Biochemistry, founded and animated for many years by Malcolm Dixon. That Commission had, and still seems to have, the noble ambition of systematizing the nomenclature of enzymes whose anarchistic state seemed to get worse and worse as each new enzyme was discovered. I was asked to participate regarding the proteolytic enzymes which, unfortunately, were among the most lacking in rational classification. In the beginning I was still rather sceptical about the actual usefulness of the Commission, an attitude which was in fact unfounded; the systematic names and reference numbers listed in a succession of thicker and thicker catalogues are currently employed in most scientific publications. However, in my eyes one of the Commission's inestimable advantages was that it was composed of a restricted number of friendly enzymologists such as A.E. Braunstein, S.P. Colowick, W.A. Engelhardt, E.F. Gale, O. Hoffman-Ostenhof, A.L. Lehninger, K. Linderstrøm-Lang, F. Lynen, and F. Egami; and that the President had a special gift for organizing meetings in particularly picturesque places.

The enzyme nomenclature reminds me, by the way, of an amusing anecdote. The neuter gender is unknown in the French language; everything must be either masculine or feminine. The first enzymologists agreed to say "un enzyme". But one fine day, an official commission announced that one should refer to "une enzyme", for reasons which I do not remember. People, who are surprisingly flexible about that type of thing, were beginning to get used to the new rule, when the same commission decided that after all it was a better idea to let everyone have the liberty of choosing the sex of their enzymes. Nobody, on the contrary, ever contested the right of the term "coenzyme" to be masculine. These at times passionate discussions, which left our French-speaking friends flabbergasted, unfortunately recall those concerning the sex of angels which diverted the intellectual elite of Byzantium at the time of its decline.

A short while later, Robert Thompson did me the great honor of proposing that I succeed him as General Secretary of the International Union. That position did not really suit me, and I only accepted the responsibility for six years. Nevertheless, it did give me multiple opportunities to work in the most perfect harmony with two eminent Presidents, Severo Ochoa and Hugo Theorell, and with the indefatigable and infinitely committed Treasurer, Elmer Stotz. In spite of much effort, we did not succeed in giving as much scientific consistency to the Union as we would have desired. But the very existence of that Union assuredly corresponded to the need men feel, whether they are scientists or not, to organize themselves into societies and to equip them with statutes foreseeing every possible situation which may occur in minute detail. My functions also awarded me the pleasure of meeting distinguished delegates from a constantly increasing number of "Adhering Bodies", and to participate more or less closely with the organization of two international congresses, one in Tokyo and the other, for various reasons, held simultaneously in three cities in Switzerland. For scientists, international contacts are thus both a pleasure and an obligation. They are more relaxed than those at the national level because the competition, with the exception of the sometimes delicate scientific priority domain, is less intense. I remember an American friend who, disembarking in Marseilles one day, said to me:

"One thing very agreeable about France is that you always seem to get along so well together."

I did not try to lift the wool from his eyes. Looking back on it, it appears that without getting angry or gesticulating, reasonable solutions can be found to most of the differences which might divide the scientific community. But there, as elsewhere, reforms, even those corresponding to evident progress, are born only after a painful struggle.

I do not want to finish without mentioning my ties with the international review *Biochimica Biophysica Acta*. Claude Fromageot had been one of the founders, and after his death, Westenbrinck and later Bill Slater asked me to be on the editorial board where I remained for almost twenty-five years. That exceptional longevity gave me the opportunity to read a number of papers at an average rate of six or seven per month. They were mainly submitted for publication in the enzymology section, but there were also some for the protein structure, membrane, and lipid sections. In the enzymology domain, I did my best to encourage the publication of thorough studies on known enzymes, to the detriment of those relating to the partial purification of an enormous mass of new enzymes isolated with uncertain motivations from various sources. For proteins, I tried, within the framework of *Biochimica et Biophysica Acta* and that of other reviews, to defend sequence works against often violent attacks. The sort of self-interested snobbery that tends to place work with directly biological conclusions above that involving molecular structures seems ridiculous to me, which is not too serious, but it is also very harmful. Indeed, everything seems to confirm that a previous knowledge of the structures is indispensable for understanding biological processes as fundamental as the mechanism of enzyme action, toxins and antibodies, the evolution of living organisms at the molecular level, the mechanisms by which proteins are excreted or inserted into membranes, the relations of viruses with infected cells, etc. The same remark naturally also applies to genetics where certain concepts had to be rather seriously modified when DNA and RNA sequences were elucidated. In the membrane section directed by L.L.M. van Deenen, I was present during the research boom concerning the essential structures of the internal life of cells and their attempt to live in peaceful harmony. The membrane domain is interesting insofar as it shows the degree of perfection and biological efficiency which can be reached by a suitable association of proteins, carbohydrates and lipids. It also shows that biochemistry techniques have currently evolved enough to permit one to surpass isolated molecule systems on behalf of supramolecular structures, thus getting closer to the real conditions of cellular life. The same evolution applies to nucleic acids and nucleoproteins.

In short, at the beginning of the century biochemistry was separated from chemistry because many chemists refused to get interested in anything but "mini-molecules" ("natural products") extracted from plants and animals. It was also separated from those biologists who claimed that by touching a living being, its specificity is destroyed, thus making any solution of the bigger problems impossible because of their excessive complexity. The ties between biochemistry and chemistry, and also those with physics, have been renewed at present thanks to the birth and spectacular performance of biophysical chemistry. The ties with biology are constantly reinforced thanks to a double awareness. Biologists are more and more willing to recognize the immense contribution to their discipline made by biochemistry, and biochemists now realize that the in vitro systems which permitted them to make such progress, are in reality only models which continually need to be perfected.

An office employee arrives every morning knowing what his day's work will be. After finishing it, he goes back home with his mind at ease. On the contrary a researcher – I mean a true researcher – finds more and more treasures while excavating his terrain; he has the impression, painful or exalting depending on his temperament, that the more work he does, the more work there will be left to do. I have sometimes met people who seem to fear that their research is progressing too rapidly, because they wonder with growing anxiety what they will do when everything is "finished". In research, nothing is ever finished. Generations come and go, but research remains and is a tireless bearer of hope for all humanity.

G. Semenza (Ed.) Selected Topics in the History of Biochemistry: Personal Recollections (Comprehensive Biochemistry Vol. 35)

Chapter 10

From α-corticotropin through β-lipotropin to β-endorphin

CHOH HAO LI

Laboratory of Molecular Endocrinology, Room 1018 HSE, University of California, San Francisco, CA 94143 (U.S.A.)

After receiving a B.Sc. degree from the University of Nanking in 1933, I was appointed as an Instructor in Chemistry to assist in teaching inorganic and physical chemistry. One of the professors, F.H. Lee, had just returned from graduate studies at Northwestern University in Evanston, Illinois. Lee asked me to perform the unfinished experiments from his Ph.D. dissertation with W.V. Evans at Northwestern. This was my first research project and resulted in a journal article entitled "The Decomposition Voltage of Grignard Reagents in Ether Solution" which appeared in the March issue of the Journal of the American Chemical Society [1]. The $1^1/_2$-page article became critically important in my career.

In the fall of 1937, I decided to obtain a higher degree in the United States and applied for post-graduate studies to the Universities of California and Michigan. I received a negative reply from California, but Michigan accepted me as a graduate student. With the Michigan certificate, I received a visa and the necessary documents to enter the U.S.A. in late September of 1938. At that time, one of my brothers was in the process of completing his Ph.D. degree in Business Administration at the University of California. I stopped for a few days at Berkeley on my way to Michigan. My brother insisted that I should see G.N. Lewis, then Dean of the College of Chemistry. Apparently, Lewis knew nothing about the University of Nanking and the College of Chemistry at the Berkeley campus had not previously accepted any graduate students from China. It happened that I had a reprint of my first journal article, and presented it to Lewis. He was somewhat surprised that a young student in China had a paper published in the Journal of the American Chemical Society, even as

Plate 14. Choh Hao Li.

a junior author. In addition, Lewis was a friend of W.V. Evans. He immediately agreed to accept me as a graduate student on probation for one semester. I received my Ph.D. degree [2] from the university in 1938 and have stayed on as a faculty member ever since.

During the period 1935–1938, it was not easy to find a part-time job as a graduate student – especially for a Chinese. Fortunately, I was able to obtain a job teaching Chinese. I first taught in the Berkeley Chinese Community Church and later in Chung-Mei Home for boys in El Cerrito. For a monthly salary of $30, I taught in the late afternoon and on Saturday mornings. The drive to the Chung-Mei Home was a terrifying experience as I had received only a one-hour driving lesson and the car was not in good shape.

In 1938, positions in academic and industrial institutions for new Ph.D.s were very limited and there were few post-doctoral fellowships available. Fortunately, I was able to obtain a temporary position in the Institute of Experimental Biology (Director, H.M. Evans). This provided me with an excellent opportunity to learn biology, especially endocrinology.

In 1949, I received a J.S. Guggenheim Memorial Foundation fellowship to study separation methods of peptides with Arne Tiselius at the Biochemical Institute in Uppsala. Before going to Sweden, I spent one month in the laboratory of F. Sanger and R.R. Porter at the University of Cambridge after the first International Congress of Biochemistry in Cambridge (August 19–25, 1949). In January 1950, I returned to Berkeley to establish a newly organized research unit separate from the Institute of Experimental Biology.

Hormone Research Laboratory

The Hormone Research Laboratory (HRL) was organized in 1950 at the suggestion of President Robert G. Sproul and officially established by the Regents in 1956 as an organized unit in the College of Letters and Science on the Berkeley Campus. In 1960, the laboratory became a unit of the School of Medicine, remaining on the Berkeley Campus until the spring of 1967 when space became available in the newly constructed Health Sciences West building at the San Francisco campus. During the period 1950–1967, the laboratory occupied space on various floors of the Life Sciences Building in Berkeley. Not until 1967 was it consolidated on a single floor in a modern laboratory on the San Francisco campus. At that time, with the help of the university architect, I designed an effective, compact laboratory in the limited space afforded me.

The laboratory is concerned with advancing our existing knowledge of the chemistry and biology of the hormones of the pituitary and its target organs. An equally important objective is the training of graduate students as well as postdoctoral investigators in the techniques and methods requisite for work in protein chemistry and experimental endocrinology. In addition, the laboratory has been a source of highly purified hormones for biological and clinical investigations.

During its early years, university funds provided all the support for the laboratory. Today the university continues to provide an annual budget for its operation, but the major support comes from the National Institutes of Health, the National Institutes of Mental Health, the American Cancer Society, private foundations and individuals.

Chemical messengers

Hormones are an important class of biological compounds that function as regulators of numerous physiological processes. They are produced by special organs called endocrine glands, and are secreted by these glands into the blood stream for transportation to various organs and tissues, where they carry out specific regulatory functions. The hormonal function is to direct aspects of the growth, development, metabolism or physiological operations of an organ or tissue. A given hormone may control more than one physiological process and a specific process may be under the control of more than one hormone. The normal development and maintenance of an organism are strictly dependent on the balanced production of hormones by the various endocrine glands, and many disease states result from inadequate or excess production of one or more of these substances.

As chemical compounds, hormones fall into three classes: (a) proteins or polypeptides, e.g., pituitary, pancreatic gastrointestinal and parathyroid hormones; (b) phenol derivatives, e.g., adrenal medulla and thyroid hormones; and (c) steroids, e.g., gonadal and adrenocortical hormones. Although hormones thus differ in chemical nature, they have two characteristics in common: they are effective in minute amounts and are present in an organism in only small quantities. These properties have made them one of the more difficult classes of biological compounds to isolate and to study in terms of their mechanism of action at the cellular level.

Among the endocrine glands, the pituitary (or hypophysis) is sometimes regarded as the "master" gland, since a number of the hormones it produces regulate the functioning of other endocrine organs. The pituitary is a small body located at the base of the brain and anatomically consists of two parts, the adenohypophysis (anterior and intermediate pituitary) and neurohypophysis (posterior pituitary). The posterior pituitary secretes two hormones, vasopressin and oxytocin, which are of physiological importance in regulating antidiuresis, uterine contraction and milk ejection. The anterior pituitary produces eleven known hormones (Table I) which help regulate the following functions: growth, lactation; reproduction; pigmentation; carbohydrate, lipid and protein metabolism; analgesia; behavior; and the production of thyroid, adrenal and gonadal hormones (Fig. 11). Thus, the pituitary gland plays an extremely important role in regulating the development and functioning of an organism.

TABLE I

The eleven hormones of the adenohypophysis

Group	Hormone		Number of amino acids	Principal function
Simple peptides	(1) Corticotropin (ACTH)		39	Stimulates the adrenal cortex to produce cortical hormones
	(2) and (3) Melanotropin			
	α-MSH	human	13	Darkening of skin (pigmentation)
	β-MSH	bovine	18	
		human	22	
	(4) and (5) Lipotropin			Fat mobilizing activity; β-LPH is the prohormone for endorphins
	β-LPH		91	
	γ-LPH		58	
	(6) β-Endorphin		31	Opiate-like activities
Simple proteins	(7) Somatotropin (growth hormone, GH)		191	General body growth
	(8) Prolactin (lactogenic hormone)		199	Development and lactation of the mammary gland
Glyco-proteins	(9) Lutropin (LH, ICSH)	ovine	215	Affects reproduction
		human	204	
	(10) Follitropin (FSH)	ovine	196	Affects reproduction
		human	210	
	(11) Thyrotropin (TSH)	bovine	209	Stimulates thyroid gland to produce thyroid hormones
		human	211	

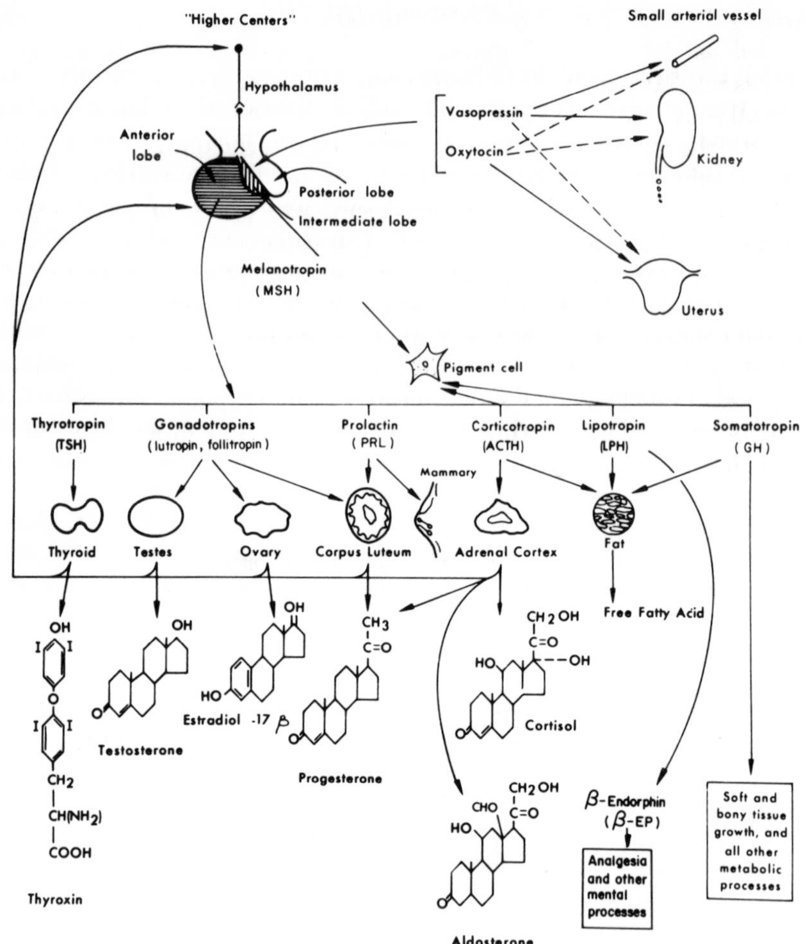

Fig. 11. Diagrammatic summary of biological properties of pituitary hormones.

It is with the anterior pituitary hormones that we have been primarily concerned since the founding of the Hormone Research Laboratory in 1950. At that time, six of these hormones had been discovered. None had been completely characterized in terms of chemical structure, and only three (lutropin, prolactin and somatotropin) had been highly purified.

α-Corticotropin (ACTH)

In 1950, the first problem to occupy our attention was the study of the exact nature of adrenocorticotropin (ACTH), the hormone that stimulates the adrenal cortex to produce such steroid hormones as cortisol, corticosterone and aldosterone. Prior to 1950, a protein having adrenal-stimulating activity had been isolated independently by two laboratories from pituitary glands of sheep and pig. However, various kinds of experimental evidence suggested that this protein was possibly a carrier or precursor of a smaller, more highly active substance. In 1953, our search for the smaller molecule led to the isolation of α-ACTH from sheep pituitary glands [3]. This highly potent adrenal-stimulating substance was found to consist of 39 amino acids linked in a single peptide chain. In 1955, we succeeded in determining the exact sequence of these amino acids [4], thus establishing the chemical structure of the molecule (Fig. 12).

In the early 1950s, three pharmaceutical laboratories (Armoir, Merck, and American Cyanamid) were also actively engaged in the isolation of ACTH from pig pituitary glands. In 1953, White in the Armour

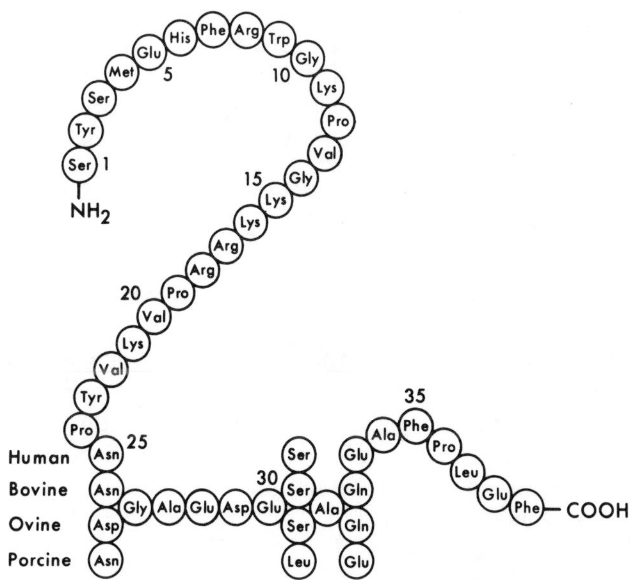

Fig. 12. Amino acid sequence of α-ACTH.

Laboratories described the preparation of an apparently pure peptide from pig glands containing 25 amino acids which was designated Corticotropin-A [5]. The following year, Bell and his colleagues [6] of the American Cyanamid Company reported the isolation of β-corticotropin from ACTH concentrates of pig pituitaries with 39 amino acids. In 1955, the American Cyanamid group published the final sequence studies of β-ACTH [7].

The period from 1950–1955 saw an exciting race to obtain a peptide with ACTH activity from pituitary extracts. We were at a disadvantage in competing with three well-equipped and aggressive pharmaceutical laboratories. In addition, two of these companies provided us with financial support. Research directors of the companies visited us regularly to learn our results, but we knew nothing concerning the work in their laboratories. I believe we were the first to isolate the ACTH peptide to homogeneity with the correct amino acid composition [3,8,9]. We had a group of talented young postdoctoral fellows and associates (A.L. Levy, J.I. Harris, I.I. Geschwind, J.O. Porath and others) who worked closely together in a collaborative spirit as an effective team. During this period, we developed useful techniques in protein chemistry. Levy [10] devised a paper chromatographic method for the quantitative estimation of amino acid. Harris [11] proposed the use of carboxypeptidase to determine the carboxyl end group analysis.

Soon after we knew the primary structure of ACTH, I contemplated the synthesis of an active fragment of the hormone, as it was known from earlier studies that the hormonal activity in ACTH endures after partial hydrolysis by pepsin or acid (for review, see [12]). In 1956, I wrote:

> "Further investigations on a combination of peptic or other enzymatic hydrolysis and acid hydrolysis may enable us to obtain the smallest active product that will still possess an ACTH activity equal to that of the native hormone". [13]

The best approach in solving this possibility was to synthesize the active fragment instead of isolating it from a mixture. In the mid 1950s, synthesis of a peptide with more than 10 amino acids was a formidable task. No one in HRL had any knowledge of peptide synthesis. I felt that if a peptide synthesis group was to be organized, I should have some knowledge of the field. In late 1956, I decided to take a 6-month sabbatical leave to learn peptide synthetic techniques. At that time, two groups of investigators in Basle, Switzerland, were very active in the synthesis of biologically active small peptides. I wrote to T. Reichstein for advice, and he suggested that I come to his Institute of Organic Chemistry at the University of Basle in the summer of 1957.

When I arrived in Basle in early August after attending the IUPAC Symposium on Protein Structure in Paris from July 25–29, 1957, Reichstein introduced me to M. Brenner, the only peptide chemist in his Institute. I told them of my desire to work in the laboratory on my own. This led to my peptide synthesis work in the laboratory of H. Kappeler in the Ciba Ltd.'s peptide group, headed by R. Schwyzer. The invitation came from A. Wettstein, then Vice-President of Ciba Ltd. in charge of research. After four months of intense work, I succeeded in synthesizing *p*-methoxyphenylazocarbobenzoxy(MZ)-His-Phe-Arg-Trp-Gly-OBzl [α-ACTH-(6–10)]. Unfortunately, I did not have time to obtain the free heptapeptide, H-Met-Glu-His-Phe-Arg-Trp-Gly-OH [α-ACTH-(4–10)], as orginally planned.

Before returning to California, Wettstein and I had a conference on the therapeutic implications of ACTH. I suggested to him the high probability of ACTH and other biologically active peptides as therapeutic drugs in clinical medicine. I also proposed that it would be profitable for Ciba to give Schwyzer a sabbatical leave to work for a few months in HRL.

Peptide synthesis

Soon after I returned to the laboratory in early February 1958, plans were made to organize a peptide group with the objective of synthesizing a corticotropically active fragment of ACTH. Since it had been known that the biological activity of ACTH resides in the NH_2-terminal segment of the molecule, and since the Lys-Lys-Arg-Arg sequence in positions 15–18 are unique for ACTH, it was decided to synthesize the first 19-amino acid sequence of the ACTH structure (see Fig. 12) as the target of our synthetic program.

In 1958, Schwyzer arrived in May, and E. Schnabel in August. By 1959, the group consisted of E. Schnabel, T.B. Lo, J. Meienhofer, J. Ramachandran and D. Chung. During this period, the *p*-toluenesulphonyl (tosyl) group was introduced for the first time to protect the guanidine group of arginine for peptide synthesis [14]. In addition, the common segment (His-Phe-Arg-Trp-Gly) occurring in ACTH and melanotropins and its analogs were synthesized and their melanocyte-stimulating activities described [14–16]. In early August 1960, we synthesized, for the first time, the ACTH-active nonadecapeptide which has an amino acid sequence identical with the first nineteen residues from the NH_2-terminus of ACTH [ACTH-(1–19)] and which possessed both corticotropic and melanotropic activities. This report appeared in the November 5, 1960, issue of the Journal of the American Chemical Society [17]. In the meantime, Schwyzer et al. submitted, on October 19, a report on the synthesis of [Gln[5]]-ACTH-(1–19) for publication in Angewandte Chemie [18].

During the period from 1960–1966, we synthesized various analogs of ACTH-(1–19) and summarized our data in a review article entitled "Structure-Activity Relationship of the Adrenocorticotropins and Melanotropins: The Synthetic Approach" [19].

In 1963, R.B. Merrifield [20] described a new method (solid-phase peptide synthesis) for synthesis of peptides by attaching a growing peptide chain to an insoluble polymeric support. This appeared to me to be the only method for synthesis of large peptides. As expected, there were many problems to be resolved before this method was applied to the synthesis of a peptide consisting of Met, Gln, His, Arg, Trp. It took J. Blake three years to succeed in the synthesis of the heptapeptide, Met-Gln-His-Phe-Arg-Trp-Gly [Gln[5]-ACTH-(4–10)] [21]. With D. Yamashiro and Blake, we made considerable improvements in the solid-phase method. In 1973, total synthesis of human ACTH, by the solid-phase method, was achieved with a yield of 3% based on Boc-Gln(Bzl)Phe-resin [22]. The identity of the synthetic product, with the natural hormone, was established by amino acid analysis, chromatography on carboxymethylcellulose, partition chromatography, optical rotation, circular dichroism spectra, electrophoresis on both paper and polyacrylamide, electrophoretic patterns of both chymotryptic and tryptic digests as well as by three biological tests.

For the last 24 years, the HRL peptide synthesis group has enjoyed very pleasant collaborations with visiting scientists, postdoctoral fellows and graduate students including D. Chung, W. Danho, R.G. Garzia, B. Gorup, B. Hemmasi, R.A. Houghten, J. Izdebski, K. Kovács, Y. Kovács-Petres, S. Lemaire, J. Meienhofer, L. Nadasdi, R.L. Noble, W. Oelofsen, J. Ramachandran, M. Rigbi, E. Schnabel, R. Schwyzer, G. Viti, K.T. Wang, H.W. Yeung, and M. Zaoral.

Lipotropins

The procedure for the isolation of ACTH as reported in 1955 [8] obtained only 12 mg pure hormone from 1 kg of whole sheep pituitaries. I suspected that the low yield was due to the oxycellulose step. Although oxycellulose is a very good adsorbent for ACTH, it is difficult to adsorb the hormone completely in a single step. I decided to omit the oxycellulose step and apply the material from the NaCl precipitation directly to chromatography on carboxymethylcellulose column. In late 1962, Y. Birk, who had just arrived from the Hebrew University, performed the experiment at my suggestion. To our surprise, a new peak appeared in the CMC chromatography before ACTH [23]. Further purification of the material in the new peak resulted in a peptide which was chemically distinct from ACTH, melanotropin and other known hormones of the pituitary gland. At that time, we were actively investigating the lipolytic activities of various synthetic fragments of ACTH and lipolytic assays were carried out almost every day. Since the new peptide behaved as a homogeneous substance, I asked A. Tanaka to assay its lipolytic potency using rabbit adipose tissue. It had considerable lipolytic activity when compared with ACTH. Thus, I suggested that the new peptide be named lipotropin (LPH) [24].

Subsequent assays indicated that the new peptide had little corticotropic activity, but that it possessed similar melanotropic activity to ACTH [23,25]. In addition, it was found to have no detectable somatotropic, gonadotropic, thyrotropic and lactogenic activities. If the new peptide had been first assayed for melanotropic activity, it might have been named γ-melanotropin (γ-MSH).

In order to ascertain that the lipotropin molecule (subsequently known as β-LPH) was a new biologically active peptide, we determined its primary structure. With L. Barnafi, M. Chrétien and D. Chung, we determined the amino acid sequence of β-LPH within one year [26] and reported to the Sixth Pan-American Congress of Endocrinology in October, 1965 [27]. Fig. 13 presents the amino acid sequence of β_S-LPH. Total synthesis of β_S-LPH by the solid-phase method was achieved in late 1977 [28]. The synthetic product has been found indistinguishable from the natural hormone in thirteen different criteria.

From the 1965 article, quote:

> "It is of great interest to note the presence in β-LPH of a core, Met-Glu-His-Phe-Arg-Try-Gly, a sequence identical with that held in common by adrenocorticotropic and melanotropic hormones from various species. In addition, the amino acid sequence 37–58 in β-LPH is identical to that of human β-MSH except that amino acid residues in positions 42 and 46 are serine and lysine instead of glutamic acid and arginine as in the case of the human hormone." [26]

It may also be mentioned that residues 41–58 in β_S-LPH (see Fig. 3) are identical to that of bovine β-MSH [29]. Thus, it was assumed at the time that β-LPH is a prohormone for β-MSH. This assumption was strengthened by the discovery of another new biologically active peptide

H-Glu-Leu-Thr-Gly-Glu-Arg-Leu-Glu-Gln-Ala-
5 10

Arg-Gly-Pro-Glu-Ala-Gln-Ala-Glu-Ser-Ala-
15 20

Ala-Ala-Arg-Ala-Glu-Leu-Glu-Tyr-Gly-Leu-
25 30

Val-Ala-Glu-Ala-Glu-Ala-Ala-Glu-Lys-Lys-
35 40

Asp-Ser-Gly-Pro-Tyr-Lys-Met-Glu-His-Phe-
45 50

Arg-Trp-Gly-Ser-Pro-Pro-Lys-Asp-Lys-Arg-
55 60

Tyr-Gly-Gly-Phe-Met-Thr-Ser-Glu-Lys-Ser-
65 70

Gln-Thr-Pro-Leu-Val-Thr-Leu-Phe-Lys-Asn-
75 80

Ala-Ile-Ile-Lys-Asn-Ala-His-Lys-Lys-Gly-Gln-OH
85 90

Fig. 13. Amino acid sequence of β_S-LPH.

(γ-LPH) from sheep pituitary glands [30].

In the course of isolating β_S-LPH, certain side fractions were obtained which were also found to possess both lipolytic and melanotropic activities. From these side fractions, Chrétien isolated a peptide (γ_S-LPH) which consisted of 58 amino acids and had comparable lipolytic and melanotropic activities of β_S-LPH. Stuctural studies indicated that it had amino acid sequences identical to the NH_2-terminal segment (residues 1–58; see Fig. 13) of β_S-LPH and that γ_S-LPH contained at its COOH-terminus the complete amino acid sequence of β-MSH [30].

It may be emphasized that the discovery of β-LPH originated from purely chemical investigation. Its biological properties were tested after it had been characterized chemically. This chemical approach for the discovery of new biologically active compounds was developed independently and extended by V. Mutt [31,32].

β-Endorphin

To explore the more specific function of β-LPH, I decided to study it in the camel, since the camel is a very lean and hardy animal. I was told that one can stick a knife in its side and the camel will not feel the pain. When Danho returned to Iraq in 1971, he collected camel pituitaries for a joint project, investigating pituitary hormones. In the summer of 1973, Danho returned to HRL with 500 dried camel pituitary glands, and succeeded in obtaining highly purified somatotropin [33] and prolactin [34].

With Chung we were unable to obtain β-LPH from the camel pituitary extract, but we did isolate a new peptide with 31 amino acids on January 22, 1974. The yield was 7.5 mg. The amino acid sequence of the new peptide was completed on June 21, 1974. It was found that the new peptide was the COOH-terminal sequence of β_S-LPH, i.e. β_S-LPH-(61–91). By this time we had only 2.0 mg of the new peptide on hand.

On Dec. 11, 1975, A. Goldstein of the Addiction Research Foundation in Palo Alto sent me a letter with a preprint of the paper by Hughes et al. [35] and asked for peptide samples related to Met-enkephalen (Met-EK) which structure is identical to β_S-LPH-(61–65) (see Fig.13). Since the first 5 amino acids of the new peptide β_S-LPH-(61–91) are the Met-EK molecule (see Fig.14), it was obvious that the new peptide might possess morphine-like activity. On December 19, I weighed out 0.30 mg of the new peptide for Goldstein. His colleague, B. Cox, showed that it had typical opioid effects in both guinea pig ileum and opiate receptor assay. The data on the isolation, sequence determination and opioid activity of β_S-LPH-(61–91) were reported in early 1976 [36,37]. This opioid peptide was named β-endorphin (β-EP) [36]. In the meantime, camel β-EP was synthesized by the solid-phase method [38].

When injected directly into the brains of mice, β-EP had about 30 times the analgesic potency of morphine [39]. It was also active as an analgesic agent when injected intravenously [40]. In addition to its analgesic potency, β-EP exhibits various behavioral activities. In the rodent, β-EP

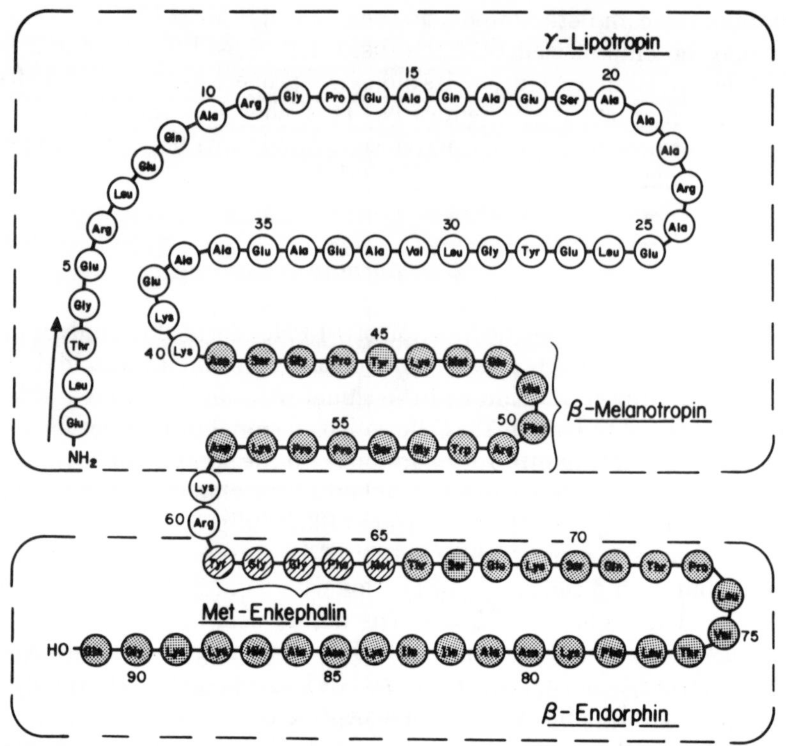

Fig. 14. Structural relationship of β-lipotropin, γ-lipotropin, β-melanotropin, β-endorphin and Met-enkephalin.

induces a pronounced sedation and a state of immobility [41,42]. A small dose causes hyperthermia [43] but at high dosages the body temperature decreases [44]. It appears that body temperature is controlled by β-EP.

In cat behavior, β-EP has a profound effect [45]. After injections, the cats showed little response to external noises or even shaking of the cage; they maintained their fixation on phantom objects with what appeared to be visual hallucinations. They did not respond even when a mouse was in view.

In another test, six monkeys were made dependent on morphine. After 14 h of acute withdrawal behavior, they were given intraventricular injections of β-EP by a special needle implanted in the brain. The β-EP proved to be twenty times more active than morphine [46].

In addition to camels, horses and various other mammals, fishes and

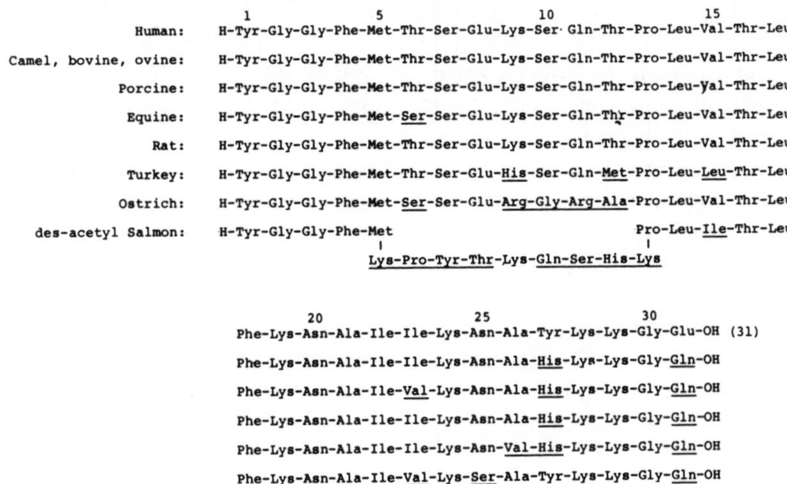

Fig. 15. Amino acid sequence of β-endorphin from various species.

birds also produce β-EP. W. Oelofsen, a former student from South Africa, provided us with ostrich pituitaries. From these glands, we obtained β-EP [47] and found that its chemical structure is similar to the mammalian hormone (see Fig. 15). Apparently, β-EP is highly conserved during evolution.

In 1976, we obtained β-EP from human pituitary glands and determined its structure [48]. Subsequently, human β-EP was synthesized [49] in sufficient amounts for studies in human subjects with strict double-blind procedures.

In Taipei, Su and his colleagues [50,51] found on two occasions the effectiveness of β-EP in suppressing the signs and symptoms of heroin addicts undergoing opiate withdrawal with relatively small intravenous doses of β-EP given rapidly. None of the subjects reported euphoria after the β-EP injection. One hour after injection, three of the subjects slept and upon awakening ate well.

In Los Angeles, Catlin and coworkers [52] studied the effect of β-EP in patients with mental disorders. Ten depressed patients received single β-EP infusions. The patients improved significantly 2 to 4 h after β-EP treatment. The typical response consists of a bright facial expression, decreased psychomotor retardation, increased social interaction, and less depressed speech content. No subject showed evidence of a hypomanic response or a rebound increase in depression. My Los Angeles colleagues have also reported the effectiveness of β-EP in patients with cancer pain. After intravenous injections, the three subjects showed good analgesic effect and mild improvement in mood.

In Tokyo, Oyama and colleagues [53] report on the profound and long-lasting analgesic effect of β-EP when injected directly into the spinal cord or intrathecally in all of 14 cancer patients with intractable pain. They also showed that rapid and prolonged analgesia was obtained in all of 14 obstetric patients who received β-EP intrathecally at the time of delivery [54]. Normal uterine contractions were maintained and all the women were fully conscious. No depression of respiration rate, cardiovascular or central nervous system was observed in any of the patients. The condition of the infants was excellent.

Concluding remarks

Probably in any language, happiness means "a state of pleasurable contentment with one's condition of life". But the relative importance of the different ingredients that produce contentment varies according to different cultures. The Chinese, with one of the oldest civilizations in the world, regard happiness as the composite of "good fortune, high position, and long life" or alternately, of "five blessings" namely, long life, riches, good health, good relations with others, and natural death. And for any given period of one's life, young or old, it also means a state of well-being "without worry or concern".

The elements that are involved in creating happiness are numerous and diverse; one can be inspired by a kind and loving individual; a simple accomplishment can give contentment. Health, talent and awareness are concomitants of happiness. Weather, surroundings and family are significant. Friends, relationships and quality of life also contribute.

True happiness usually involves, in one way or another, conscious effort and a high degree of "awareness". Euphoria, on the other hand, connotes passive enjoyment and is exclusively controlled by the senses. Physical fatigue often accompanies and sometimes creates euphoric reactions.

How much a person's individual body chemistry contributes to his or her happiness is difficult to say, but there must be a relationship.

Morphine, heroin and codeine are opiates. They reduce pain and induce happy moods. Morphine is used medically for its pain-killing (analgesic)

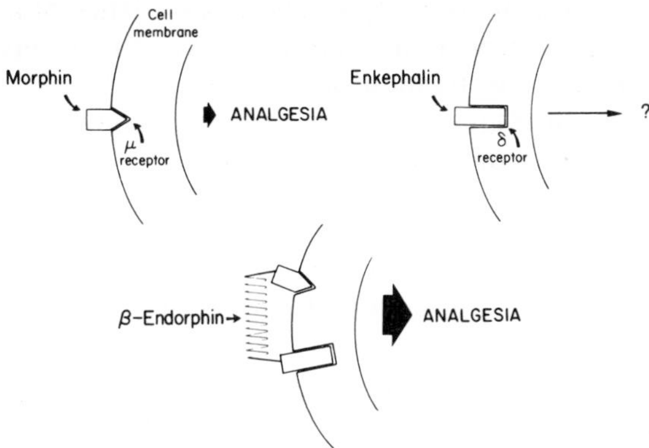

Fig. 16. Opioid receptors in the brain for morphine, enkephalin and β-endorphin.

action. Heroin is known for its euphoric effect. Morphine and other opiates work by binding brain cells specialized to handle the body's pain messages and also to turn on the body's sensation of pleasure. These binding sites in brain cells are called receptors (Fig. 16). The recent discovery of these opiate receptors in human brains [55,56] led to the search for endogenous morphine-like substances (enkephalins (EKs) and β-EP).

People who are addicted to drugs, depressed or in pain are not happy. When injected intravenously or intrathecally, β-EP can decrease pain, eliminate depression and reduce craving for drugs. There is other evidence indicating that β-EP is related to the happy state of mind and body.

Acupuncture has been used as a medical tool in China for over 5000 years. In recent years, acupuncture has been effective as an analgesic in surgical operations. In early 1971, electrical stimulation of the brain was shown to elicit analgesia in rats [57]. This work was extended to human studies in 1977 [58,59]. Patients suffering from chronic intractable pain derived long-term relief from electrical stimulation of the brain. Both acupuncture and electrical brain stimulation in human subjects for pain relief have been shown to cause a rise of the β-EP level in the circulating blood and cerebrospinal fluid.

It has been reported that persons who exercise regularly produce high levels of β-EP. Men and women who jog regularly and are physically fit produce β-EP more rapidly and in far greater amounts than those who are doing little exercise. After the activity is stopped, β-EP levels return to normal; it is generally released in response to physical stimulation and enhances the feeling of well-being.

In normal human subjects, the circulation level of β-EP is very low, but pregnant women have very high levels of β-EP [60]. It is generally agreed that pregnant women are usually happy women.

From these studies, it is evident that happiness, euphoria and optimism of an individual are enhanced by his own β-EP, secreted from the pituitary gland.

Acknowledgment

This essay is based on investigations supported in part by grants (GM-2907, MH-30245) from the U.S. Public Health Service and the Hormone Research Foundation.

REFERENCES

1 W.V. Evans, F.H. Lee and C.H. Li, J. Am. Chem. Soc., 56 (1935) 486.
2 T.D. Stewart and C.H. Li, J. Am. Chem. Soc., 60 (1938) 2782.
3 C.H. Li, I.I. Geschwind, A.L. Levy, J.I. Harris, J.S. Dixon, N.G. Pon and J.O. Porath, Nature, 173 (1954) 251.
4 C.H. Li, I.I. Geschwind, R.D. Cole, I.D. Raacke, J.I. Harris and J.S. Dixon, Nature 176 (1955) 687.
5 W.F. White, J. Am. Chem. Soc., 75 (1953) 503.
6 P.H. Bell, J. Am. Chem. Soc., 76 (1954) 5565.
7 K.S. Howard, R.G. Shepherd, E.A. Eigner, D.S. Davis and P.H. Bell, J. Am. Chem. Soc., 77 (1955) 3419.
8 C.H. Li, I.I. Geschwind, J.S. Dixon, A.L. Levy and J.I. Harris, J. Biol. Chem., 213 (1955) 171.
9 A.L. Levy, I.I. Geschwind and C.H. Li, J. Biol. Chem., 213 (1955) 187.
10 A.L. Levy, Nature, 174 (1954) 126.
11 J.I. Harris, in D. Glick (Ed.), Methods of Biochemical Analysis, Vol II. Interscience, New York, 1955, 397.
12 C.H. Li, in M.L. Anson, K. Bailey and J.T. Edsall (Eds.), Advances in Protein Chemistry, Vol XI, Academic Press, New York, 1956, 101.
13 C.H. Li, in M.L. Anson, K. Bailey and J.T. Edsall (Eds.), Advances in Protein Chemistry, Vol XI, Academic Press, New York, 1956, 165.
14 R. Schwyzer and C.H. Li, Nature, 182 (1958) 1669.
15 C.H. Li, E. Schnabel and D. Chung, J. Am. Chem. Soc., 82 (1960) 2062.
16 E. Schnabel and C.H. Li, J. Biol. Chem., 235 (1960) 2010.
17 C.H. Li, J. Meienhofer, E. Schnabel, D. Chung, T-B. Lo and J. Ramachandran, J. Am. Chem. Soc., 82 (1960) 5760.
18 R. Schwyzer, W. Rittle, M. Kappler and B. Iselin, Angew. Chem., 23 (1960) 915.
19 J. Ramachandran and C.H. Li, Adv. Enzymol., 29 (1967) 391.
20 R.B. Merrifield, J. Am. Chem. Soc., 85 (1963) 2149.
21 J. Blake and C.H. Li, J. Am. Chem. Soc., 90 (1968) 5882.
22 D. Yamashiro and C.H. Li, J. Am. Chem. Soc., 95 (1973) 1310.
23 Y. Birk and C.H. Li, J. Biol Chem., 239 (1964) 1048.
24 C.H. Li, Nature, 201 (1964) 924.
25 P. Lohmar and C.H. Li, Endocrinology, 82 (1968) 898.
26 C.H. Li, L. Barnafi, M. Chrétien and D. Chung, Nature, 208 (1965) 1093.
27 C.H. Li, L. Barnafi, M. Chrétien and D. Chung, Excerpta Medica Int. Congr. Ser., 112 (1966) 349.
28 D. Yamashiro and C.H. Li, J. Am. Chem. Soc., 100 (1978) 5174.
29 I.I. Geschwind, C.H. Li and L. Barnafi, J. Am. Chem. Soc., 79 (1957) 6394.
30 M. Chrétien and C.H. Li, Canad. J. Biochem., 45 (1967) 1163.
31 K. Tatemoto and V. Mutt, Proc. Natl. Acad. Sci. USA, 75 (1978) 4115.
32 K. Tatemoto and V. Mutt, Proc. Natl. Acad. Sci. USA, 78 (1981) 6603.
33 W.O. Danho and N.S. Al-Khidair, J. Fac. Med. Baghdad Iraq, 15 (1973) 141.
34 W.O. Danho, J. Fac. Med. Baghdad Iraq, 15 (1973) 57.
35 J. Hughes, T.W. Smith, H.W. Kosterlitz, L.A. Fothergil, B.A. Morgan and H.R. Morris, Nature, 258 (1975) 577.

36 C.H. Li and D. Chung, Proc. Natl. Acad. Sci. USA, 73 (1976) 1145.
37 B.M. Cox, A. Goldstein and C.H. Li, Proc. Natl. Acad. Sci. USA, 73 (1976) 1821.
38 C.H. Li, S. Lemaire, D. Yamashiro and B.A. Doneen, Biochem. Biophys. Res. Commun., 71 (1976) 19.
39 H.H. Loh, L-F. Tseng, E. Wei and C.H. Li, Proc. Natl. Acad. Sci. USA, 73 (1976) 2895.
40 L-F. Tseng, H.H. Loh and C.H. Li, Nature, 263 (1976) 239.
41 Y. Jacquet, N. Marks and C.H. Li, in H. Kosterlitz (Ed.), Opiates and Endogenous Opioid Peptides, Elsevier/North-Holland Biomedical Press, Amsterdam, 1976, p. 411; Y.F. Jacquet and H. Marks, Science, 194 (1976) 632.
42 F. Bloom, D. Segal, N. Ling and R. Guillemin, Science, 194 (1976) 630.
43 L-F. Tseng and C.H. Li, Int. J. Peptide Protein Res., 15 (1980) 471.
44 L-F. Tseng, H.H. Loh and C.H. Li, Biochem. Biophys. Res. Commun., 74 (1977) 390.
45 M. Meglio, Y. Hosobuchi, H.H. Loh, J.E. Adams and C.H. Li, Proc. Natl. Acad. Sci. USA, 74 (1977) 774.
46 E.F. Domino and C.H. Li, Soc. Neurosci. Abstracts, Vol. 6 (1980).
47 R.J. Naudé, D. Chung, C.H. Li and W. Oelofsen, Biochem. Biophys. Res. Commun., 98 (1981) 108.
48 C.H. Li, D. Chung and B.A. Doneen, Biochem. Biophys. Res. Commun., 72 (1976) 1542.
49 C.H. Li, D. Yamashiro, L-F. Tseng and H.H. Loh, J. Med. Chem., 20 (1977) 325.
50 C-Y. Su, S-H. Lin, Y-T. Wang, C.H. Li, L.H. Hung, C.S. Lin and B.C. Lin, J. Formosan Med. Assoc., 77 (1978) 133.
51 C-Y. Su, C.S. Lin, C. Peng, C.S. Cheng, H.H. Loh, C.H. Lin and E.L. Way, in E. Costa and M. Trabucchi (Eds.), Neural Peptides and Neuronal Communication. Advances in Biochemical Psychopharmacology, Vol. 22, Raven Press, New York, 1980, p. 503.
52 R.H. Gerner, D.H. Catlin, D.A. Gorelick, K.K. Hui and C.H. Li, Arch. Gen. Psych., 37 (1980) 642.
53 T. Oyama, T. Jin, R. Yamaya, N. Ling and R. Guillemin, Lancet, Jan. 19 (1980) 122.
54 T. Oyama, A. Matsuki, T. Taneichi, N. Ling and R. Guillemin, Am. J. Obstet. Gynecol., 137 (1980) 613.
55 J.M. Hiller, J. Pearson and E.J. Simon, Res. Commun. Chem. Pathol. Pharmacol., 6 (1973) 1052.
56 M.J. Kuhar, C.B. Pert and S.H. Snyder, Nature, 245 (1973) 447.
57 D.J. Mayer, T.L. Wolfe, H. Akil, B. Carder and J.C. Liebeskind, Science, 174 (1971) 1351.
58 D.E. Richardson and H. Akil, J. Neurosurg., 47 (1977) 178.
59 Y. Hosobuchi, T.F. Adams and R. Linchitz, Science, 197 (1977) 183.
60 H. Akil, S.J. Watson, J.D. Barchas and C.H. Li, Life Sci., 24 (1979) 1659.

G. Semenza (Ed.) Selected Topics in the History of Biochemistry: Personal Recollections (Comprehensive Biochemistry Vol. 35)

Chapter 11

A.N. Bach, Founder of Soviet School of Biochemistry

W.L. KRETOVICH

A.N. Bach Institute of Biochemistry, Leninsky prospect, 33, Moscow, B-71 (U.S.S.R.)

Short biography

Alexei Nikolaevich Bach was born on March 17, 1857 in Zolotonosha in the Ukraine to a wine distillery technician's family. In 1875 he graduated from a gymnasium in Kiev and joined the Physico-Mathematical Department of Kiev University. However, in 1878 he was expelled for taking part in student disturbances and joined the *Narodnaya Volya* (People's Freedom) revolutionary party which was active against the tsarist regime [1], was arrested and exiled to Belozersk, from where he returned to Kiev in December 1881 to continue his revolutionary activity [2]. As from 1883 Bach went underground to live in Kharkov, Yaroslavl, Kazan and Rostov. During that period, he wrote his celebrated revolutionary book *Tsar Hunger*, which played an important role in spreading the ideas of scientific socialism in Russia.

In 1885, after *Narodnaya Volya* was crushed, Bach emigrated to Paris to engage in scientific and literary work in the *Moniteur Scientifique* journal. In 1890, he started experimental studies in the laboratory of Professor Paul Schutzenberger, a well-known chemist at the Collège de France in Paris. In 1894, Bach moved to Geneva, where he worked in his private chemical laboratory and did joint research with Professor R. Chodat of Geneva University, a well-known Swiss scientist. It was precisely during the Geneva period of his life that Bach developed his peroxide theory of respiratory processes and took active part in Swiss scientific activities.

Swiss scientists showed respect for Bach: the Geneva Society of Physical and Natural Sciences elected him its chairman for the year 1916. In a letter to his daughter, L.A. Bach dated April 14, 1916, he wrote:

Plate 15. Alexei Nikolaevich Bach (1857–1946).

"I am fairly chairing the Société de Physique et d'Histoire naturelle, and as its representative attend funerals of men of science and other ceremonies."

In his report on the work of the Society in the year 1916 to its annual meeting on January 20, 1917, he noted with satisfaction that

"Despite the terrible crisis caused by the World War the number of scientific communications has increased in comparison with the number of pre-war publications. Possibly this was due to the fact that in time of catastrophes people turn to permanent values, and science is one of the most reliable ones."

In 1916, two review papers were delivered at Society meetings, one by R. Chodat titled *Modern concepts of genesis and heredity*, and the other by Ch.E. Guy called *The principle of evolution in physical and chemical sciences*. Among other communications by Society members on their experimental work, they noted Bach's *On reactions of peroxidase purified by ultrafiltration*.

In early 1917, the University of Lausanne, conferred on Bach the degree of doctor honoris causa for all his scientific publications combined. In a letter dated February 9, 1917, Professor A. Chavanne, Rector of Lausanne University, wrote Bach:

"Dear Sir, I have the privilege to advise you that, on the occasion of the twentieth anniversary of its founding, Lausanne University has ruled at the proposal of its Department of Natural Sciences to bestow upon you the degree of doctor honoris causa. In this way, the University expresses to you its admiration and gratitude for your innovatory research which opened the way to theoretical studies of respiration and, in particular, for your outstanding works on oxidative and reductive enzymes."

This is what Bach wrote to his daughter, L.A. Bach, on February 19, 1917 about the ceremony at which he was awarded the degree:

"Went to Lausanne on a special invitation of the University Senate to attend a grand meeting devoted to proclaiming of doctors honoris causa. You naturally realize that the outward aspect of all this is very insignificant to me. However, I am pleased that competent men of science have recognized my work as having significance for science to thus justify my existence before mankind, so to speak."

The diploma of doctor honoris causa of Lausanne University was presented to Bach on February 15, 1917, one month before he turned sixty.

After the Russian Revolution in 1917, Bach returned to Russia to begin a new and highly fruitful period of his scientific, organisational and public activity.

In 1918, he organized the Central Chemical Laboratory of the USSR Supreme Economic Council. Later, it was reorganized into the L.Ya. Karpov Physico-Chemical Institute, now a major research institute in the USSR, and Bach was appointed its first director. Bach's particular interest in biochemistry showed in that, as early as in 1921, he organized a biochemical institute in Moscow under the People's Commissariat of Health. He was appointed its first director and the institute, where outstanding Soviet biochemists V.A. Engelgardt, A.E. Braunstein, B.I. Zbarsky, D.M. Mikhlin, A.I. Oparin, and others initially worked, played a major role in the development of biochemistry in the USSR. In 1929, Bach was elected full member of the USSR Academy of Sciences. In 1935, together with A.I. Oparin, he organized in Moscow the Institute of Biochemistry, USSR Academy of Sciences, of which he was director to the last days of his life; the Institute is now named after Bach. Also in 1935, Bach founded the Soviet scientific journal *Biochemistry* and was elected President of the D.I. Mendeleev All-Union Chemical Society. In 1939, he was elected Academician-Secretary, Division of Chemical Sciences, USSR Academy of Sciences. In addition to being active in research and scientific organization, Bach was very active in public affairs: he was a member of the USSR Central Executive Committee and Deputy of the USSR Supreme Soviet.

Research

Bach's initial works in the 1890's were devoted to the chemical mechanism of assimilation of carbon dioxide by green plants. At that time, Baeyer's concept was already prevalent in science. Proceeding from Butlerov's discovery that sugars form from formaldehyde under the effect of alkalis, Baeyer suggested that formaldehyde, which in condensing yields sugar, was the primary product of photosynthesis. In his works devoted to the biochemistry of photosynthesis, Bach, while agreeing with Baeyer in regard to the role of formaldehyde in the forming of sugar, gave a somewhat different interpretation of the mechanism itself. He wrote:

> "If the basic carbon dioxide assimilation principles proposed by Beier's hypothesis do not evoke any doubts, the explanation it provides for the chemical mechanism of that phenomenon is utterly unsatisfactory."

In his hypothesis, Bach proceeded from the fact established by O. Loew that, under the effect of sun rays, sulphurous acid transforms into sulphuric acid to release sulphur and yield water. Bach assumed that carbon dioxide is involved in photosynthesis in the form of H_2CO_3, which is similar to sulphurous acid. In this view, the effect of light results in decomposition of H_2CO_3, with concomitant forming of percarbonic acid and formaldehyde:

$$3H_2CO_3 \rightarrow 2H_2CO_4 + HCHO$$

The percarbonic acid then decomposes in accord with the following equation:

$$2H_2CO_4 \rightarrow 2CO_2 + 2H_2O_2$$
$$2H_2O_2 \rightarrow 2H_2O + O_2.$$

Hence, Bach assumed that carbon dioxide assimilation by green plants is based on a conjugated redox reaction that involves water [3, 4].

Although Baeyer's hypothesis has presently been abandoned and the molecular mechanism of photosynthesis is considered from an absolutely different angle, Bach's idea that photosynthesis represents a conjugated redox process nonetheless proved absolutely correct.

In developing Bach's interest for the mechanism of photosynthesis, the Institute of Biochemistry, USSR Academy of Sciences, is presently conducting studies on the biochemistry and photobiochemistry of chlorophyll and other pigments, on the mechanism of initial photosynthesis reactions, and on the enzymology of ribulose diphosphate carboxylase.

Bach's attention was attracted to nitrate reduction, concerning which he wrote:

> "The nitrogen included in nitrogenous compounds that form in plants is known to originate largely, if not wholly, from soil nitrates that reduce in the plant organism. This nitrate reduction is one of the most interesting phenomena in plant chemistry. It is interesting for physiologists, since its direct result is the forming of protein substances, whose existence is an essential prerequisite for any vital process." [5]

In examining various concepts on the nitrate reduction mechanism, e.g. the hypotheses of Gauthier, Loew, Meyer and Schultze, Bach arrived at the conclusion that nitrates transform into nitrites and further into hydroxylamine, which, in interacting with formaldehyde, yields formaldoxime and then formamide. Although the idea that hydroxylamine, formaldoxime and formamide form during nitrate reduction has been abandoned, the fact that nitrites form in that process has nonetheless been unquestionably proven. At present, nitrate reduction has been studied in detail; nitrate reductases and nitrite reductases have been studied to show the involvement in those enzymes of flavins and various metals; nitrate photoreduction reactions have also been studied in detail. Nitrate and nitrite reductases of bacteroids from nodules of leguminous plants have been studied at the A.N. Bach Institute of Biochemistry.

Bach's works on the chemistry of respiration date back to the period of his stay in Geneva. In 1897, Professor Paul Schutzenberger reported to the Paris Academy of Sciences Bach's work titled *On the role of peroxides in slow oxidation processes*, a paper that was published in the journal of the Russian Physico-Chemical Society in 1897 [5a]. As in subsequent publications, the basic idea developed by Bach in that work was that slow oxidation and respiratory processes are essentially based on the forming of peroxides. In his view, a prerequisite of any oxidation by means of free, molecular oxygen is in eliminating its inert state. Conversion of oxygen from inert to active state is possible only with cleavage or attenuation of bonds with which the oxygen atoms are retained in its molecules. Bach maintained that the oxidation reaction entails cleavage of one of two bonds in the oxygen molecule, and that peroxides form at attachment to the oxidised substances; being highly unstable and chemically active compounds, they further oxidise the substrates. According to Bach, the forming of peroxides at biological oxidation temperature is due to the fact that in nature there exist substances whose valencies are not fully saturated. Such substances are capable of deforming the oxygen molecule, cleaving one of the two bonds in its molecule, and attaching it to themselves. Bach termed the attachment product a "peroxide" in the sense that it readily donates one or both of the attached oxygen atoms to another molecule of the same or another oxidised substance. Bach corroborated the correctness of his theory both by facts known before him and those obtained by himself experimentally. It should be noted that the German chemist Engler had developed the concept of the important role of peroxides in biological oxidation processes simultaneously with Bach.

Basing on his ideas on the role of peroxides, Bach published a whole series of model experimental studies on the role of peroxide compounds in oxidative reactions. These investigations were, for instance, devoted to the mechanism of action of hydrogen peroxide on permanganic acid, and so on. Proceeding from his idea on the role of peroxides in biological oxidation, Bach, during his work in Geneva, and later after his return to Russia, gave major attention to oxidative enzymes, viz. oxidases, peroxidase and catalase. During his research in Geneva, Bach conducted and published some relevant investigations together with Chodat [6–14]. These works, published mostly in the Transactions of the German Chemical Society (Berichte der Deutschen Chemischen Gesellschaft) were devoted to proving the presence of oxidative enzymes in various plants, and in higher and mouldy mushrooms. Bach paid particular attention to peroxidase, which he obtained from horseradish and for which, together with Chodat, he developed a quantitative method for estimating its activity, a technique based on due regard for purpurogallin, a pyrogallol oxidation product. In her memoirs, Bach's daughter, L.A. Bach, wrote that since her father had no assistants in Geneva,

"He, already a well-known scientist, joked and shed tears while rubbing the horseradish on a grater of impressive size."

Bach devoted a number of his works to the study of catalase: its relationship with peroxidase, its effect on alcoholic fermentation, its content in yeast, and so on [8, 15, 16].

In 1896, Bourguelot and Bertrand discovered tyrosinase in some mushrooms. This enzyme attracted the attention of Bach, who studied its action mechanism, the effect on it of various metal ions, and the mechanisms of tyrosine conversion under its influence. In his experiments, Bach clearly showed tyrosinase to be an oxidative enzyme to thus disprove Honnerman's hypothesis which assumed tyrosinase action to be connected with hydrolysis [17–19]. During Bach's stay in Geneva, many scientists studied laccase and, since that enzyme is an oxidoreductase, Bach also gave it major attention [20, 21]. His concepts on direct participation of molecular oxygen in biological redox processes proved correct for a number of enzymic reactions. The subsubclass 1.13, oxygenases in modern enzyme nomenclature, includes a whole series of enzymes that catalyze introduction of molecular oxygen into the oxidised substrate, e.g. polyphenols, amino acids, ascorbic acid, non-saturated fatty acids, and carotinoids (Enzyme Nomenclature [22]).

The redox reactions of microsomal oxidation, the hydroxylase and reductase reactions occurring in the membranes of the endoplasmic reticulum, the reactions of peroxide oxidation of non-saturated fatty acids, formations of oxygen peroxide, and reactions of desaturation and transfer of electrons in redox chains all have exceedingly important significance. In the light of contemporary experimental evidence on microsomal oxidation, the Bach-postulated "non-saturated readily oxidisable substance, an enzyme capable of attaching –O–O– groups and then transferring its active oxygen to hardly oxidisable substances" is actually none other than the principal component of microsomal oxidation, cytochrome P-450. Unlike with mitochondrial oxidation, in which dehydration reactions play a leading role and molecular oxygen is utilised solely as a hydrogen acceptor to form water, in microsomal oxidation processes activated oxygen is introduced directly into the oxidised substance. The functional roles of mitochondrial and microsomal oxidation reactions in the cell are distinctly different. Mitochondrial oxidation is a mechanism whereby oxygen is utilized in bioenergetic processes, whereas microsomal oxidation is a mechanism for utilizing oxygen for plastic purposes. The peroxidase, polyphenoloxidase, tyrosinase and laccase studied by Bach play the same role of reactions ensuring the plastic needs of the cell and organism. These enzymes play a highly important role in plant metabolism.

In recent years, it became clear that catalase, together with superoxide dismutase, has an important function in rendering harmless superoxide-anion, a highly toxic radical that forms with numerous redox reactions involving oxygen. In this case, the reaction sequence is as follows:

$$O_2 + O^{\circ}_2 + 2H^{\circ} \xrightarrow{\text{superoxide dismutase}} H_2O_2 + O_2$$
$$2H_2O_2 \xrightarrow{\text{catalase}} 2H_2O + O_2.$$

Bach continued to study redox enzymes after returning home to Russia. A large group of young researchers, who worked enthusiastically under the guidance of the well-known scientist that Bach already was in those years, formed a nucleus around him at the L.Ya. Karpov Chemical Institute and the Biochemical Institute, People's Commissariat of Health; both Institutes were located in Moscow. Together with B.I. Zbarsky, V.A. Engelgardt, D.M. Mikhlin, K.A. Nikolaev, E.N. Ivanovsky and A.I. Oparin, they worked with various enzymes [23].

Already in the nineteen-twenties, immediately after his return from Geneva, Bach began a series of works devoted to quantitative enzyme indices in connection with various states of the organism. In those studies, methods were developed for quantitating peroxidase, catalase, esterase and protease activities in a blood drop to show that they differed in animals and humans [23]. Research performed jointly with A.I. Oparin studied enzymes in germinating and maturing wheat grains, and also in various cultivars of wheat grains [24–26]. They showed that catalase, peroxidase, amylase and protease activities sharply grow at germination of wheat and sunflower seeds to attain maximum values and then decline.

Contrariwise, enzyme activities were found to gradually decrease with seed maturation to then attain certain constant levels.

A characteristic feature in Bach's research was that he sought to link theoretical studies in chemistry and biochemistry with practical needs. He always liked to repeat the words of the great Pasteur that there is no such thing as theoretical and applied science, but simply science and its applications.

The study of blood quantitative indices was ultimately designed for use in medicine to diagnose various diseases. Relevant works by Bach played a pioneer role in this, and diagnosis of diseases based on enzymic tests and indices is presently known to be a major chapter in clinical biochemistry.

The study of enzymes in maturing, latent and germinating wheat grains was directed at developing the biochemical indices of the bread-baking quality of flour with a view to rationally organizing production involving mechanized and automated processes. Bach consistently emphasised that the baking qualities of flour depend not only and not so much on its chemical composition, but on the enzymic processes in it. Keeping in mind the tendency to increasingly mechanize and automate bread baking, Bach wrote:

"In conditions of automated bread baking, it is highly important to know the biochemical processes involved in kneading, arranging and baking the dough, and today it may be said with complete certainty that without that knowledge rational management of production would be impossible." [27]

Bach's idea that enzymic indices are a major factor in grain quality was further developed in the works of the Institute of Biochemistry, USSR Academy of Sciences, where he was director, and also in those of K. Myrbäck in Sweden; they showed that various barley cultivars and batches from different regions sharply differ in the content and activity of the amylolytical complex [28–30].

In the thirties, organization of tea production was given major attention in the USSR, and vast tea plantations appeared on the Black Sea coast of Georgia. However, tea production was then essentially empirical. Bach became very interested in the processing of tea leaves, assuming that redox processes should essentially play the principal role. Proceeding from that thought, he suggested that his collaborators at the Institute of Biochemistry start studying what happens with the tea leaf when it is turned into a ready product. The work was conducted under A.I. Oparin in very close contact with tea factories and tea industry workers. As a result, Bach's collaborators created the biochemical foundations of tea production and developed relevant biochemical control techniques [31].

Studies in the field of applied biochemistry were performed at the Institute of Biochemistry and other Soviet scientific institutions and published in special volumes of collected articles, namely in *The Biochemistry of Bread Baking* [28] (3 volumes), *The Biochemistry of Grain* [29] (7 volumes), *The Biochemistry of Tea Production* [31] (9 volumes), and *The Biochemistry of Winemaking* [32] (7 volumes). As a result, today the Institute of Biochemistry is the nucleus around which scientists engaged in applied biochemistry unite.

Bach was always attracted by the problem of molecular nitrogen fixation and the forming of ammonia in that process. Proceeding from the idea that, like fermentation, fixation of molecular nitrogen results from the action of specific enzymes, he and his coworkers completed the first experiments for obtaining from Azotobacter cell-free enzymic preparations that catalyze nitrogen fixation and ammonia formation. Bach wrote in this connection:

"Similar to how theoretical studies of the mechanism of bird flight led to the construction of a flying apparatus heavier than air, we hope by theoretically studying the conjugate effect of biological redox catalysts causing the bonding of atmospheric nitrogen by bacteria to reveal the most favourable conditions for commercial synthesis of ammonia." [33]

Since then, the enzymology of nitrogen fixation has been studied quite extensively to show that the nitrogenase enzymic complex represents an oligomer consisting of two proteins, a molybdoferredoxin tetramer and an azoferredoxin dimer; molybdenum was found to play a major role in the process. Thanks to new methods involving the use of ^{15}N and the acetylene technique, the list of nitrogen-fixing microorganisms is now considerably longer.

At present, the A.N. Bach Institute of Biochemistry conducts diverse biochemical studies devoted to fixation of molecular nitrogen by bacteria; it also investigates the role of other molybdoenzymes, viz. nitrate reductase and xanthine oxidase. Other investigations involve enzymes that catalyze ammonium-binding reactions by bacteroids. A metlegoglobin reductase system, due to which legoglobin is maintained in reduced state to be capable of binding oxygen, has been discovered and the metabolism of the polymer of β-hydroxybutyric acid was shown to be closely related with nitrogen fixation in symbiotrophic microorganisms.

In developing Bach's ideas on the enzymic nature of molecular nitrogen fixation, the Institute of Biochemistry conducts special Bach colloquia on the biochemistry and enzymology of nitrogen fixation, in which scientists from various Soviet research institutions take part. In 1980, the Sixth Colloquium was held in Chernigov to discuss the most urgent problems of nitrogen fixation.

REFERENCES

1 L.A. Bach and A.I. Oparin, Alexei Nikolaevich Bach. Biographical sketch. USSR Academy of Sciences Publishing House, Moscow (in Russian), 1957.
2 A.N. Bach, Notes by Member of the People's Freedom Party. Molodaya Gvardiya Publishers, Moscow (in Russian), 1929.
3 A.N. Bach, Moniteur Scientifique, Serie 4, 7 (1893) 669.
4 A.N. Bach, Arch. Sci. Phys. Nat., 4, 5 (1898) 401–415.
5 A.N. Bach, Moniteur Scientifique, Serie 4, 11 (1897) 5–11.
5a A.N. Bach, Zhurnal Russkogo Fiziko-Khimicheskogo Obshchestva, 29 (1897) 373–398.
6 A.N. Bach and R. Chodat, Ber. Deut. Chem. Ges., 35 (1902) 2466–2470.
7 A.N. Bach and R. Chodat, Ber. Deut. Chem. Ges., 36 (1903) 600–605.
8 A.N. Bach and R. Chodat, Ber. Deut. Chem. Ges., 36 (1903) 1756–1761.
9 A.N. Bach and R. Chodat, Ber. Deut. Chem. Ges., 37 (1904) 1342–1348.
10 A.N. Bach and R. Chodat, Ber. Deut. Chem. Ges., 37 (1904) 2434–2440.
11 R. Chodat and A.N. Bach, Ber. Deut. Chem. Ges., 35 (1902) 1275–1297.
12 R. Chodat and A.N. Bach, Ber. Deut. Chem. Ges., 35 (1902) 3943–3946.
13 R. Chodat and A.N. Bach, Ber. Deut. Chem. Ges., 36 (1903) 606–608.
14 R. Chodat and A.N. Bach, Ber. Deut. Chem. Ges., 37 (1904) 36–43.
15 A.N. Bach, Ber. Deut. Chem. Ges., 38 (1905) 1878–1885.
16 A.N. Bach, Ber. Deut. Chem. Ges., 39 (1906) 1669–1670; 1670–1672.
17 A.N. Bach, Ber. Deut. Chem. Ges., 41 (1908) 221–225.
18 A.N. Bach, Ber. Deut. Chem. Ges., 42 (1909) 594–601.
19 A.N. Bach, Biochem. Z., 60 (1914) 221–230.
20 A.N. Bach and B.I. Zbarsky, Biochem, Z., 34 (1911) 473–480.
21 A.N. Bach and V. Marianowitsch, Biochem. Z., 42 (1912) 417–431.
22 Enzyme Nomenclature, Academic Press, New York, 1979.
23 A.N. Bach, Collection of Works on Chemistry and Biochemistry, USSR Academy of Sciences Publishing House, Moscow (in Russian), 1950.
24 A.N. Bach and A.I. Oparin, Biochem. Z., 134 (1923) 183–189.
25 A.N. Bach, A.I. Oparin and R.A. Wähner, Biochem. Z., 34 (1927) 473–480.
26 A.I. Oparin and A.N. Bach, Biochem. Z., 148 (1924) 476–481.
27 A.N. Bach, Khlebopekarnaya Promyshlennostj, No. 1 (1939) 4–5 (in Russian).
28 The Biochemistry of Bread Baking, USSR Academy of Sciences Publishing House, Moscow, 1938, 1941, 1942.
29 The Biochemistry of Grain, USSR Academy of Sciences Publishing House, Moscow, 1951, 1954, 1956, 1958, 1960.
30 The Biochemistry of Grain and Bread Baking, USSR Academy of Sciences Publishing House, Moscow, 1960, 1964.
31 The Biochemistry of Tea Production, USSR Academy of Sciences Publishing House, Moscow, 1935, 1936, 1937, 1940, 1946, 1950, 1959, 1960, 1962.
32 The Biochemistry of Wine-making, USSR Academy of Sciences Publishing House, Moscow, 1947, 1948, 1950, 1953, 1957, 1960, 1963.
33 A.N. Bach, Z.V. Ermolieva and E.P Stepanian, Dokl. Akad. Nauk. SSSR, 1 (1934) 22–24 (in Russian).

G. Semenza (Ed.) Selected Topics in the History of Biochemistry: Personal Recollections (Comprehensive Biochemistry Vol. 35)

Chapter 12

Sergei E. Severin: life and scientific activity

S.E. SEVERIN

Department of Biochemistry, Moscow State University, 117234 Moscow B-234 (U.S.S.R)

I was born in Moscow on December 21, 1901. Having now reached an age which is rather impressive, evidencing the years of a long and intense life, I believe it is the proper time to look back upon the events and circumstances which caused me to become a scientist and a university professor.

To begin with, I should probably relate the most important facts and landmarks of my biography [1].

I entered the Medical Faculty of Moscow University in 1918 having completed by the time a 7-year course in a gymnasium. In 1918 the last, 8th form of gymnasium was cancelled and the admittance to the university was announced free, i.e. without entrance examinations. Over 6000 students were enrolled that year to the Medical Faculty of Moscow State University (MSU). Naturally many of them could not cope with the difficulties of the first year, failed in many subjects and gradually withdrew. Of those admitted in 1918 only about 10% completed a 5-year course.

During the first year I took up a course in human anatomy under the guidance of Professor E.O. Greylikh, a remarkable teacher who aroused in me a deep interest in biological science and who was teaching far beyond the curriculum of a medical college. I completed a two-year course in anatomy in one year, which took a great deal of effort and time to the detriment of other important subjects. As a result, in due time I found myself unprepared to take examinations in these subjects and with much persuasion on my part I got permission to remain for another year in the second course. Later on it turned out to be one of the best moves in my life.

Plate 16. S.E. Severin

That year I had rather a lot of spare time and could devote myself with enthusiasm to studying some optional subjects. I attended lectures in higher mathematics, chemistry, anthropology and embryology, went through practical courses in physiology, embryology and histology and read several fundamental works on these subjects. Of particular interest to me was an optional theoretical course in analytical chemistry given by Professor A.M. Berkengeim, a remarkable teacher, whose lectures were always listened to with great interest and unfailing attention. At the same time I received an opportunity to attend for the second time the lectures of Professor V.S. Gulevich, Head of the Chair of Biological Chemistry of MSU, later Active Member of the USSR Academy of Sciences. The high scientific level of his lectures, the masterful way of presentation of the lecture material and perfect logics, all that accompanied by excellent demonstrations aroused great interest of the audience and encouraged the students to specialize in biochemistry. And, last but not least, the personal charm of Professor Gulevich won love and respect of all those who happened to come into contact with him.

In the autumn of 1920, when I was in my second year, I asked Professor Gulevich for permission to study biological chemistry under his supervision and was given a place in his laboratory. There I could do some practical work on analytical, organic, physical and biological chemistry and, later on, carry out experiments on the isolation of nitrous compounds of skeletal muscle extract by electrodialysis.

In 1923 I married V.A. Kafieva, my fellow-student, with whom I have walked hand in hand through life.

In 1924 I graduated from the University and started working, first as a research assistant and a little later as a postgraduate at the Department of Biological Chemistry of MSU headed by Professor V.S. Gulevich.

The twenties and especially the late twenties were marked by an extensive cultural program launched in our country. Many research institutes, universities and laboratories were being opened, the already existing ones were enlarged and modernized. There appeared to be a great need to fill the arising new vacancies with qualified personnel.

At the beginning of 1925, parallel to my research work in the laboratory of Professor Gulevich, I was offered a place in a physiological laboratory set up by Professor I.P. Razenkov at the Institute of Occupational Diseases, where I worked till 1933. That year (1925) was also marked by another event in my personal life – the birth of my daughter Irina. Now she is a Doctor of Biological Sciences and is successfully working in the field of biochemistry.

In 1929 I finished my postgraduate course at the Department of Biological Chemistry of MSU but continued to work as a teacher of biochemistry for second-year students at the Medical Faculty. At the end of that year I became Assistant Professor at the Department of Animal Physiology of the Biological Faculty of MSU. My duty was to deliver lectures and to supervise practical work on the biochemistry of blood, in particular, on the respiratory function of blood, for fourth-year students. I knew the subject very well having dealt with it during my stay at the Institute of Occupational Diseases. So my pedagogical and administrative work at the Biological Faculty of MSU began in 1929 and, I am pleased to say, is continuing up to the present day.

In 1931 I became the head of the Department of Physical and Biological Chemistry at a newly established Third Moscow Medical Institute. For this reason I had to give up my work at the Institute of Occupational Diseases and at the Department of Professor V.S. Gulevich and to spend a great deal of time and energy for organizing normal work in the empty premises I got there. Simultaneously, I continued my activities at the Biological Faculty of MSU* as an Associate Professor, supervising lectures and practical courses on animal biochemistry for the students specializing in physiology and some aspects of zoology.

In 1934 my son Evgenii was born. At present he is a professor, Doctor of Chemistry, head of a laboratory and Deputy Director of the Institute of Molecular Biology of the USSR Academy of Sciences (Director–Academician V.A. Engelgardt).

In 1935 I was nominated head of a newly organized laboratory of Animal Biochemistry at the Biological Faculty of MSU. At my disposal were a research worker, a laboratory assistant, and a few rooms which, frankly speaking, were hardly suitable for experimental work. However, despite all inconveniences the three of us started to work out a manual for the students specializing in animal biochemistry. We were full of energy and ideas and took up with enthusiasm every new task that seemed interesting to us. Beside scientific, pedagogical and administrative work in the laboratory of biochemistry of MSU and in the Third Medical Institute, I was also the head of a biochemical laboratory at the Institute of Hematology and Blood Transfusion where I worked for about 20 years with some intervals, in particular, when the Institute was evacuated from Moscow during World War II.

---

* In 1930 all medical faculties at the universities were abolished and higher medical education became subordinate to the Ministry of Public Health.

In the same year, 1935, I got the degree of Doctor of Biological Sciences and became a professor.

In 1939 the Laboratory of Animal Biochemistry was raised to the rank of a department, of which I was elected chairman and I am still holding this chair (later on the word "animal" was omitted from its name). Since the day of its foundation the department has been considerably enlarged; the number of graduates who are now successfully working in different research institutions and laboratories of USSR and abroad is more than 1000. The experimental work which is carried out in the laboratories of this department has always been up to the present-day standards and employs the most advanced methods of investigation. About 160 young biochemists defended their candidate dissertations there. Ten of them continued to work further and after having defended their doctor dissertations became the heads of various research laboratories and institutes.

In 1939 our family acquired a new member: Andrei, the son of my sister-in-law who died soon after childbirth. Now he is a research worker specializing in cryosurgery at the Oncological Research Center of the USSR Academy of Medical Sciences.

In 1941 the work was interrupted because of the war and only in 1943 could we return to Moscow, where I resumed my work in the same institutions.

In 1945 I was elected Corresponding Member of the USSR Academy of Medical Sciences and was appointed Director of the Institute of Nutrition. In 1948 I got a new assignment as Director of the Institute of Medical and Biological Chemistry. At the end of 1948 I was elected Active Member of the USSR Academy of Medical Sciences and Academician-Secretary for the Medical-Biological Department of this Academy. Because of my new duties I had to resign from all the other posts; the only one I retained was the Biochemical Department of MSU.

In 1953 I was elected Corresponding Member of the USSR Academy of Sciences, but this event did not add much to my current duties. In 1954 Moscow University moved to a new building in Lenin Hills. There, our department got spacious premises and modern equipment, while we could increase our staff. The research facilities improved considerably. The assistant professors N.P. Meshkova and A.V. Golubtsova helped me in many ways and shared with me the heavy burden of planning the laboratory rooms, ordering and purchasing the necessary equipment and chemicals, installing laboratory furniture, etc.

In 1957 my term of duty as Academician-Secretary of the Medical

Academy expired and I could accept another post as a chief of a biochemical laboratory at the Institute of Pharmacology, where I worked till 1973.

In 1968 I became Active Member of the USSR Academy of Sciences. In 1970 I was elected Foreign Member of the Polish Academy of Sciences and in 1971 I was admitted to the German Academy of Naturalists "Leopoldina". In 1977 I was awarded the title of an Honorary Doctor of Medicine of the Karl Marx University of Leipzig.

My public activities have always been connected with scientific societies. From 1930 till 1959 I took an active part in the work of a biochemical section of Moscow Physiological Society. In 1959 the All-Biochemical Society of the USSR was established, which was at first headed by Academician A.I. Oparin and from 1964 by Academician A.V. Palladin.

In 1964 I received an invitation to act as a plenary speaker at the Sixth International Biochemical Congress which was held in New York City.

In 1969 I was elected President of the All-Union Biochemical Society, was twice reelected and still continue to perform this function. Simultaneously I was entitled to act as Chairman of the Scientific Council for Problems of Human and Animal Biochemistry under the auspices of the USSR Academy of Sciences. In 1973 I was elected Honorary Member of the Biochemical Society of GDR. From 1955 till 1967 I was Editor-in-Chief of the journal *Voprosy meditsinskoi khimii* (Problems of Medical Chemistry); starting in 1967 up till now I have been Editor-in-Chief of the journal *Biokhimiya.* I have also been a member of editorial boards of a number of other biochemical journals, both federal and foreign ones.

In 1971 I was awarded the honorary title of the Hero of Socialist Labour. In April 1982 I was awarded the National Lenin Prize for fundamental research in the field of muscle biochemistry. In the USSR this is the most honorable public recognition of one's scientific merits.

This is practically all I wished to relate from my life. One last thing I should like to mention is the great bereavement that befell our family in 1976, the death of my wife, V.A. Severina, Professor of Biochemistry at the Second Moscow Medical Institute.

Now I should like to focus on the problems that interested me at different steps of my scientific career.

One of the main research tasks in the laboratories headed by me was to further the investigation of nitrous compounds of skeletal muscle and their two natural histidine dipeptides, which was started by my teacher, Professor V.S. Gulevich.

The first of these compounds, carnosine (β-alanyl-L-histidine) was discovered by V.S. Gulevich and S. Amiradzhibi in 1900 in a meat extract (the so-called Liebig extract) [2]; the second, anserine (β-alanyl-*N*-methylhistidine) was discovered in 1928 independently by Ackermann in Würzburg in goose muscle [3] (hence its name) and by N.F. Tolkachevskaya from Gulevich's laboratory in chicken muscle [4].

The biological significance of the dipeptides, their formation and role in muscle activity were ambiguous at the time. The problem of localization of enzymatic synthesis and degradation, the time of their emergence in muscles and their occurrence in nature as well as their participation in skeletal muscle metabolism became the main subject of our experimental work.

The dipeptides occur in large amounts (varying from 150–1000 mg% of wet weight) only in skeletal muscles of vertebrates with the exception of a few fish species [5], which do not contain the dipeptides but contain their constituent amino acids, β-alanine and histidine. In parenchymatous organs and blood only trace amounts of the dipeptides (approx. 1–2 mg%) were detected; their content in heart muscle is 1–10 mg% of wet weight. In skeletal muscles of invertebrates the dipeptides are absent. We demonstrated that carnosine is the first to appear in skeletal muscles: in chicken by the 12–13th day [6], in duck by the 18–20th day [7] and in rabbit by the 15–16th day [8] of embryogenesis. When, in addition to carnosine, the muscle tissue of adult animals contains anserine, it is discovered somewhat later, by the 4–5th day of postnatal development depending on the animal species. Simultaneously with the emergence of anserine in muscle tissues the amount of carnosine in them decreases. The muscle tissue [9, 10] and, possibly, liver [11] are the sites of carnosine synthesis. However, in parenchymatous organs and blood the enzymatic hydrolytic degradation of the dipeptides is effected by an enzyme called carnosinase, although its action is not strictly specific [12]. Degradation of the dipeptides also takes place in the intestines of many mammalian species (e.g. cats, dogs); human or pig intestine does not contain carnosinase. In muscle tissue carnosine and anserine splitting is very weakly pronounced; it may be revealed only in in vitro experiments using glycerol extracts of skeletal muscles [13] and prolonged incubation.

Undoubtedly, both processes, degradation and synthesis of the dipeptides, occur in muscles during life; it was confirmed by our experiments with the injection of [^{14}C]β-alanine into chicken. Firstly the label was detected in carnosine, where its content gradually decreased giving rise to the increased content of anserine [14].

In 1938 phosphocarnosine was synthesized by carnosine treatment with phosphorus oxychloride ($POCl_3$). Under variable conditions mono-, di- and triphosphocarnosines were obtained. Two phosphoryl residues are bound to N_1 and N_3 of the imidazole ring and one to the amino group of β-alanine. The N–P links in phosphocarnosine and phosphoanserine proved to be energy-rich bonds. According to the calorimetric data about 12 kcal/mol of phosphate are released during hydrolysis of one N–P bond. Nevertheless, all our attempts to detect phosphocarnosine in muscle tissue have failed so far.

We may thus conclude that carnosine appears in muscles in the course of embryogenesis; in rabbit muscles its appearance coincides with the time of a reflectory arc completion, i.e. when muscle contraction occurs in response to nerve excitation [15]. It is of interest that simultaneously the formation of an actomyosin complex, which responds by superstriction to adenosine triphosphate addition, takes place.

The above-mentioned steps of embryogenesis are generally associated with substantial changes in many parameters reflecting the composition and functional activity of skeletal muscles.

It should also be noted that dyskinesia of muscles, caused by tenotomy and, moreover, by denervation or deefferentation results in a gradual decrease of the carnosine content [16]. The reduction of carnosine content in a denervated muscle may be due to its increased degradation, decreased synthesis and release from the muscle caused by increased membrane permeability. The increased carnosine degradation during denervation was proved experimentally. Conceivably, carnosine synthesis induced by innervation is also blocked or markedly inhibited by denervation. Upon nerve regeneration and renewal of innervation the carnosine content shows another rise [17]. The enhanced release of carnosine from the denervated muscle of the frog was demonstrated by perfusion with Ringer's solution.

What is the role and specificity of dipeptides in skeletal muscle metabolism?

During glycolysis in rabbit muscle homogenates the formation of triphosphoglycerate increases upon addition of carnosine [18], apparently as a result of accelerated glycolytic oxidoreduction and a phosphoryl group transfer to ADP. The effect of carnosine on glyceraldehyde phosphate dehydrogenase is not specific. Histidine exerts similar effects; however, in mammalian tissues it is present in negligible amounts, while the carnosine and anserine content is high enough to produce a stimulating effect. The muscle tissue of some fish species (e.g. crucian, carp, bonito, etc.) is rich in histidine [5] and is characterized by a low rate of oxidative metabolism.

A more noticeable effect of carnosine on oxidative phosphorylation is observed in the case of skeletal muscle mitochondria. Carnosine and, to a still greater extent, anserine when added to Ringer's solution (0°C, 1 h) containing "ageing" mitochondria prevent their swelling. At the same time the coupling between respiration and phosphorylation is changed but only insignificantly. In control samples the P/O ratio during "ageing" of mitochondria shows a marked decrease within 1 h; in the experimental samples the P/O fall is insignificant. In other words, the dipeptides prevent mitochondria against the uncoupling between respiration and phosphorylation [19]. The dipeptides have a very slight effect on oxygen consumption by mitochondria; however, at a higher level the formation of energy-rich phosphorus compounds continues [18, 20]. This effect was repeatedly reproduced in experiments with phosphate acceptance by creatine in the presence of creatine kinase or glucose and hexokinase. Successive experiments involved a short-term incubation (2, 4, 6, and 8 min) of mitochondria which consumed oxygen in the absence of accepting enzyme systems followed by a direct assay of the ATP formed in the presence of the dipeptides, using α-ketoglutarate and succinate as substrates. In the presence of the dipeptides the rise in ATP content was markedly increased already after 2 min. In experiments with α-ketoglutarate this increase was about 30–40%, in the case of succinate about 25–35%. The increased production of ATP continued and could readily be determined after 4, 6, and 8 min. These results are strictly specific for skeletal muscles but are not observed when gruels, homogenates, and mitochondria of myocardium are used.

Concerning localization of carnosine in muscle cells, preliminary autoradiographic data indicated that the dipeptides are absent in myofibrils, mitochondria and cell nuclei but are detected in sarcoplasmic reticulum vesicles.

When studying the dipeptide action on oxidative phosphorylation, it should be emphasized that excessive formation of ATP is strictly paralleled with carnosine penetration inside the mitochondria isolated from pigeon breast muscle, special care being taken to preserve them in a native state. It is of interest that the dipeptides do not only maintain the coupling between respiration and phosphorylation but also facilitate Ca^{2+} accumulation in sarcoplasmic reticulum vesicles of skeletal muscle by increasing the Ca/ATP ratio [21]. Na^+ and K^+ transport through sarcolemma mediated by ATPase is enhanced in the presence of the dipeptides [22]. And, finally, we made an attempt to directly check the effect of the dipeptides added to Ringer's solution washing an isolated frog muscle on its working capacity both upon direct excitation and during nerve stimulation. It was shown that the dipeptides restore the working capacity of a fatigued muscle irrespective of the mode of stimulation [23, 24]. When the impulsation is transmitted from a nerve to a muscle, imidazole eliminates the effects of myorelaxants of the cholinolytic type but enhances that of cholinomimetics. Carnosine does not possess this property but increases the end-plate miniature potential frequency.

Summing up the results of the first line of our research one may draw some preliminary conclusions.

Carnosine biosynthesis occurs primarily in skeletal muscles beginning with embryogenesis which coincides with the reflectory arc completion. Conceivably, the penetration of nerve endings into the muscle tissue is a prerequisite for carnosine synthesis; the more abundantly innervated segments of muscles contain more dipeptides [25]. Deefferentation, that is, disturbances of motor (and trophic) innervation, results in a drop of carnosine content in the muscle. We are inclined to regard this phenomenon as a result of enhanced hydrolytic splitting, increased membrane permeability and reduced synthesis of carnosine.

Quite obviously, the dipeptides play an essential role in the maintenance of constant pH in muscle tissue, since they produce a distinct buffering action which was reported by Bate-Smith as early as 1938 [26].

At the same time the dipeptides also promote glycolysis, preserve the structure and function of mitochondria, prevent the uncoupling between respiration and phosphorylation, i.e. proton "leakage" through the mitochondrial membrane, which otherwise would lead to the above-mentioned disturbances, etc. The action on the membranes of sarcoplasmic reticulum vesicles prevents them from Ca^{2+} leakage. And, finally, their action on sarcolemma ensures rapid elimination of depolarization and restores the potential difference on both sides of the plasma membrane typical of skeletal muscle. As a result, the participation of the dipeptides in all these metabolic pathways of skeletal muscle ensures their prolonged and more effective functioning. The mechanism by which the dipeptides are involved in the maintenance of membrane structure properties and functions, still remains unknown and awaits further elucidation.

The second, enzymological, trend in our research was originally closely connected with the first one. On the one hand, the study of the role of dipeptide in skeletal muscle metabolism demanded a more detailed knowledge of certain enzymes and possible effects of the dipeptides on their activity. On the other hand, the physiological studies based on extensive investigation of blood, particularly of the respiratory function of blood, drew our attention to oxidative processes occurring in the organism, to their diversity and biological significance. Chronologically our research was developing as follows. The starting point was the study of the acid-base equilibrium in the organism under variable physiological and pathophysiological conditions, e.g. muscle exercise, pain stimulation, overheating, intoxication, etc. Naturally, questions arose as to the role of blood buffer systems in the maintenance of constant pH, erythrocyte function, position of oxyhemoglobin dissociation curves, blood ability to bind carbon dioxide, ratios of bases, chlorides and phosphates in erythrocytes and blood plasma, etc.

All the described events took place in 1925–1928, that is, at the time when blood was regarded as a physicochemical system. Hence it was considered possible to determine the amounts of components on the basis of the already known ones, using Henderson–Van Slyke nomograms.

Those who are acquainted with the blood analysis procedure for drawing Henderson nomograms [27] are aware how tedious and time-consuming this process is. First of all it is necessary to plot oxyhemoglobin dissociation curves at CO_2 partial pressures typical for arterial and venous blood. This dictates the necessity of determining a curve for blood plasma and total blood binding of CO_2 at partial pressure of O_2 characteristic for arterial and venous blood.

Once hemoglobin dissociation curves and CO_2 binding curves have been determined, one needs to know the actual content of O_2 and CO_2 in arterial and venous blood in order to determine partial pressure of respective blood gases (we did not use the Krog procedure with a gas bubble). On the basis of these data it is rather easy to calculate pH values for arterial and venous blood. The results thus obtained are expected to be consistent with experimental pH values from an electrometric analysis. It is also necessary to estimate the total amount of bases, chlorides and water in erythrocytes and plasma. Having all these data on hand, we drew nomograms. Yet the values of different blood and plasma components calculated from the nomograms were inconsistent with the experimental data. The differences were especially noticeable under conditions of experimental pathology, e.g. overheating. We supposed that either the principles of plotting nomograms needed revision, or the methods employed were inadequate and thus excluded the accuracy of later steps.

We dropped this line of investigation after another attempt of our own and the appearance of some literature which convinced us that the basic principles of nomographic representation of blood analysis data were oversimplified. Indeed, such important events as carbohemoglobin formation, active ion transport into erythrocytes, participation of lipids, but not exclusively proteins, in binding of bases, etc. could not be taken into consideration when plotting nomograms. Hence the nomograms failed to give an adequate presentation of interrelationship between the components of an intricate and variable blood system.

Having thoroughly studied the respiratory function of blood, we undertook a new task to follow the further fate of blood-transported oxygen in tissues. It was 1928, the year when tissue respiration was almost exclusively investigated by the method of Warburg devised in 1925 and when H. Krebs and B. Kisch were chiefly concerned with the study of oxidative deamination of amino acids. The first question we tried to answer was whether carnosine was subjected to oxidative deamination by kidney sections or gruels.

We were immediately faced with a strange phenomenon. The increased oxygen consumption in the presence of carnosine and ammonia production was preceded by a "lag-period" (approx. 1 h). These data were interpreted only some time later when enzymatic degradation of carnosine in parenchymatous organs to individual amino acids was demonstrated and O_2 adsorption and NH_3 production by the sections and gruels of kidney cortical layer after addition of histidine and β-alanine were investigated. A delayed, enhanced O_2 consumption and NH_3 production (like in the case of carnosine) occurred only when β-alanine was added. By the time, 1937, A.E. Braunshtein and M.G. Kritsman described the process of transamination of amino acids, it had been originally believed, however, that β-amino acids were not involved in transamination [28]. In fact, β-alanine is not subjected to deamination but is involved in transamination in liver and kidney with pyruvic acid to form α-alanine. Consequently, NH_3 production and enhanced O_2 uptake by kidney sections and gruels at the expense of the carnosine added are caused by a sequence of reactions, namely, hydrolytic splitting of carnosine, transamination of β-alanine to yield α-alanine and then glutamic acid, oxidative deamination of α-amino acids resulting in additional consumption of O_2 and NH_3 production.

In 1928 K. Lohman [29] discovered ATP in the composition of skeletal muscles. In 1933 G. Embden proposed a new scheme of glycolysis [30] which with certain modifications got the name of the Embden–Meyerhof scheme. The pathway of glucose conversions according to this scheme made clear the biological significance of glycolysis which led to ATP production under anaerobic conditions. In 1930 V.A. Engelhardt described ATP formation with respect to oxygen consumption [31, 32] and perceived in this fact the biological meaning of respiration. The data of V.A. Engelhardt were supplemented by the results obtained by H.M. Kalckar [33] and V.A. Belitser [34] who introduced the concept of a phosphorylation coefficient. The pathway of aerobic formation of ATP got the name of oxidative or respiratory phosphorylation. With the time ATP was accepted to be a universal source of energy used in a great variety of vital processes and first and foremost in the processes of muscle activity. That was to a great extent promoted by the discovery by V.A. Engelhardt and M.N. Lyubimova of the adenosine triphosphatase activity of myosin [35].

Beginning with the postwar period we started a new, purely enzymological line of investigation which, in a broad sense, included study of enzymatic reactions directly or indirectly involved in formation and utilization of energy-rich phosphorous compounds.

The enzymes of glycolysis, glycogenolysis, glycogen and pentose phosphate synthesis and oxidative metabolism of skeletal muscles, e.g. pigeon breast muscle, and myocardium have been studied in more detail.

Initially all the topics of would-be investigations were proposed and further supervised by myself. But in 1965–1966 several research groups were set up in order to study some particular problems or enzymes within the framework of a general project of the laboratory. Each group consisting of 2 to 9 investigators was headed by a senior supervisor who was supposed to have sufficient experience in a given field. As a rule the investigations were directed and some of the experiments continuously supervised by me. The general progress of the work was totally dependent on collective efforts and efficiency of the head of the group.

The general project in enzymology included the following steps: isolation of an enzyme in a homogeneous state, study of its physicochemical properties including definition of its functional groups of the active center and elucidation of the mechanism of its catalytic acts. Special emphasis was laid on enzymes with a complex quaternary structure as well as on principles and specificity of regulation of their enzymatic activity. Gradually the number of enzymes under study and that of research groups has become considerably increased. Our present-day investigations comprise the enzymes of the anaerobic pathway of carbohydrate metabolism (hexokinase, glyceraldehyde 3-phosphate dehydrogenase, lactate dehydrogenase, glycogen phosphorylase, phosphorylase b kinase, transketolase) and some other enzymes of the anaerobic part of the pentose phosphate pathway. Special reference was given to the study of NAD kinase, creatine kinase of hyaloplasm and mitochondria, myosin adenosine triphosphatase, cAMP and cGMP-producing cyclase system, phosphodiesterase and protein kinases both dependent and independent of cyclic nucleotides. Among the enzymes of aerobic metabolism the Krebs cycle enzymes (dehydrogenases of pyruvate, ketoglutarate and succinate), succinyl-CoA synthetase and the respiratory chain and oxidative phosphorylation enzymes have been studied.

(I) Study of hexokinase interaction in rat skeletal muscle hyaloplasm showed that there are no special mitochondrial, microsomal or hyaloplasmic forms of the enzyme. Under variable environmental conditions the enzyme is either bound to subcellular structure membranes (pH $\geqslant$ 7.4 + glucose + Mg^{2+}) or completely and reversibly solubilized (pH $\leqslant$ 7.4) in the presence of adenyl nucleotides, glucose 6-phosphate, K^+, and phosphate. The membrane-bound hexokinase of mitochondria (only isozyme II) has a higher V_{max} and a lower K_m value; hence this form is more active than the free one. When interacting with sarcoplasmic reticulum membranes, both isozymes I and II change their properties which manifests itself in a reversible increase of V_{max}.

These data may serve as a model of an adsorption mechanism of regulation of enzyme activity and can form a basis for a simple method of extraction and purification of muscle hexokinase [36, 37].

(II) Entirely different results were obtained from a comparative study of cytoplasmic and mitochondrial forms of creatine kinase. Much attention has been paid recently to the mitochondrial enzyme which has been studied comparatively little. A simple procedure for isolation of mitochondrial creatine kinase in a homogeneous state from bovine heart has been developed. This protein has a pI of 9.6 (that for the cytoplasmic enzyme is 6.3), which can be explained by a high content of amino acids, arginine and lysine. According to our data the enzyme fixation in mitochondria can be due to electrostatic interactions of the positively charged amino acid residues of the enzyme with the negatively charged lipid groups of the mitochondrial membranes. The M_r value of the enzyme subunit after incubation in 8 M urea is 41 000–43 000. The enzyme usually exists in an aggregated form, i.e. in the form of a hexamer (M_r = 240 000–260 000). The degree of aggregation depends on the enzyme concentration; the dimer with M_r = 80 000 is evidently the main functional unit involved in the formation of polymers. Of great significance for the enzyme activity are two arginine residues. Inactivation of the enzyme with butanedione proceeds in two stages: rapid inactivation (by 50%) is achieved by adding butanedione (2–20 mM); further addition of the inhibitor causes a slow inactivation to a complete loss of activity.

Adenyl nucleotides protect the enzyme against inactivation; phosphocreatine does not exert such protective effect. A mixture of adenyl nucleotides and phosphocreatine protects the enzyme against butanedione-induced inactivation only by 35%.

Sulfhydryl groups are also very essential for the creatine kinase activity. The active center of mitochondrial creatine kinase is likely to contain one cysteine and two arginine residues; one of the latter binds adenyl nucleotides, while the other is responsible for maintaining the active conformation of the enzyme [38, 39].

(III) The active form of glyceraldehyde phosphate dehydrogenase was considered to have a tetrameric structure consisting of four identical subunits. A method has been developed which allows to obtain monomeric, dimeric and trimeric forms possessing the same specific activity as the tetramer. If the total activity of the tetramer be taken for 100% (with respect to protein) the activity of the trimer, dimer and monomer amounts to 75, 50 and 25%, respectively. Lactate dehydrogenase having a specific activity characteristic of this enzyme in its tetrameric form was also obtained in monomeric form possessing an equal specific activity. The method used is based on covalent immobilization of the enzyme on Sepharose or adsorption on specific antibodies which does not affect the enzyme activity.

Our experimental results allow to conclude that the role of intersubunit association in the molecule of NAD-dependent dehydrogenase is not to maintain the catalytically active conformation but to ensure the interaction of the enzyme active centers which determines the marked cooperativity and the ability to respond to regulatory influences [40].

In order to analyze conformational changes in the enzyme-active site, we used fluorescent probes of anionic (e.g. aniline naphthalenesulfonate) and cationic (auramine) nature. The former probe was used to localize the site of coenzyme binding; the latter the site of substrate binding. This approach allowed us to achieve stabilization of the conformational state of the enzyme under the effect directed to a certain part of the active site [41]. Moreover, some new and unexpected results were obtained in experiments that evidenced the involvement of one of the arginine residues in the formation of a productive ternary enzyme-NAD-substrate complex essential for a catalytic act [42, 43]. The results obtained suggest that arginine residues are necessary for the formation of a catalytically active structure of some other enzymes, e.g. phosphorylase, α-ketoglutarate dehydrogenase, transketolase, creatine kinase, etc.

(IV) When speaking about the role of aggregation and deaggregation in the regulation of enzymatic activity, one has to mention such enzymes as α-ketoglutarate dehydrogenase, NAD-kinase, glycogen synthetase, in which changes in their aggregational state produce marked alterations in enzymatic activity. Here it seems appropriate to present some data obtained by hydrolysis of phosphorylase b kinase with subtilisin.

(V) As is known, this kinase is a complex made up of four subunits, α, β, γ and δ, differing in their structure and M_r values. Each of the subunits consists, in its turn, of four subunits; therefore, the kinase molecule may be represented by a formula $\alpha_4\beta_4\gamma_4\delta_4$ or $(\alpha\beta\gamma\delta)_4$. Which of the four subunits plays a key role in phosphorylation of phosphorylase b kinase, is still unclear. Our preliminary data suggest that this role belongs to the β-subunit [44].

Hydrolysis of phosphorylase b kinase with subtilisin results in a number of fragments with a relatively low M_r value; one of them possesses high kinase activity. The initial complex protein conglomerate is not responsible for the catalytic act proper, but determines the enzyme ability to respond to regulatory influences [45].

(VI) The results of numerous investigations on skeletal muscle phosphorylases a and b have been published more than once and will be mentioned here only in connection with the original analogs of pyridoxal phosphate and AMP synthesized in our laboratory and widely used for the study of catalytic mechanism of the phosphorylase reaction. The use of these analogs as irreversible inhibitors allowed us to obtain the enzyme-inhibitor complexes; the corresponding peptide fragments were isolated from these complexes and characterized after hydrolysis [46].

(VII) In the studies on protein kinases from skeletal muscles (e.g. pigeon breast muscle) it was shown that the protein isolated and purified by us consisted of two regulatory and one catalytic subunits; each of the subunits was obtained in a homogeneous state. To investigate the mechanism of enzymatic activity we carried out an inhibitory analysis using synthesized original analogs of cAMP and cGMP [47, 48].

In collaboration with the Laboratory of Enzyme Regulation of Cell Activity from the Institute of Molecular Biology, USSR Academy of Sciences, we studied phosphorylation of histone H1 by protein kinases dependent and independent on cyclic AMP and cyclic GMP.

It appeared that strictly depending on the type of protein kinase, phosphorylation occurs on serine residues located in different sites of the histone H1 molecule, which affects the cell cycle by making it longer or shorter. As the object of study, the lower eukaryote *Physarum polycepha-*

lum was chosen, which is a unicellular multinuclear organism characterized by synchronous division of cell nuclei [49].

(VIII) Cyclic nucleotide-independent phosphorylation at N-terminal serine was reported for troponin T. The respective kinase detected in our laboratory [50] can be related to a casein-type kinase. However, the enzyme phosphorylates troponin T much stronger than casein and is thus believed to be specific for troponin T [51].

(IX) Of the wide variety of multienzyme high-M_r complexes, pyruvate dehydrogenase has been studied in more detail. The enzyme was prepared from pigeon breast muscle in a homogeneous state and was subjected to electron microscopy and kinetic and sedimentation analyses. The first decarboxylating component of this complex has been studied more thoroughly. The study included enzyme purification to homogeneity, inactivation by phosphorylation with a respective protein kinase, dephosphorylation with phosphatase up to reconstitution of the enzymatic activity and separation into two subunits in accordance with the formula $\alpha_2\beta_2$. It was found that only the α-subunit undergoes phosphorylation [52]. Using original thiamine pyrophosphate analogs synthesized in the laboratory of Professor A. Schellenberger (Halle, GDR), we examined coenzyme binding in the active center of the pyruvate dehydrogenase decarboxylating component. The molecule of the latter $(\alpha,\beta)_2$, was found to contain two cysteine and two histidine residues essential for the catalytic act [53].

The last step of conversions catalysed by the multienzyme high-M_r pyruvate dehydrogenase complex consists in oxidation of the reduced form of lipoic acid. NAD, a coenzyme for many dehydrogenases, is an indispensable participant of this reaction. In some separate cases the phosphorylated form of NAD (NADP) performs this function.

Then a question arose as to successive steps and localization in the cell of biosynthesis of these compounds and, in the first place, of phosphoribose.

(X) Muscle tissue of rabbit heart was used in our investigations. We succeeded to establish that the enzymes of synthesis were distributed between numerous cell organelles, i.e. nucleus, mitochondria, microsomes and cytosol [54].

(XI) The activity of glucohydrolase in myocardium as one of the enzymes causing degradation of NAD and NADP was studied in detail. A comparison of the enzyme activity with respect to NAD and NADP showed that decomposition of NAD is more intensive and that inhibition with nicotinamide is more expressed in the case of NAD. It should be added that the reduced form of NAD (NADH) is a powerful inhibitor of NAD kinase –

it actually stops the formation of the phosphorylated form of the coenzyme. The above data are sufficient in evaluating the numerous possibilities of heart muscle to control the NAD/NADP ratio [54].

The synthesis of nicotinamide coenzymes is known to begin with phosphoribose.

Then the question arose as to the sources of pentose phosphates and the possibility of their biosynthesis in heart muscle not only via oxidation of glucose 6-phosphate with involvement of two dehydrogenases and lactonase, but also under anaerobic conditions using the intermediate products of glycolysis.

(XII) A new, experimentally substantiated scheme of carbohydrate metabolism in heart has been proposed. According to this scheme the enzymes and substrates involved in the anaerobic part of the pentose phosphate pathway and in initial steps of glycolysis and the products formed are common for both processes and constitute a single system. The pentose phosphate formation from glycolytic products or intermediate glycolytic product formation from pentose phosphates depends on the state of a cell as well as on the set and concentration of substrates. The crucial role in these conversions belongs to erythrose 4-phosphate which, in its turn, may be formed both from glycolytic products and pentose phosphates.

Hence the key role in the regulation of the direction and rate of these reactions can be attributed to the concentration ratio of the reaction substrates, pentose phosphates and hexose phosphates [55].

(XIII) Another example of a reaction, the direction of which is controlled by substrate concentration, is the succinyl-CoA-synthetase. The enzyme was isolated from pigeon breast muscle in a homogeneous state and characterized in terms of physicochemical properties and kinetic behaviour [52].

We investigated the formation of (a) succinyl-CoA when the enzyme, ATP, Mg^{2+}, succinate, CoA and P_i were present in the incubation mixture, and (b) ATP, succinate and free CoA when the incubation sample contained the enzyme, succinyl-CoA, Mg^{2+}, P_i and ADP*. The formation of intermediate phosphates took place in both cases. When the reaction proceeds in the direction of succinyl-CoA synthesis, four phosphoryl groups are joined to the protein, two of which are replaced by the succinyl

---

* In pigeon breast muscle (as it was shown by us [52] and was confirmed by M.L. Hamilton and D.H. Ottaway [56]) adenyl but not guanyl nucleotides participate in the succinyl-CoA-synthetase reaction.

residues immediately before succinyl-CoA formation. Within the protein the phosphoryl groups are joined to the histidine residues; they are stable in alkaline media and, under special conditions after completion of protein hydrolysis, may occur as phosphohistidine as was earlier described by Boyer.

When the reaction is directed towards ATP formation, i.e. when succinyl-CoA is the initial compound, the reaction intermediate is a phosphoryl derivative of the protein containing four phosphoryl residues. These residues are readily split off from the protein in alkaline media and, consequently, are not attached to the histidine residues of the enzyme. The investigation of the mechanism and intermediate products of succinyl-CoA-synthetase is under way in our laboratory.

The above described role of substrate concentration in regulation of the direction and rate of reversibility enzymatic processes is only one of the possible regulatory influences.

Apparently the secondary messengers of hormonal effects, cyclic nucleotides (cAMP and cGMP), play a very essential and varied role in the regulation of substance and energy metabolism. These cyclic nucleotides are formed from corresponding triphosphates under the action of adenylate cyclase and guanylate cyclase. cAMP activates protein kinases which phosphorylate proteins and causes their modification. Phosphorylation sharply changes the enzyme properties, sometimes by stimulating (e.g. phosphorylase b) or inhibiting (e.g. pyruvate dehydrogenase) the enzymatic activity. The structure and mechanism of action of protein kinases have been investigated extensively in our laboratory.

(XVI) Adenylate cyclase is a typical membrane enzyme. Our studies demonstrated that different types of receptors after having interacted with corresponding hormones produce an activating effect of various intensity on the enzyme activity. A combined action of, say, two hormones which form two different hormone-receptor complexes leads to a competition between them for the same cyclase molecules. As a result the degree of activation of the cyclase attains some average value with the account of the effects produced by the two aforementioned different hormone-receptor complexes [57]. One of the specific features of adenylate cyclase from heart is its ability to undergo desensitization, i.e. to become insensitive to hormones, for example catecholamines. The decreased number of adrenergic receptors and the fall in the activating effect of catecholamines on adenylate cyclase constitute the essence of this phenomenon [58].

(XV) As regards diesterase, which also belongs to the cyclase system, among the numerous detected forms of the enzyme two forms may be readily distinguished from each other. One of them increases the catalytic activity in the presence of Ca^{2+} and calmodulin, while the other is insensitive to these effects. All the diesterase forms detectable in heart muscle (i.e. fractions eluted from the column) are capable of splitting both cAMP and cGMP at different rates depending on the experimental conditions [59].

(XVI) Another example of a membrane enzyme is adenosine triphosphatase responsible for active ion transport, namely, that of Na^+ and K^+ through sarcolemma [60] or Ca^{2+} transport across sarcoplasmic reticulum membranes. These enzymes have a complex structure and form complexes made up of 2 to 4 protomers. The transport function seems to be due to cooperative interactions of Na^+, K^+-ATPase protomers which can be revealed only during ATP or GTP hydrolysis. The enzymatic degradation of other nucleotide triphosphates does not reveal any cooperativity, nor does it provide for ion transport [61].

The rearrangements in the membrane lipid bilayer at critical temperatures as revealed by the Arrhenius plots consist in a change of viscosity indicating some structural disturbances in the enzyme molecule. As a result, the enzyme affinity for the substrate, activation energy and cooperativity of interaction of protòmers are altered. Study of this group of enzymes isolated from different sources demonstrated that the protomer interactions in the oligomeric forms of adenosine triphosphatases is a necessary condition for operating a control of active ion transport.

Active ion transport is dependent on specific enzymes of ATP hydrolysis. So we undertook a detailed study of ATP formation, primarily via a pathway coupled with cell respiration, i.e. oxidative phosphorylation. This line of investigations in our chair has its history. The first step in this direction was an elucidation of the role of imidazole-containing dipeptides of skeletal muscle in the formation of energy-rich phosphorus compounds. Further progress of the work required a detailed study of phosphorylating processes which take place in mitochondria isolated from muscle tissues and in phosphorylating particles obtained from these mitochondria. In the fifties these investigations were carried on under my direct day-by-day guidance but by 1960 a very active group of young scientists with V.P. Skulachev at the head was established. At that time Skulachev was a postgraduate and later on a research assistant at our department.

Some time later Skulachev continued these studies on his own as the head of an independent Division of Bioenergetics of MSU. In our department, beginning in the late sixties this line of research was carried on by a research team headed by Assistant Professor Dr. A.D. Vinogradov.

(XVII) This research team focussed their attention on the study of two enzymatic systems of heart mitochondria, namely: succinate oxidase and ATP synthetase. Succinate oxidase has been examined as an enzyme possessing its own dehydrogenating function and electron transport activity. These studies demonstrated the significant role of the sulfhydryl group of the protein in the catalytic activity of the latter. A model for binding of the natural competitive inhibitor, oxaloacetate, to the active site of the enzyme by a thiosemiacetal bond was proposed [62]. The discovery of a new catalytic effect of succinate dehydrogenase, that is, the succinate ferricyanide reductase activity, was a good prerequisite for investigation of the electron transport function of the enzyme. The intramembrane localization of the electron transport site of succinate dehydrogenase was proved and the important role of the iron-sulfur complexes in its activity was demonstrated. Of great importance for elucidating the mechanisms of electron transfer from reduced dehydrogenase to the natural acceptor ubiquinone appeared to be the isolation of a protein with $M_r = 13\,000$ which made the soluble enzyme reactive with respect to ubiquinone [63].

The next task of this research team – investigating the ATP-synthetase complex – consisted primarily in a comparative kinetic analysis of the ATP hydrolase and ATP synthetase reactions catalyzed by submitochondrial particles. It was suggested that the kinetic mechanisms of ATP synthesis and hydrolysis are different. The coupling factor F_1 exists in two slowly interconvertible conformations: the equilibrium between them is controlled by the ATP/ADP ratio. The conditions were elaborated when the enzyme is fully inactive as ATP hydrolase but fully active as ATP synthetase. A hypothesis on different pathways of ATP synthesis and hydrolysis suggests entirely new approaches to the study of the mechanisms of oxidative phosphorylation [64].

Summarizing the results of our enzymological studies described here, the following conclusions can be made. Some data concerning the structure and properties of a number of enzymes have been partly established for the first time and partly obtained in support of previously reported results; the ways of regulating the direction and rate of the reactions catalyzed by these enzymes have been elucidated.

In the simplest cases, in reversible processes, the direction of a reaction is mainly determined by the concentrations of substrates and reaction products (e.g. the pentose pathway and glycolysis); in more complicated cases a special significance acquires the adsorption of an enzyme on the surface of cell organelles (e.g. hexokinase). One of the ways for regulating the enzyme activity appears the formation of composite protein complexes from more simple components, which, depending on conditions, are capable either of forming still larger aggregates or of dissociating into dimers or monomers. Each of the resulting forms is characterized by a definite enzymatic activity (e.g. α-ketoglutarate dehydrogenase, NAD-kinase).

Studies on various enzymes of a polymeric structure provided experimental evidence for a very important statement about the determining role of the quaternary structure in cooperativity between the subunits, the capacity to receive regulatory effects, though (in many cases) without affecting the specific activity of an enzyme, which remains constant both in the case of a polymer and of a monomer (GAPD, lactate dehydrogenase, transketolase).

The high-M_r multienzyme complexes were used as a convenient object for demonstrating a special role of the component initiating the reaction. The maintenance of its native state or, on the contrary, its modification, determines the rate of the whole cycle of processes characteristic of the multienzyme complex in general (e.g. pyruvate dehydrogenase). Numerous examples are known of protein modification due to their phosphorylation, which sharply affects the enzyme activity by increasing it in one case and inhibiting it in the other (e.g. phosphorylase, phosphorylase kinase, glycogen synthetase). Moreover, phosphorylation of serine residues in various parts of the histone H1 molecule changes the timing of the cell cycle causing a delay or an advancement of certain stages of preparation for cell division. The study on the membrane-bound enzymes which, as a rule, have the form of proteolipid conglomerates showed a particular significance of the dependence of the structure of these complex formations on the environment. Due to some environmental changes the reaction can start to proceed in the opposite direction; however, not by a simple reversal but independently, through other intermediates with different kinetic parameters (e.g. succinyl-CoA-synthetase, ATP synthetase).

Thus, the general outline of our studies can be given as follows: our group described two previously unknown enzymes, found new ways of regulation, distinguished between the ATP-hydrolase (F_1) and ATP-synthetase activities, developed a number of convenient methods for isolation and purification of enzymes and proposed several new schemes illustrating the succession of basic metabolic reactions.

At this point I should like to finish the account of investigations undertaken by our group in recent years and of those in progress now. Certainly the description of the work performed is very superficial and lacks strict logics in presentation. Some directions of our work are entirely omitted, for instance, those concerning blood preservation, biochemical pharmacology, nutritional problems, experimental oncology, etc. However, it was not the aim of this essay to give a detailed description of the whole lot of our studies. The basic idea, as I saw it, was to acquaint the reader with the organization of the Biochemical Department of MSU, with the pedagogical activity there and, in particular, with the general trend of our scientific interests.

Naturally, a question may arise as to whether the enzymes under study are not too numerous and diverse in nature. The point is that the choice of enzymes for our work was made gradually and in a natural way. Partly, this was conditioned by the necessity to comprehend one or another phenomenon observed and, partly, by new ideas and considerations which appeared as a result of careful analysis of the literature. It should be noted that the subject chosen for experimental study always corresponded to the personal inclinations and wish of each individual research worker, whether the theme was offered by the chief or whether he suggested the problem himself and obtained the consent for its exploration. Of course, there were a number of instances, though fortunately very few, where inside the group disagreement appeared as to the future tendencies in the development of experimental work. In these cases, either the research worker pursued his work, with my consent, in an individual order or moved to another department.

Coming back to the question about the expediency of investigating a large number of enzymes in our group, it should be first of all taken into account that all these enzymes are involved in the processes connected with the formation or utilization of energy-rich phosphorous compounds and, consequently, one general problem unite these studies. A high variety of methodological approaches utilized in the studies on different enzymes permit a wide exchange of experience, enable us to modify, improve and invent more and more complicated tasks, which we offer the students during their practical work as well as to make these tasks individual and applicable to each specialization. From the point of view of scientific utility, one should not forget that the analysis of properties and mechanisms of action of various enzymes as well as the studies in the regulation of their activity reveal general principles of the catalytic action, similarity of the structural organization of the active sites. A comparison of the results obtained in different groups in the enzyme studies invariably appeared fruitful and useful for future work. Nevertheless we can expect a question (if this question can be considered justified), whether such an organization of work contributes to the formation of "a scientific school"? Our group has become larger, it unites scientists of different generations; at present the number of middle-aged and young research workers gradually tends to become predominant. From my point of view, "a scientific school" should not only and so much be characterized by a narrow topical plan of work. Rather, the scientific direction of the group can and must undergo changes parallel to the increased experience and qualification of the research workers and the progress of science. The "school" is formed on the basis of the whole complex of requirements for the level of research, attitude to work, the ethical criteria, etc. The atmosphere of consideration and benevolence, mutual respect and assistance should be created within such a group of scientists, with the freedom for initiation and reasonable quest. At the same time, one should make scrupulous demands of the accuracy and reliability of the reported facts, avoid speculativity in the conclusions drawn and be careful in theoretical formulations. I dare to believe that our group does its utmost to meet all these requirements and look forward to our fruitful pedagogical and scientific work in the future.

And the last question: Am I satisfied with the years gone by?

Like every scientist I certainly wish I could have done more and better. However, I greatly appreciate my relations with my coworkers. The mutual understanding which I have always enjoyed allowed us to smoothen the differences in opinions in those rare cases when they appeared. With great satisfaction and gratitude I think about the attitude to my research of official instances who always showed sympathy and aid for my ideas. And, finally, I have been very happy in my family life; my relationship with my wife and later on, with my grown-up children has always been very warm, friendly and full of mutual understanding. Boredom is unknown to me; I have always had so much to do, so many problems and so many duties to attend to, most of which I fulfilled with pleasure, that I have no reason to complain or to think that I have not been lucky in my life.

REFERENCES

1 S.E. Severin, Vopr. Med. Khimi., 17 (1971) 564–573 (in Russian).
2 V.S. Gullevich and S. Amiradzhibi, Ber. Deut. Chem. Ges., 33 (1900) 1902–1903.
3 D. Ackermann, O. Timpe and K. Poller, Z. Physiol. Chem., 183 (1929) 1–10.
4 N.F. Tolkachevskaya, Z. Physiol. Chem., 185 (1929) 28–32.
5 P.L.Vulfson, Usp. Biol. Khim., 4 (1962) 81–92 (in Russian).
6 S.E. Severin and V.N. Fedorova, Dokl. Akad. Nauk SSSR, 82 (1952) 443–446 (in Russian).
7 P.L. Vulfson, Biokhimiya, 23 (1958) 300–306 (in Russian).
8 S.E. Severin and N.A. Yudaev, Biokhimiya, 16 (1951) 286–291 (in Russian).
9 L.G. Razina, Dokl. Akad. Nauk SSSR, 111 (1956) 161–164 (in Russian).
10 T. Winnik and R.E. Winnik, Biochim. Biophys. Acta, 23 (1957) 649–650.
11 A.N. Parshin, T.A. Goryukhina, E.A. Sherstnev and L.M. Vdovichenko, Dokl. Akad. Nauk SSSR, 141 (1961) 233–235 (in Russian).
12 S.E. Severin and E.F. Georgievskaya, Biokhimiya, 3 (1938) 148–162 (in Russian).
13 N.P. Meshkova, Biokhimiya, 5 (1940) 151–161 (in Russian).
14 O.V. Dobrynina, Biokhimiya, 32 (1967) 612–617 (in Russian).
15 A.A. Volokhov, in Regularities of Ontogenesis of the Nervous System in the Light of Evolutionary Theory, USSR Acad. of Sci. Publishers, Moscow-Leningrad, 1951, p. 178 (in Russian).
16 L.V. Slozhenikina and S.E. Severin, Dokl. Akad. Nauk SSSR, 113 (1957) 399–402 (in Russian).
17 N.A. Yudaev, M.I. Smirnov and L.G. Razina, Biokhimiya, 18 (1953) 732–738 (in Russian).
18 N.P. Meshkova, Usp. Biol. Khim., 6 (1964) 86–107 (in Russian).
19 S.E. Severin and Yu Shu Yui, Biokhimiya, 23 (1958) 862–869 (in Russian).
20 N.P. Meshkova, Biokhimiya, 24 (1959) 323–328 (in Russian).
21 A.A. Boldyrev, A.V. Lebedev and V.B. Ritov, Biokhimiya, 34 (1969) 119–124 (in Russian).
22 A.A. Boldyrev, Biokhimiya, 36 (1971) 826–832 (in Russian).
23 S.E. Severin, Vestnik MGU (Biology, Pedology series) No. 1 (1972) 3–18 (in Russian).
24 S.E. Severin, Proceedings of the Plenary Sessions of the Sixth International Congress of Biochemistry, Vol. 33, New York City, 1964, pp.45–61.
25 S.E. Severin, P.L. Vulfson and L.L. Trandofilova, Dokl. Akad. Nauk SSSR, 145 (1962) 215–217 (in Russian).
26 E.C. Bate-Smith, J. Physiol., 92 (1938) 336–343.
27 L. Henderson, Blood. A Study in General Physiology, Yale University Press, New Haven, 1928.
28 A.E. Braunshtein, in Biochemistry of Amino Acid Metabolism, USSR Acad. Med. Sci. Publishers, Moscow, 1949, p. 67 (in Russian).
29 K. Lohman, Biochem., 203 (1928) 164–171; 172–207.
30 G. Embden, H.J. Deuticke and G. Kraft, Klin. Wochenschr., No. 12 (1933) 213–217.
31 V.A. Engelhardt, Biochem. Z., 227 (1930) 16–38.
32 V.A. Engelhardt, Kazansky Med. Zh., 27 (1931) 4–5; 499–502 (in Russian).
33 H.M. Kalckar, Enzymologia, 2 (1937) 47–52.
34 V.A. Belitser and E.T. Tsybakova, Biokhimiya, 4 (1939) 516–535 (in Russian).

35 V.A. Engelhardt and M.N. Lyubimova, Biokhimiya, 7 (1942) 205–231 (in Russian).
36 L.N. Shcherbatykh, N.Yu. Goncharova and N.V. Aleksakhina, Biokhimiya, 42 (1977) 1408–1418 (in Russian).
37 N.Yu. Goncharova, L.N. Shcherbatykh, V.I. Tsupak and N.V. Aleksakhina, Biokhimiya, 46 (1981) 628–634 (in Russian).
38 L.V. Betlousova, T.Yu. Lipskaya, V.I. Temple and A.P. Rostovtsev, J. Mol. Cell. Cardiol., 12/1 (1980) 15.
39 T.Yu. Lipskaya, V.I. Temple, L.V. Belousova, E.V. Molokova and I.V. Rybina, Biokhimiya, 45 (1980) 1155–1166 (in Russian).
40 N.K. Nagradova, L.I. Ashmarina, R.A. Asriyants, T.V. Cherednikova, T.O. Golovina and V.I. Muronets, Adv. Enzyme Reg., 19 (1981) 171–204.
41 S.V. Ermolin, A.A. Kost, M.V. Ivanov and N.K. Nagradova, Dokl. Akad. Nauk SSSR, 238 (1978) 245–248 (in Russian).
42 L.I. Ashmarina, V.I. Muronets and N.K. Nagradova, Biochem. Int., 1 (1980) 47–54.
43 L.I. Ashmarina, V.I. Muronets and N.K. Nagradova, FEBS Lett., 128 (1981) 22–26.
44 N.V. Gulyaeva, P.L. Vulfson and E.S. Severin, Biokhimiya, 43 (1978) 373–382 (in Russian).
45 S.E. Severin, S.A. Shur, A.N. Pegova, L.K. Skolysheva and P.L. Vulfson, Biochem. Int., 3 (1981) 125–130.
46 L.I. Mikhailova, L.K. Skolysheva, P.L. Vulfson, M.B. Agalarova and E.S. Severin, Biokhimiya, 43 (1978) 2016–2021 (in Russian).
47 I.A. Grivennikov, T.V. Bulargina, Yu. V. Khropov, N.N. Gulyaev and S.E. Severin, Biokhimiya, 44 (1979) 771–780.
48 E.S. Severin, N.N. Gulyaev, T.V. Bulargina and M.N. Kochetkova, Adv. Enzyme Reg., 17 (1979) 251–282.
49 I.N. Trakht, I.D. Grozdova, N.N. Gulyaev, E.S. Severin and N.V. Gnuchev, Biokhimiya, 45 (1980) 788–793 (in Russian).
50 N.B. Gusev, A.B. Dobrovolsky and S.E. Severin, Biochem. J., 189 (1980) 219–226.
51 V.V. Risnik and N.B. Gusev, Biochem. Int., 3 (1981) 131–138.
52 S.E. Severin and M.M. Feigina, Adv. Enzyme Reg., 15 (1977) 1–22.
53 L.S. Khailova and O.V. Lisova, Dokl. Akad. Nauk SSSR, 259 (1981) 506–508 (in Russian).
54 S.E. Severin and L.A. Tseitlin, Circul. Res., Suppl. III, 34/35 (1974) 121–127.
55 S.E. Severin and N.G. Stepanova, Adv. Enzyme Reg., 19 (1981) 235–255.
56 M.L. Hamilton and J.H. Ottaway, FEBS Lett., 123 (1981) 252–254.
57 V.A. Tkachuk and G.N. Baldenkov, Biokhimiya, 43 (1978) 1097–1110 (in Russian).
58 V.A. Tkachuk and S.E. Severin, Proc. Fourth Soviet-American Symp. on Myocardial Metabolism, Vol. XVII, Tashkent, 1979, pp.1–25.
59 V.A. Tkachuk, V.G. Lazarevich, M.Yu. Menshikov and S.E. Severin, Biokhimiya, 43 (1978) 1622–1630 (in Russian).
60 S.E. Severin and A.A. Boldyrev, Dokl. Akad. Nauk SSSR, 188 (1969) 1415–1417.
61 A.A. Boldyrev, Usp. Fiziol. Nauk, 12/2 (1981) 91–130 (in Russian).
62 A.D. Vinogradov, E.V. Gavrikova and V.V. Zuevsky, Eur. J. Biochem., 63 (1976) 365–371.
63 A.D. Vinogradov, V.G. Gavrikov and E.V. Gavrikova, Biochim. Biophys. Acta, 592 (1980) 13–27.
64 I.B. Minkov, E.A. Vasilyeva, A.F. Fitin and A.D. Vinogradov, Biochem. Int., 1 (1980) 478–485.

Name Index